Is Apartheid Really Dead?

Is Apartheid Really Dead?

Pan-Africanist Working-Class
Cultural Critical Perspectives

Julian Kunnie

Westview Press
A Member of the Perseus Books Group

This book is dedicated to the memory of Steve Biko and my sister Lorna, whose lives of struggle were taken away prematurely, and in honor of all those matriots and patriots who sacrificed their lives for a free Azania and a liberated Africa.

Map on page xvi from, *Regionalism in the New South Africa*, p. 46, courtesy of the University of Natal, Durban, South Africa.

Copyright © 2000 by Westview Press, A Member of the Perseus Books Group

Published in 2000 in the United States of America by Westview Press, 5500 Central Avenue, Boulder, Colorado 80301-2877, and in the United Kingdom by Westview Press, 12 Hid's Copse Road, Cumnor Hill, Oxford OX2 9JJ

Find us on the World Wide Web at www.westviewpress.com

A CIP catalog record for this book is available from the Library of Congress.
ISBN 0-8133-3758-5

The paper used in this publication meets the requirements of the American National Standard for Permanence of Paper for Printed Library Materials Z39.48-1984.

PERSEUS
POD
ON DEMAND   10   9   8   7   6   5   4   3   2   1

Contents

Acronyms

AALAE	African Association for Literacy and Education
ANC	African National Congress
AZACTU	Azanian Congress of Trade Unions
AZAPO	Azanian People's Organization
AZAWU	Azanian Workers Union
BAMCWU	Black Allied Mining and Construction Workers Union
BCM	Black Consciousness Movement
BPC	Black People's Convention
BWF	Black Women's Federation
CNETU	Council of Non-European Trade Unions
CODESA	Convention for a Democratic South Africa
COSATU	Congress of South African Trade Unions
ECOMOG	Economic and Monitoring Group of West African States
ESAF	Enhanced Structural Adjustment Facility
EU	European Union
FSAW	Federation of South African Women
FEDSAL	Federation of South African Labor
GEAR	Growth, Employment, and Redistribution Program
IAWUSA	Insurance and Assurance Workers Union of South Africa
ICU	Industrial and Commercial Workers Union
IMF	International Monetary Fund
JCI	Johannesburg Consolidated Investments
NACTU	National Council of Trade Unions
NAIL	New African Investments Limited
NECC	National Educational Crisis Committee
NIC	Natal Indian Congress

NUM	National Union of Miners
NUMSA	National Union of Metalworkers of South Africa
OAU	Organization of African Unity
PAC	Pan-Africanist Congress of Azania
PEBCO	Port Elizabeth Civic Organization
RUF	Revolutionary United Front
SACP	South African Communist Party
SADCC	Southern African Development Coordination Conference
SADF	South African Defense Force
SASO	South African Students Organization
SOPA	Socialist Party of Azania
UDF	United Democratic Front
UPC	Cameroon People's Union
VAT	valued-added tax
WOSA	Workers Organization of South Africa
ZANU	Zimbabwe African National Union

Preface

There has been much discussion within media and political circles, particularly in Europe and North America, about a "post-apartheid" society being imminent in South Africa. This book raises some fundamental questions on whether the notion of "post-apartheid" has any tangible basis in reality, considering the historical, contemporary, and near-future actuality of the South African context. In responding to the issue of "post-apartheid," this discourse will explore some of the corollary questions that emerge from such a subject, namely, politics, power, and liberation from the tenacity of colonialism, economic exploitation, and social oppression.

The questions that loomed constantly in my mind as I traversed the highways and byways of South Africa's townships, towns, and villages in the summers of 1994, 1997, 1998, and 1999 were multifarious, but one rang constantly in my ears: Are we Black South Africans a defeated people? Others pondered the euphoria that so many people abroad, including many academic "specialists," had expressed about "post-apartheid."[1] Where was the post-apartheid society that was being hailed in so many political circles, since all that I saw and experienced was the transmogrification of apartheid, its accompanying violence, and the totality of white power in every social, political, and economic sphere? Are Black people in South Africa in a more advantageous position today when it comes to assessing liberation probabilities, or are we more powerless today than we were a decade or two ago in terms of critical consciousness and resolve to eradicate oppression? From the vantage point of many whites in South Africa and globally, the country may have reached a precipice anticipating a post-apartheid society; from the perspective of many Black people, particularly working-class and poor people, South Africa is still a nightmare of colonial oppression and structural violence that is not anywhere near surrendering power to veritable democratic Black rule. Mandla Tshabalala, a social-work professor at the University of Cape Town, raised the question of the function of changes in South Africa in the late 1980s:

> [W]ho is this reform supposed to benefit? Is this reform supposed to benefit the white community and the Black community? Or is it meant to benefit the Black community on a large scale and the white community on a lesser scale, that is my immediate question? If this is change, is it for people who are more privileged? Or maybe it is threatening for them to see some more changes which are going to threaten their privileged position. . . .

. . . And for them therefore this is reform. But to me it is a burning issue if you are at Crossroads, Khayelitsha, and all these kind of things are not resolved.[2]

Several auxiliary questions that issue from the ones raised above warrant some form of sensible response and explanation to weed out the seeds of ideological confusion and obfuscation sown by the colonial regime of South Africa and its allies in the capitalist and imperialist world.

One of the distinctive features of this book is that it describes the South African situation both historically and contemporaneously from the vantage point of the oppressed Black people of South Africa and explores issues of transformation from an indigenous perspective. Specifically, it emanates from someone who has been historically involved and continues to be involved in that country's struggle for liberation, self-determination, and independence via the Black Consciousness Movement (BCM). The book does attempt to articulate the more radical views of the Black South African working class, views that the bourgeois media generally ignores because the media function more in the interests of racist ruling classes than the ruled.

A curious question that ought to be pondered by all freedom-loving people the world over is: Why is it that the Black Consciousness Movement is either never or seldom mentioned by the many critics scrutinizing the South African reality, even though it was this same movement that propelled the Black struggle for freedom into a new era through the intensified resistance movements of Black youth and workers during the late 1960s and early 1970s, culminating with the historic insurrection of 1976?[3] One probable explanation for this sin of omission is the fact that organizations like the Azanian People's Organization (AZAPO) and the newly formed Socialist Party of Azania (SOPA) are Black-led, and uncompromisingly socialist in orientation.[4] Many people today, both Black and white, inside and outside South Africa, buy into the argument that the Black Consciousness Movement is an organization marginal to the South African struggle and engaged in regressive, reactionary politics because it is idealistic in its objectives and is essentialist in outlook. As Nkosinathi Biko, the twenty-six-year-old son of Steve Biko, admonished:

> It was politically opportune in the late 1980s for some propagandists to spread confusion about by claiming that Black Consciousness had become outdated, redundant, and thus irrelevant. But we are all now aware that was part of a calculated political mis-education campaign that has resulted in many of our people being emptied of the essence of their identity.[5]

This argument is one indication of how deep racist ideology runs in the world and how successful the imperialist world is in defining which organizations represent the legitimate aspirations of the oppressed Black people of South Africa and then dictating the contents of liberation so that they are considered both "realistic and pragmatic."

A countervailing question that needs to be raised from the vantage point of many colonized and oppressed people is: Where has "realism, pragmatism, and expediency" landed Black people in South Africa's struggle for national liberation? For the Black Consciousness Movement to unabashedly declare that the South African liberation struggle is a Black struggle that ought to be led by the working class is neither racist nor chauvinistic; it is merely stating the truth: It is Black working-class people who are oppressed and colonized in South Africa and it is Black working-class people who will ultimately liberate themselves and everyone else from oppression.

Steve Biko, the founder of the Black Consciousness Movement of Azania, dedicated his life to the elimination of ideological obfuscation, political confusion, and epistemological brainwashing, for the purpose of clarifying the situation for ordinary people and telling the South African truth even though many people were not ready for hearing such truth.[6] It is in the wake of his insights into the South African reality during the late 1960s and 1970s and according to his lucid analysis and perspicacious vision, then, that we are compelled to elevate these questions now.

Why was it that the racist minority regime of South Africa found it timely to discuss negotiations about a "new South Africa" in early 1990? Why was it that Black people in South Africa were led to believe that our historical enemies were now our friends and many began to laud the erstwhile oppressors as liberators? Why have Black people so quickly forgotten our history of struggle, especially the history over the past three decades? Why are questions of indigenous people's freedom from colonization in South Africa hardly considered, when South Africa still signifies a settler-colonial republic in the Southern part of Africa? Why have so many dismissed the philosophy of Black Consciousness cultivated by Steve Biko as anachronistic and outdated for our contemporary age of politics and resistance? How is the rest of Africa connected with recent changes in South Africa? Where do working-class women and the rural community feature within the dynamic of the movement for decolonization and liberation? This book will attempt to respond to these questions and sketch the complexity of the principal issues in an intelligible manner to help elucidate what is at stake in South Africa and the rest of Africa as current political developments unfold in the country and abroad.

I also bring my experiences as a Black theologian involved with the liberation struggle in South Africa and the United States (Turtle Island, in the words of the indigenous people of North America) to bear in the articulation of this book. I will thus illumine issues of Black religio-culture that I consider relevant to the Black struggle in South Africa, or Azania (as those within the Black Consciousness Movement and Pan-Africanist Movement refer to South Africa), a subject that few "post-apartheid specialists" have deemed worthy of discussion. Notwithstanding the aforesaid presuppositions, this book will endeavor to be optimally objective and maximally analytical in the presentation of the subject matter, utilizing the dialectical approach of question and challenge to describe the South African reality as it actually exists, while being cognizant of the plethora of factors that function as obstacles to the realization of an authentically liberated South Africa. I am fully aware that such a view itself can be subject to further critical scrutiny.

Julian Kunnie

Acknowledgments

At the heart of the African and all indigenous traditions is the collective definition of individuality. This book represents the culmination of a nine-year study, during which I discussed the manuscript with scores of individuals whose input was invaluable in the final writing of the manuscript.

I want to thank my colleagues Babatunde Agiri and Nomalungelo Goduka from Norfolk State University and Central Michigan University, respectively, for their suggestions, as well as Percy Hintzen, chair of African American Studies at the University of California, Berkeley, for his scholarly insights in African political economy that inspired me to embark on this project. In Azania, the insights of people with whom I have been in regular communication have been extremely helpful: Sam and Laura Abrahams, Barney Mokgatle and the Mokgatle family, Lybon Mabasa and all members of the Socialist Party of Azania, Patrick and Thandi Mkhize, Colin Geoffries, Willy Mathys, Sonwabo Hoyi, Xolisi Mfazwe, and the Kunnie family.

I appreciate the support of the dean of the College of Humanities, Charles Tatum, for supporting my travel to Northwestern University and to Southern Africa to conduct research for this book. I am also grateful to the Office of the Provost of the University of Arizona, Paul Sypherd, for an Author's Support grant. Nnamdi Elleh and Reinessa Neuhalfen at Northwestern University were helpful in identifying research materials for my work during my visits to Northwestern. The staff of Africana Studies at the University of Arizona, Eleanor Navarro and Chantele Carr, provided much-needed administrative support during my crafting of this book, assisting me with making copies of various drafts.

I am especially thankful to Jennifer Chen, associate editor at Westview Press, for her active interest in this book project; to Michelle Trader for assistance in the layout; and to Kim Kunnie for proofreading the final manuscript.

Finally, I need to acknowledge a profound debt to my dearest wife, Kim, and our two beautiful sons, Mandla and Sibusiso, for their unflinching support throughout the process of the writing of this manuscript. They have embodied the true meaning of family in their coping with my exhaustive research, travel, and writing schedules.

All these familial contributions have made the publication of this book possible. For any shortcomings and oversights in its writing, I am fully responsible.

J. K.

Notes

1. One example would be Timothy Sisk in his work *Democratization in South Africa: The Elusive Social Contract* (Princeton: Princeton University Press, 1995), especially in opening statements in the preface.

2. Cited in Eric Wainright and Brigette Wakabayashi, eds., *South Africa: Reform or Revolution,* papers from a conference on the South African Professors World Peace Academy, held in Sandton, South Africa, April 7–10, 1988.

3. See, for instance, Gail Gerhart's *Black Power in South Africa* (Berkeley: University of California Press, 1978), and Ernest Harsch, *White Rule, Black Revolt* (New York: Monad Press, 1980).

4. *Tribute*, March 1993.

5. *City Press,* Johannesburg, September 13, 1998.

6. For a detailed explication of an analysis of the South African situation from the perspective of Black-consciousness philosophy as developed by Steve Biko, see Steve Biko, *I Write What I Like* (San Francisco: Harper and Row, 1978).

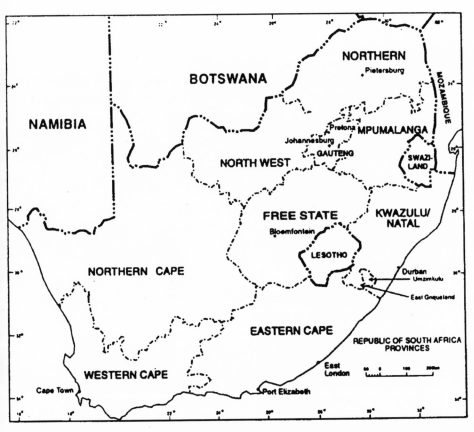

The New Provincial Map of South Africa

1

A Comprehensive History of the South African Struggle

It has been about a decade since the racist white minority regime of South Africa unbanned major political organizations such as the African National Congress (ANC), the Pan-Africanist Congress of Azania (PAC), and the South African Communist Party (SACP), and released ANC leader Nelson Mandela and other ANC members from prison. It has been 347 years since the first European intruders landed on South African shores, marking the beginning of one of the longest colonial eras of recent history.[1] South African history, notwithstanding its distinctive colonial past, must be considered within the purview of African history, since regardless of its Westernized outlook, it is still an intrinsic part of Africa.

This chapter will furnish a detailed and comprehensive history of the various trajectories of Black resistance in South Africa in order to shed light on the current wave of changes observed in post-apartheid South Africa, including the watershed events that led to the negotiated settlement and the first post-apartheid national elections of 1994. Such a delineated history needs recounting to further explain the tenacity of colonial occupation and racial capitalism entrenched in modern-day South Africa.

From the perspective of many among the oppressed Black majority, a radical critical social and historical analysis of the South African situation reveals that the South African entity represents a combination of European settler-colonialism and racial capitalism within the context of the Western European subjugation of Africa. Apartheid connotes a socioeconomic system of settler-colonialism where racism is the ruling ideology and monopolistic capitalism is the economic philosophy.[2] Settler-colonialism signifies a situation in which Europeans initially colonized the indigenous populace through wars of attrition, expropriated the lands of the latter through military conquest, and later decided to settle in the country and establish European-controlled republics. In the South African case, this process transpired following the discovery of diamonds in 1867 and gold in the Witwatersrand region in 1886. (This colonial process follows similar patterns of European settler-colonialism extant in the Americas, Australia, New Zealand [called Ao Te Roa by the indigenous Maoris], and Palestine.) Apartheid must therefore be perceived in terms of its socioeconomic underpinnings, with race functioning as the visible ideological mechanism for the pursuit of capital accumulation via a system of the exploitation of forced and cheap Black labor. Bernard Magubane, a South African political economist, contends:

> While colonialism has an ancient history, the colonialism of the last five centuries is closely associated with the birth and maturation of the capitalist socio-economic system. The pursuit and acquisition of colonies accompanied the mercantile revolution and the founding of capitalism. To study the development of capitalism is thus the best way to study race inequality, for to do so places socio-economic relationships at the heart of the problem, and shows how underdevelopment and racial inequalities developed together.[3]

Analysis depicting South Africa as a "settler-colonial" republic is not popular in the Western world, and understandably so, because it delegitimates the right of European colonial settlers to rightful occupation of indigenous peoples' lands and challenges the fundamental right of such regimes and populations to determine the destiny of indigenous peoples. Patently, too, this kind of perspective asserts the primary right of self-determination of indigenous peoples based on their own histories, cultures, and experiences. In the South African context, this implies the refusal by indigenous Black people to recognize the occupation of their country as legitimate, though the Western capitalist world demands such acceptance as bona fide and normative nation-state reality.

This view of history warrants some form of explication because it certainly does not accord with the mainstream view of African history, or global history, for that matter, held by both bourgeois and even some left-leaning Eurocentric historians. The question of perspective in most historical narratives of colonized regions such as South Africa is crucial in the discourse on contemporary socioeconomic analysis. Histories of most places in the world, particularly histories of those continents and areas where Europeans invaded and violently conquered the indigenous populations through protracted wars of attrition, forcing the subjugated populations to work for the colonizers, have generally been written by white hands, from the vantage point of the settler-colonialists. The American continent, Africa, Australia, and New Zealand are places that come to mind.[4]

With regard to South Africa, historical discussions have almost without exception viewed the intrusion of Europeans into Southern African shores as a given, as an inevitable course of natural historical evolution. From the vantage point of the indigenous African population, however, the presence and invasion of Southern

Africa constituted a brutal fracturing of the independent historical evolution of the region.

The Indigenous African Struggle Against Colonialism and Black Working-Class Resistance to Industrial Capitalism

Early Years

In examining the history of the Black resistance struggle in South Africa, one discovers that it has always been a resilient formation in the South African experience. The earliest indigenous African cultures extant in the southwestern tip of Southern Africa were those of the San (or Abathwa) and the Khoi Khoi. The Abathwa were hunter-gatherers whose history reaches to 60,000 years ago. Their political organization was rudimentary, as W. M. Tsotsi notes.[5]

The Khoi Khoi were pastoral people who depended on cattle and sheep farming for subsistence. They resided on the land from the "Keiskama River to the Cape Point northward along the Atlantic coast past the Olifants River."[6] Unlike the Abathwa, the Khoi Khoi accumulated cattle and sheep and traded with their neighbors, the Nguni, who lived in the eastern region of Southern Africa, from Delagoa Bay in the North to Algoa Bay in the South. The roots of the Khoi Khoi are most probably in contemporary Botswana, and they are believed to have been present in the region for at least a thousand years prior to the Portuguese invasion in 1488.[7] They are credited for introducing pottery and sheep farming to the Cape anywhere between 200 and 100 B.C.E. The Khoi Khoi were responsible for initiating Southern Africa's first indigenous resistance movement, repelling the attempted Portuguese intrusion by killing many of the marauders, including their leader Viceroy Francisco D'Almeida, in 1510.[8] Although the Portuguese were armed with swords, lances, and crossbows and numbered 150 men, the Khoi Khoi were successful in staving off the Portuguese invasion, and after the exchange, sixty-five invaders lay dead.[9] The myth that Africans in Southern Africa volitionally gave their lands to Europeans is shattered with the recounting of such incidents.

The Sotho were another adjacent indigenous nation that practiced cattle farming and agriculture, occupying the area north of the Orange and Vaal Rivers. They engaged in iron smelting and mined copper and tin at Dithakong in the Western Cape. Iron smelting sites have been found in the Northern Transvaal dating to the third century and in the Eastern Transvaal, traced to the fifth and sixth centuries.[10]

These diverse indigenous groups lived in relative propinquity to each other, and though there were cases of serfdom of the Abathwa and the Khoi Khoi, there were frequent intermarriages among the various communities and even mutual linguistic influences.[11] The Nguni groups practiced cattle herding and the cultivation of crops like sorghum, calabashes, beans, coco yams, and groundnuts, and the Sotho grew hemp and tobacco, resulting in a thriving trade in the region. This emergence of commercial activity eventually led to a gender division of labor between men and women, with women cultivating crops and men farming with cattle and hunting. Although agriculturalism led to forms of early accumulation by the Nguni, at no time was land isolated for privatized individual use; families had usufruct rights within a system of communal land tenure.

It was this ethos of indigenous African social stability and evolution that European colonizers invaded, establishing a legacy of genocide and systems of exploitation that Africa had never witnessed in its vast, far-reaching history. In 1652, when the Dutch East India Company initially decided to establish a "refreshment" station at the Cape of Good Hope in South Africa as a halfway point between Europe and the East Indies, its representatives were allotted the status of "temporary guests" by the indigenous people, in a manner that would similarly apply to the hypothetical situation of Africans attempting to construct a trade mission station in Liverpool, England. On realizing the fertility of the land, the beauty of the landscape, and the vastness of the natural resources, the Dutch (and later the British) decided that they should own South Africa, setting these European groups in conflict with the indigenous Black population. The Black people considered Azania (South Africa) their national homeland since this was the land they had lived on and with for countless generations, the same land where their historical ancestors had been buried for centuries.

The Dutch East India Company represented the first major intruders, led by Jan Van Riebeeck, who referred to the Khoi Khoi as "black stinking dogs" and "dull, stupid and odorous," akin to the way Christopher Columbus disparagingly referred to the indigenous peoples of the Caribbean and the Americas when he first encountered them.[12] This nascent anti-Black racism impelled Van Riebeeck to introduce Black slaves from Guinea and Angola into South Africa as early as 1658. The Dutch colonizers attempted to enslave the Khoi Khoi and the Abathwa, but a smallpox epidemic erupted in 1713, which, coupled with the subsequent actions by Dutch military commandos, resulted in the virtual extermination of these indigenous African nations, much as Native peoples in the Americas were almost totally eliminated. The colonials then imported slaves, many of whom were skilled artisans, from Madagascar, East Africa, Delagoa Bay, the Bay of Bengal, Indonesia, and Malaya.

From the time of the earliest European intrusion, the Khoi Khoi resisted, from 1658 to 1677 defending the sovereignty of their land under the courageous leadership of their leader, Gounema. The Khoi Khoi struggled to repossess their traditional grazing grounds in the Liesbeeck Valley, but the Dutch relentlessly encroached on all of the best land. Van Riebeeck contended that "if their (the Khoi Khoi) lands were restored there would not be enough grazing for both nations," referring to the Dutch colonizers.[13] The Khoi Khoi realized that the European invaders were land greedy, and following decades of fierce military defense, they were physically exhausted as an indigenous nation. Van Riebeeck stolidly declared that the land had been taken from the Khoi Khoi through war and was now the possession of the

Dutch, totally rejecting the Khoi Khoi's claim to the land as the first and indigenous inhabitants. This war over land would continue for centuries, well into the twentieth century, as the European colonizers who defeated the indigenous African people forcefully imposed their illegitimate acquisition of the Southern African land on the subjugated Black majority via the apartheid system. At the heart of the Black struggle against white domination in South Africa is the struggle to regain the land that was stolen by European invaders.

The Dutch launched a series of wars in the seventeenth and eighteenth centuries against the Xhosa-speaking people of the Cape. How ironic it is to describe these colonial wars as "Xhosa Wars" when it was actually the Dutch who declared war on the Africans! The Dutch were committed to eliminating the indigenous people because they viewed them as obstacles to their frenzied obsession with acquiring the land. In 1850 and 1877–1878, Xhosa communities like the Ngqika, Gcaleka, and Thembu, together with Griqua communities in the Cape, launched attacks against the British and the Dutch, which in effect were indigenous wars of resistance for the preservation of African lands and the right to self-governance in the wake of European colonial violence and slavery.[14]

These African communities strongly protested against inhumane and repressive anti-proletarian legislation such as the Masters and Servants Act of 1856, and though not successful in getting such colonial laws repealed, they were able to gain some limited reforms that protected servants employed on colonial farms. Saul Solomon was one such working-class resister who campaigned vigorously for the defense of Black workers' rights at a time when the only law governing Africans was that of legalized European colonial terrorism.[15]

The Dutch endeavored to force the Xhosa nation off their land at the end of the seventeenth century, particularly in the area on the west bank of the Fish River and all the way to the Gamtoos River, but were unsuccessful. The Xhosa communities resisted, fearlessly protecting their ancestral lands. It is worth noting that the Xhosa nation and the Dutch coexisted as neighbors for fifty years, with only three conflicts, a fact of which apartheid's architects were very much aware.[16]

With regard to the Abathwa, they were hospitable to decent and civil passersby but rejected the intrusive presence of the Europeans who were bent on occupying their hunting grounds and water holes, which they viewed as ancestral. The Europeans dispersed their own herds of cattle that they had received through trade with the Khoi Khoi and Xhosa peoples, occupying traditional lands of the Abathwa. The result was open conflict between the indigenous Abathwa and the European colonizers.[17] The Dutch established commando bands, the likes of which would be employed two centuries later under apartheid rule, to attack Abathwa communities, using the assistance of co-opted Khoi Khoi individuals, to launch these war operations. These military incursions continued throughout the eighteenth century and lasted until 1827. Abathwa communities were decimated, and their children were enslaved and scattered. In 1774, a commando killed 500 Abathwa and captured 239. One Dutch commando unconscionably described having killed or captured 3,600

Abathwa over a six-year period, and another participated in violent attacks that ended up killing 2,600 people.[18]

In the Cape, the Dutch invaders were followed by the British colonizers, who formally annexed the Cape in 1806. The Dutch subsequently left the Cape in 1836 to penetrate the Southern African hinterland and encountered the Zulu nation in the North, under the rule of Dingaan, the brother of the great king Shaka.

Shaka had consolidated a formidable Zulu kingdom and, through distinguished military prowess and intellectual genius, was able to establish a Zulu empire from 1818 to 1828. Smaller ethnic groups living in the vicinity of Zululand were forced to be assimilated into Shaka's expanding empire. Shaka invented the small spear, making the weapon more effective in internecine warfare, solidified earnest discipline within his army, and transformed the sociopolitical terrain of Zulu society through a more centralized ruling and administrative structure. The military establishment of the Zulu empire under Shaka had dire consequences for smaller groups in the region such as the Fingo, resulting in a massive dispersal of thousands of people, in what came to be known as the Mfecane.

In a repudiation of Eurocentric versions of Zulu national history that attempt to associate Shaka's monarchical emergence with reaction to European colonization of the area, it is important to consider ecological, climactic, and social factors in the rising of the Zulu kingdom. James Gump, for instance, argues that the Madhlatule famine, the increasing population, and the exhaustion of natural resources such as cereals may have functioned as catalysts to the Mfecane phenomenon, since material conditions in the early 1800s were conducive to the geopolitics of territorial and national expansion.[19] However, the reality is that the British did conduct slave raids into the interior, because England had outlawed the slave trade in 1807. The British needed to conceal their own slaving operations and thus construed the Zulu expansion as a decoy.[20] In 1828, for instance, British colonial leader Henry Somerset penetrated the Transkei and conducted massive slave raids, claiming to defend the Xhosa ruling establishments against Shaka's onslaught. The increase of the formation of centralized states such as Shaka's Zulu kingdom and Moshoeshoe's Sotho kingdom may have been in response to the slave raids and practices of plunder by the Portuguese at Delagoa Bay and the British at Port Natal and the Cape.

The outcome of the Mfecane for the various neighboring smaller ethnic groups and clans was retreat to the Eastern Cape. One of the most important Xhosa leaders to resist repeatedly deceptive British colonialism in the Cape was Maqoma, whose life history has been distorted by Eurocentric historians who describe him as a "cattle-rustler," a bloody "raider," and an alcoholic. These characterizations of Maqoma—and indeed of most African resisters of colonialism and European frontier expansion—do not do justice to the heroic role that he played in defending his nation against the savage forced evictions of Xhosa people from their ancestral lands, such as during the Sixth Frontier War of 1834–1835.[21] During the "War of Mlanjeni," which lasted from 1850 to 1853, Maqoma launched a massive guerrilla campaign through the hills, mountains, and forests of the Waterkloof area of the Cape, confounding the British colonial commandos. He was finally subjugated by British military scorched-earth policies that decimated Xhosa communities and was convicted by a British martial court; he was exiled to Robben Island and released, but was then returned to die there in 1873.

This wave of resistance by the Zulu nation continued. In 1835 and again in 1879, notwithstanding formidable military odds, under the adept leadership of Cetshwayo kaMpande, a unified army of 30,000 defeated the most powerful colonial army in the world at the time, the British, at the Battle of Isandlwana.[22] The British invasion of the Zululand-Transvaal region, which had intended the destruction and fragmentation of the Zulu kingdom, ironically effectuated the unification of the Zulu nation in its struggle for preservation of ancestral lands and independence.

Black people had no other option but to defend their territorial sovereignty, for it meant their very national survival. After two and a half centuries, the violence of European firepower (gunpowder through rifles and cannons, which ironically originated in Africa and was taken to Europe)[23] was finally able to overcome the military strategies of the African armies, led by leaders like Shaka and Cetshwayo. The African nations fought courageously and relentlessly in defense of their motherland. Had the Africans been able to resort to similar methods of warfare through firepower, the outcome of the conflict would most probably have been quite different.

Africans were conquered in their own land by an alien and acquisitive people, a conquest that was never accepted as morally predicated because of its violation of national independence and territorial sovereignty that all "civilized" nations upheld. The stronger European armies had proved "mightier" than the Africans, using weapons of self-defense. Yet this colonial conquest was neither moral nor just. Europeans did not and do not possess the right to claim South Africa or any country originally populated by Black—or brown or red people, for that matter—as theirs, ostensibly on the grounds that they had won the battle against the Africans. The Darwinian precept that claims that the strongest (the colonizers) survive and the Platonic logic that avers that this group deserves to rule over the weakest (the colonized) holds no moral sway because it subscribes to principles of barbarity and uncivilization. If we accept the dispossession of indigenous peoples and the subsequent genocide of such peoples in Africa, Asia, and the Americas, then we have no moral basis to establish internationally acceptable principles of law and justice. There must be room for discussion of this point of morality and justice in analyzing African and other indigenous political histories and economies.

In examining the contours of the South African historical narrative from the viewpoint of the resistance of the indigenous Black people, two salient points stand in sharp relief to any other, the consequences and contradictions of which are manifest symptomatically in the socioeconomic structures currently extant in the country. The first of these is land dispossession, a culmination of two and a half centuries of European wars of dispossession of African ancestral lands, institutionalized with the Land Act of 1913, which formally declared that Africans could only own 7 percent of the total land area of South Africa. The second element is monopoly capitalism, dominated by the mining industry, which led to the cheap labor and migrant-labor system so notorious during the days of apartheid and still an intrinsic element of the fabric of South African society today.[24]

The "discoveries" by European colonial settlers of diamonds in 1867 and of gold in the Witwatersrand region in 1886 were the catalytic events that led to the decision by the intruding population that devious ways somehow had to be found to facilitate expropriation of the precious minerals and to subjugate the indigenous people living on land that contained such precious stones so as to provide easy access to such wealth (since Western Europe was in the process of entering the industrial era, in which gold was a principal yardstick of financial and commercial value). Further, cheap labor was needed to extract the vast quantities of gold and diamonds from the earth, and the local African population was the closest to providing such labor. It was following the realization that the "discovery" of gold and diamonds would be a boon to the establishment of industrial capitalism that the Europeans were determined to possess South Africa at all cost. Frank Molteno confirms this decisive historical fact:

> With the extraction of minerals that were first discovered in 1867, capitalist production and industrialization in South Africa were established. But the capitalist mode of production depends upon the existence of a class of laborers who are "sufficiently" free of the means to production to force them ("freely") to offer their labor power as a commodity on the market. Such a class of "free" workers had to be brought into being. The mere territorial conquest of land does not in and of itself secure labor power.[25]

Subsequently, the British, upon militarily defeating the African people, forced the Africans off their ancestral lands and imposed a land tax for Africans residing on what illegally became "European-owned" land. Since the monetary economy as conceived by the British was alien to the African nations, Africans could only obtain the monies to pay such tax by working in the mines that were opened by the European invaders.

In the Gauteng (Transvaal) region where the Sotho-speaking nations historically lived, the people had produced vast surpluses in trading with other African nations between 1830 and 1865.[26] Moshoeshoe, the leader of the Sotho nation during that period, led a heroic battle of resistance in the vicious wars waged by the European colonists but was finally forced to concede defeat owing to the massive firepower of the British forces in 1869.[27] The result of this tragic defeat was the loss of productive grazing lands for Africans to the Europeans and the progressive impoverishment of the indigenous people, forcing many to seek work in the mines to earn additional wages since the indigenous economies were devastated. Realizing a progressive weakening of the Sotho-speaking people, the British imposed a home tax on the "Basutoland Protectorate" in 1870, precipitating a massive migrant-labor system that became one of the cornerstones of the apartheid economy and that still exists in the post-apartheid society of South Africa today. Molteno explains:

> The separation of the conquered from the conquered land demands full use of further coercive measures. The poll tax was one such measure. The Reserves too played a key role. Prior to conquest, production in South Africa had been predominantly according

to a communal mode which depended upon effectively unlimited land . . . Under the conditions of colonial conquest and an emergent capitalist mode of production, as population numbers rose, the contradiction sharpened and even more so once immediate producers were pushed into the market where they [had] to offer their sole remaining resource, their labor power, for sale, alongside those already deprived of any means of subsistence by the direct appropriation of their land.[28]

After African land had been expropriated, African labor was confiscated by the colonial-capitalist occupiers, so that ten years after the discovery of diamonds in 1867, over 300,000 African people were working in all of the European-controlled mines.[29] Similar patterns followed the discovery of gold in 1886. Stanley Greenberg verifies this assertion:

Capitalist development has gone a long way toward remaking South Africa. The Witwatersrand, whose mining camps housed perhaps 3,000 workers in 1887, encouraged in excess of 100,000 people by the turn of the century, and by the 1960s, included 2.5 million people and a sprawling manufacturing and mining complex, stretching from Witbank in the east to Krugersdorp in the west.[30]

Capitalist development in South Africa from the time of its inception in the 1880s implied the underdevelopment of African sociocultural patterns and indigenous economic modes of production. Black people were viewed as appendages of labor and as objects to be exploited for the benefit of white economic power and the monopoly of white social privilege. Walter Rodney, the late political scientist, explains:

In the Union of South Africa, African laborers worked deep underground under inhuman conditions which would not have been tolerated in Europe. Consequently, black South African workers recovered gold from deposits which elsewhere would be regarded as non-commercial. And yet it is the white section of the working class which received whatever benefits were available in terms of wages and salaries. Officials admitted that the mining companies could pay whites higher than miners in any other part of the world because of the superprofits made by paying blacks a mere pittance.

In the final analysis, the shareholders of the mining companies were the ones who benefited most of all. They remained in Europe and North America and collected fabulous dividends every year from the gold, diamonds, manganese, uranium, etc., which were brought out of the South African subsoil by African labor. For years, the capitalist press praised Southern Africa as an investment outlet returning superprofits on capital invested. From the very beginning of the scramble for Africa, huge fortunes were made from gold and diamonds in Southern Africa by people like Cecil Rhodes.[31]

This historical pattern was the consequence of colonial interventions by European nations into Africa and the Americas, resulting in the contemporary situation of the northern metropole countries constituting the core of the global economy, while relegating most of the southern countries to the periphery, South Africa being a classic case in point. It is unequivocal that the development of industrial capitalism lay at the heart of the rationale for the subsequent emergence of the apartheid system in the late 1940s:

> In the South African case and in others, the empirical evidence suggests, that in the words of Blumen, "the apparatus and operation introduced by industrialization invariably adjust and conform to the pattern of race relations in the given society." Furthermore, this perspective contends, white supremacy has not been eroded but reinforced by the process of South Africa's economic development, from which whites alone prosper. Any dissonances are more than compensated for by the enforcement which keeps African labor cheap. . . . the specific form of apartheid policy characteristic of the postwar years has to a great extent been shaped by the dictates of capitalist development in South Africa, in particular by the requisite of transition from an economy built around primary extractive industry to one in which manufacturing is the most dynamic sector.[32]

Yet Africans were never acquiescent in this onslaught of economic exploitation of Black labor by the demands of European industrial capitalism in South Africa, even in the aftermath of military subjugation by the colonial forces. Desertion by Black miners in the Transvaal was rampant as early as 1869, shortly after diamonds were first discovered in Hopetown in 1867, as Black workers sought to "secure better wages and living conditions for themselves."[33] In 1889, the first strike against European ownership of the gold and diamond mines was organized by Black workers, resulting in six workers being killed by police and over 300 fired.[34] Black men were compelled to work in the European-run mines and factories out of a sheer need to survive and support their families, which is still a historical trait of Black working-class culture today.

Mining (together with agriculture) represented the largest employment sector for Black people, solely driven by cheap labor. Wages paid to Black workers in other industries such as manufacturing were always determined according to the rate paid to mine workers. Wage gaps between industry and mining could not be too large, for that would have implied a shortage of labor for the mines, since higher wages in industry would attract Black workers away from the desperately worker-needy mines.[35]

Indigenous resistance also featured among the African Christian churches, which sought autonomy from white ecclesial formations. Nehemiah Tile established the Tembu Church in 1884; Mangena Mokone founded the Ethiopian Church in 1892, and subsequently, with James Dwane, forged the formation of the African Methodist Episcopal Church under the inspiration of African American bishop Henry M. Turner. Pambani Mzimba split from the Presbyterian Free Church of Scotland Mission and founded the Bantu Presbyterian Church. At the core of these religious movements was the desire for political autonomy and religious independence, free from the paternalistic tutelage of European missionaries.[36] The symbolism of Ethiopia, derived from Psalm 68:31 in the Bible, was invoked and referred to Africa and its need for unity and independence from the stranglehold of European colonial subjugation.

The Early 1900s to 1960

Through the combined armies of the British and the Dutch colonial forces, with the British vying for superiority over their rival European counterparts, the African population was finally defeated by this onslaught of firepower and military might. The South African War of 1899–1902, saw the British emerge victorious as a colonial power in South Africa. Typical of colonial morality, both intruding imperialist forces wrestled over African land over which they basically possessed no moral or political right.

The same spirit of resistance demonstrated by Shaka in the early 1800s continued in the revolt led by Bambhata, the leader of a small community near Greytown in the Zululand region in the early 1900s.[37] Bambhata attacked the British colonizers for imposing a poll tax on Africans and surrogate collaborator leaders on the Zulu nation. After organizing a massive guerrilla campaign of attacking the colonial enemies and their African stooges and evading capture for a number of years, Bambhata retreated to Nkandla, where he was finally captured by British troops and killed, together with 500 of his soldiers. Bambhata's rebellion marked the last major uprising in the region until the Pondoland insurrection of the 1960s.

Determined to wrest South Africa away because of its vast holdings of gold and diamonds, the British declared that South Africa was a union in 1910, essentially a union of white privilege and domination.[38] May 31, 1910, will always be remembered among Blacks as the day on which white supremacy was formally established, when it was tacitly accepted by whites that South Africa was a white man's country.[39] Albert Luthuli, the Nobel Peace Prize winner in 1960 and a stalwart of the anticolonial resistance, opined:

> The Act of Union virtually handed the whole of South Africa over to a minority of whites, lock, stock and barrel. English Natal sided on this issue quite happily with the Boer republics. Only in the Cape was there a little effective opposition, and this was lost in the compromises which had to be effected in order to achieve union. As far as whites were concerned the matter was settled: they became the exclusive owners of the new state. The members of other races who found themselves handed over officially, entirely without their consent, were the livestock which went with the estate, objects rather than subjects.
>
> I do not think that the African opposition to this situation brought about by the Act of Union is surprising, and I do not think that it has so far been extreme. We regard it as an act of piracy, in which the lives and strength of ten million Africans are part of the loot.[40]

The promulgation of overtly racist strictures such as the Native Labour Regulation Act; the Mines and Works Act of 1911 (the so-called Color Bar Act), which protected whites and further marginalized Blacks; and the Defense Act of 1912, which declared that only whites were full citizens, provoked the conceptualization of a unifying Black organization that would champion the cause of Black rights and foster unity among dispersed and disparate African communities.[41]

In 1912, the African National Congress was formed under the adept leadership of such individuals as John Dube, Pixley ka Isaka Seme, A. P. Xuma, and Selby Msimang, together with other women and men who envisioned a racial restructuring of the existing society. Originally called the South African Native National Congress, it changed its name to the African National Congress in 1923. The ANC sought the abolition of white monopoly of land possession and racist domination by the European minority, aspiring toward participatory, nonracial democracy. It had its roots in the African nationalist organization Imbumba Yama Afrika, which was formed in the Eastern Cape in 1882 and published the first African newspaper in 1884.[42] It became involved in wide-scale national protest against the repressive and racist legislation that emerged after 1912, particularly with the passing of the Land Act of 1913, a colonial law that confined Africans to owning 7 percent of the total land area of South Africa, which was then extended to 13 percent of the land by the Native Trust and Land Act of 1936.[43] Solomon Plaatje, the first secretary of the South African Native National Congress, described the essence of the Land Act of 1913 as follows: "Awakening on Friday morning, June 20, 1913, the South African native found himself, not actually a slave, but a pariah in the land of his birth."[44]

Parceling out the most unproductive lands to Blacks and the most fertile lands to whites was part of a genocidal policy conceived by the colonial state to force the Black residents of the 7 percent land area to kill themselves through starvation. In reflecting on the work done by W. M. Macmillan in *Complex South,* in which statistical surveys were conducted to understand the connection among land-utilization, land-ownership, and land-occupation, and the relationship of land and labor, Ralph Horwitz writes:

> The encouragement given by land-acquisition to basic environmental factors of soil and water made extensive agriculture that entrenched white tradition; the compulsion exerted by white land-acquisition on the Africans was to force the adoption of such intensive, non-traditional, agricultural practice within the limited tribal territories so as to result in a degree of malnutrition that sometimes approximated starvation.[45]

The effect of the land-dispossession policy was genocide, subsequently perpetuated in many rural Bantustan areas, where up to half of the Black children under five would suffer from malnutrition and starvation throughout the 1970s and 1980s.

In 1914, a delegation consisting of John Dube, the first chairperson of the ANC, Solomon Plaatje, A. Msane, T. M. Mapikela, and Dr. W. B. Rubusana traveled to London to appeal to the British government to repeal the British government law, a request that was refused, as was the ANC delegation request to the local white authorities in 1912.[46] The ANC's objective was the circumventing of ethnically based strategies of challenging white domination, and to that end the ANC averred, "We [the African people] are one."[47] Leaders like Pixley ka Isaka Seme contended that divisions among Africans had to be dissolved to attain justice and equal rights for Africans:

The demon of racialism, the aberrations of the Xhosa-Fingo feud, the animosity that exists between the Zulus and the Tongas, between the Basutos and every other nation must be buried and forgotten; it has shed among us sufficient blood! We are one people. These divisions, these jealousies, are the cause of all our woes and all of our backwardness and ignorance.[48]

Initially, the ANC was led by and catered to a small class of the African bourgeoisie, consisting of intellectuals and traditional chieftains. It subscribed to an appeals approach in pursuing Black rights and "had *not* attempted to wrest power from whites."[49] It was this bourgeois element that conditioned the moderate accommodationist stance of the ANC, combined with the supplicatory element of traditional African leadership character, according to Francis Meli.[50] The ANC embarked on a campaign of limited resistance, launching peaceful protests against such laws as passbooks for women and discriminatory legislation restricting Africans.[51]

World War I compelled the beginning of changes in resistance strategy, when the ANC learned that there was no value in placing its lot with Britain. The imperialist power continued to abrogate Black rights, even after Africans had died in the war on behalf of Britain. In 1918, over 100,000 mine workers went on strike for livable wages. In 1919, over 700 ANC members and associates were arrested for protesting against the notorious passbook laws, sparked by the ANC membership in the Transvaal and signifying a more radical outlook than the organization's usually accommodationist stance.[52]

In October 1920, a riot broke out in Port Elizabeth after an African trade union leader was arrested, and other Africans rallied to his support. Twenty-three Africans and one white person were killed by police and white vigilantes.[53] The ANC supported massive worker strikes, as with the dockworkers' strike organized by the Industrial and Commercial Workers Union (ICU) in 1919 under the leadership of Clements Kadalie, and the mine workers strike in 1921. The ICU grew in membership to about 86,000 in 1928, under the influence of Thomas Mbeki in the Transvaal and A.W.G. Champion in Kwazulu-Natal. This trade union movement represented the first mass organization of Black workers of its kind. Unfortunately, due to financial problems faced by Kadalie, conflicts between Kadalie and Champion over the issue of control over Kwazulu-Natal, and Champion's raising the question of undue Communist influence in the ICU, the movement gradually dissipated.[54]

In 1921, a massacre of 163 Africans, members of the Israelites, a millenarian religious sect founded by Enoch Mgijima, occurred at Bulhoek in the Cape, as the ruthless white authorities attempted to suppress the group that had refused to move from indigenous lands demanded by the colonials.[55] The confluence of religious sentiment and political aspirations was evident in this tragic account of indigenous resistance, a scenario that the ruling regime was determined to erase.

Interestingly, the ANC described itself as a "Pan-African Association" in 1919, heralding the organization's incipient consciousness of the need for Pan-African unity across colonial boundaries imposed on the continent. In 1920, it appealed to the ostensible philanthropic ideals of freedom and democracy espoused by Western Europeans, dispatching a delegation to the League of Nations in Versailles, where the organization called for international intervention in the ending of white domination of South Africa, but to no avail. The ANC had always sought to function within the framework of the "Western European democratic system" that existed in South Africa, anxiously hoping that the ideals of "democracy" that the British espoused in their homeland would also eventually be practiced in Africa. The organization had yearned for an extension of the white-dominated parliamentary system that solely favored whites and anticipated that it could be modified and expanded so that it would come to include Black people. This historically accommodationist posture of the ANC vis-à-vis the apartheid system has cost the organization dearly because it was bound to progressively compromise its objectives of liberation, evidenced in the structures and policies of the current post-apartheid government.

A landmark mine workers' strike of 1920 involved 70,000 workers and lasted twelve days. The Chamber of Mines reported that the strike had paralyzed the industry. It was sparked by two workers who had been arrested for moving from room to room in their hostel, urging workers to strike for higher pay. The next day, 2,500 workers went on strike, demanding the reinstatement of the two workers, an increase of 3 shillings a day, and other improvements in the living and working conditions. For six continuous days, 30,000 mine workers went on strike. Twenty-one of the 35 mines were brought to a standstill. The mining management ordered the workers to return to work; they refused, prompting the management to bring in the army. Three Black miners were killed and forty were injured.[56] In 1921, over 40,000 mine workers struck for more humanitarian conditions and workers' rights in the mining industry—forty people were shot dead by the armed forces.[57] Yet the struggle continued.

In 1922, white miners rioted to protest the en masse influx of Africans in the mining industry, a move supported by the South African Communist Party. The strikers "demonstrated under the slogan of a 'white South Africa' and sang the 'Red Flag.'"[58] At the end of the strike, fifty-six Africans had been killed. The white regime responded differently than in its reaction to Black protest. Steps were taken to address white workers' needs, such as the passing of the Mine Works Act of 1926 (also known as the Color Bar Act) to curtail the hiring of Africans in certain sectors of the mining industry. Many African miners were dislocated and replaced by whites. The Communist Party, formed in 1921, had injudiciously defended this discriminatory act and failed to acknowledge the decisive contours of race within structures of class in South Africa, placing "skin color above class."[59]

The racism of the white establishment toward Black workers and the justification of the ideology of white supremacy were classically illustrated in the 1925 Report of the Economic and Wage Commission by "Native commissioners" Mills, Clay, and Martin, as they prepared to construct labor and economic policies for Blacks, an extract of which reads as follows:

If the natives are this dependent on European industry for employment, equally European industry is dependent on the native population for a large part of its labour supply. And not only industry, mining, transport, and agriculture; domestic house-keeping has come to be organized on a basis of native labour. We have seen in our analysis of South African wages that the relatively high wages of white artizans are due to and dependent on the employment of large numbers of unskilled native labourers; and in this the artizan is typical of the whole White community, who are enabled to maintain a standard of life approximating rather to that of America than to that of Europe, in a country that is poorer than most of the countries of Western Europe, solely because they have at their disposal these masses of docile, low-paid native labourers.[60]

This statement captures the quintessence of the oppression of Black people by whites in South Africa and explains the reason that white workers have consistently opposed the interests of Black workers and maintained Blacks in unskilled positions: so that their own income levels would remain high. I. B. Tabata, a former president of the Unity Movement of South Africa and the African People's Democratic Union of Southern Africa, cogently explained this phenomenon when he described how the white worker was a "junior beneficiary" of the Anglo-Boer financial and commercial raping of South Africa's wealth:

As for the white worker, he becomes a junior beneficiary of the Company (the "Christian Trusteeship Proprietary Co. Ltd"), receiving a small portion of the dividends.

This fact explains the peculiar position and outlook of the White worker in South Africa. Like his counterpart in ancient history, the Roman proletariat, he lives at the expense of voiceless, rightless labour. In ancient Rome, the proletarian, though he was the lowest and poorest of its citizens, scorned manual labour as being beneath his dignity and lived off the crumbs of the rich slave-owner's table. He, too, battened on slave labour. Similarly, the White worker in South Africa today scorns manual labour as fit only for Black hands to perform. He rejects anything less than the white-collar job, which he guards as his exclusive prerogative by the divine right of his birth and for which he receives a wage out of all proportion to its value. This is possible because the White worker's bill of wages is subsidised from that of the Black worker. In other words, the White worker lives on the back of the voteless, oppressed Black masses. That is why the White worker takes his place on the side of the White boss and so enthusiastically supports the plans of his herrenvolk. That is why he is so willing to join the White bloc.[61]

This situation of white workers benefiting may be somewhat modified today, due to changing demographic and economic factors since the early 1920s, but nevertheless obtains in substance.

In 1927, Clements Kadalie, head of the Industrial and Commercial Workers Union, applied to register the ICU with the South African Trade Union Congress, a white organization.[62] He was turned down because the leadership of TUC was afraid of the anti-Black sentiments expressed by rank-and-file white workers. This decision indicates the historic deep racial divide that existed between Black and white workers, a legacy still evident today, since white workers have always viewed themselves as superior to Black workers solely on the basis of being white.

When the ICU began to fade in the late 1920s after the expulsion of numerous Communist leaders, several of these expelled workers began to organize workers in the laundry, baking, clothing, and furniture industries in the Reef. By the end of 1928, over 10,000 Black workers constituted the Non-European Trade Union Federation, unrecognized by the white state. The Native Clothing Workers Union was formed by Gana Makabeni, and the African Mine Workers Union was led by John Marks in 1941.[63]

Peaceful Black protest persisted, with coal miners in Kwazulu-Natal sporadically striking in 1927. In the late 1920s, Black women workers in the garment industry were involved in 100 strikes, two of which finally caused the clothing industry to collapse for that period. In 1928, dockworkers downed tools, and in June that year, 30,000 diamond workers remonstrated against reduction in earnings. In each instance, the police counteracted belligerently and with beatings. In 1929, many Africans were killed by gunfire in Durban after demanding the abolition of the passbook system.

As repression increased, the struggle saw a burgeoning of political groups: the African People's Organization (formed in 1902), the Non-European Unity Movement (founded in 1938), and the African Democratic Party. The ANC worked on a fraternal level with the Communist Party during that period, continued in collaborative relations between the two organizations today.

In 1934, the white "prime minister" J.M.B. Hertzog introduced certain segregation bills that were designed to annul the practice of African voter registration in the Cape by launching a quisling organization, the Natives Representative Council. Opposition from the ranks of the Black community was immediate and swift—nexuses of unity were built among major organizations such as the ANC, the Industrial and Commercial Workers Union, and the Communist Party. Such leaders as Selby Msimang, J. B. Marks, Clements Kadalie, and A. P. Xuma were instrumental in fostering this spirit of cohesiveness. The National Liberation Legacy, founded in 1935, and the Non-European United Front sought to bridge the chasms existing among the three oppressed groups, the Africans, the "Coloreds," and the "Indians." Unions from the various Black populations were united in the Council of Non-European Trade Unions, founded at a historic conference called by Moses Kotane, the new secretary of the Communist Party of South Africa in 1941.[64] In 1945, the "CNETU claimed 158,000 members from 119 unions—a figure not reached until the 1980s."[65] The formation of CNETU marked a watershed in Black worker politics.

Strikes among workers became widespread. Between 1942 and 1944, municipal workers, railway and dock laborers, milk distributors, and coal miners engaged in extensive strikes and stay-aways, resulting in scores being shot and killed and wounded. In a single incident near Pretoria, sixteen Africans were gunned down in cold blood by police. The War Measure Act No. 115 was invoked to proscribe the gathering of workers for either protest or strike reasons.

In 1943 and 1944, Africans boycotted municipal buses to voice their frustration at fare increases in Johannesburg. In 1944, 20,000 Africans occupied "white land" areas in Johannesburg, and in the same year, thousands of African teachers marched in protest for higher salaries. Again, in the same year, 20,000 Africans protested against the notorious pass laws in that city. Many protesters were injured by police gunfire. These events made Africans realize that exploitation along class lines coincided with that of race and that appealing to whites for change of their racist attitudes was futile and had deadly consequences, as had been evidenced by the bloodshed of resisters since the beginning of the century.

In August 1946, workers of the African Mine Workers Union went on strike, subsequently involving 76,000 out of a total workforce of 308,000. Twelve workers were killed and over 1,200 injured by 1,600 police who were called in. Two months before the strike, the Council of Non-European Trade Unions had resolved to support the mine workers financially and politically. However, because the leadership of the Black miners was under armed guard, it was cut off from communicating with the leadership of the CNETU, a contingency that workers' organizations need to anticipate in strike activity.[66] Between 1930 and 1950, African and "Indian" workers coalesced in a united wave of strikes that erupted following World War II, 304 in all, involving some 58,000 Black men and women from all segments of the Black population.

Later in 1946, the African Mine Workers Union staged its most powerful strike, with about 72,000 gold miners from the Witwatersrand, underlined by the demand for a livable minimum wage (10 shillings at that time). Between August 12 and 19, the mining industry, which engaged some 100,000 workers, was virtually paralyzed by the stoppage. The response of the mining industry was horrific: Two thousand armed police were summoned to quell the strike and to force the workers to return to work at bayonet point. The workers would not be intimidated and refused. The police opened fire, killing nine defenseless and unarmed strikers.[67] Black anger swelled throughout the country, and several organizations became involved in the vortex of the liberation struggle as a result of the strike, strengthening the position of radicals within the ANC who desired more militant resistance action.[68]

Under A. P. Xuma's leadership, the ANC began to move from its moderate reformist tendency toward a populist-based liberation position in the 1940s, culminating in the formation of the Congress Youth League in 1943. The inspiration for this emergence was Anton Lembede, the pioneer of Black nationalist thought in South Africa. It was the charisma and dynamism of Lembede extolling the principles of African pride, self-determination, and autochthonous rights that galvanized huge segments of Africans who were becoming disgruntled about the gradualist approach of the ANC. Lembede stressed African nationhood, unity, and the infusion of a new spirit of "Africanism," in contradistinction to the erstwhile ANC accommodationist policies concerning the legitimacy of the white occupation of South Africa.[69]

With the growing sentiment of the need for more confrontation strategies of resistance, the Youth League Program of 1949—the Program of Action—was launched. The call was for self-determination of the African nation and open defiance of the colonial status quo through mass boycotts, strikes, protests, and mass campaigns of civil disobedience. The philosophical bases for the Program of Action were the principles of self-determination, nationalist unity, African self-pride, and an outright rejection of white domination, coupled with mass action against racial injustice. This program sparked the Defiance Campaign, which consisted of open violation of racist and colonialist laws. Peaceful protest and agitation followed, with several thousand people getting arrested for defying the statutes of apartheid, introduced as law in 1948. Hundreds of Blacks were killed and wounded, propelling South Africa into a full-scale anticolonial conflict, last seen in the Bambhata-led war of the early 1900s. Sacrifice and boldness were the principal characteristics the ANC demanded of resisters to the oppressive system.

In 1946, the Natal Indian Congress (NIC) spearheaded a passive resistance offensive against the racist discriminatory statutes of the regime. Thousands participated in mass protests and land occupation of "white" areas. In that year, 50,000 miners in the Rand area engaged in a coordinated work stoppage, echoing the demand for more humanitarian wages. Several policemen swooped on the gathering, wielding clubs and coercing workers to return to work at gunpoint. Police used gunfire to disperse the striking workers. This series of protests by the NIC and the workers continued till 1949, building new bridges of solidarity between Africans and "Indians."

In 1948, the *apartheid* charter formally came into being, with the ascent to power by the white Nationalist Party, signifying an extension of the Hertzog segregation bills of the 1920s and 1930s. With the inception of formal apartheid, a myriad of racist and draconian laws came into being. The first was the Prohibition of the Mixed Marriages Act, which barred marriage between "Coloreds" and whites, since marriages between Africans and whites had already been outlawed in 1923, followed by the Immorality Act of 1950 which prohibited sexual union between Blacks and whites.

Many cornerstone apartheid laws were decreed in 1950, such as the Group Areas Act, which segregated people in residence, and the Population Registration Act, which classified all South Africans by race. The Suppression of Communism Act was also promulgated in 1950, with communism interpreted as any action rejecting the formal governmental policy of apartheid

which aims at bringing about any political, industrial, social or economic change within the Union by the promotion of disturbance or disorder, by unlawful acts or omissions, or means which include the promotion of disturbance or disorder . . . and which aims at the encouragement of feelings of hostility between the European and the Non-European races of the Union, the consequences of which are calculated to further the achievement of any object referred to earlier in the definition.[70]

This law would be repeatedly invoked to arrest, detain, and torture opponents of the apartheid system during the 1960s and 1970s, particularly activists from the Black Consciousness Movement and Black workers organizations.

In 1951, the Prevention of Illegal Squatters Act granted apartheid officials the right to remove "surplus natives," who were Blacks "living without permission on white land."[71] The Native Abolition of Passes Act, euphemistically named, tightened existing pass laws and forced all Africans to carry passbooks of identification, to be produced on demand; failure to produce them would result in arrest and imprisonment with hard-labor sentences. The Separate Representation of Voters Act of 1951 removed "Coloreds" from parliamentary representation, and the Native Laws Amendment Act of 1952 reinforced Hertzog's 1937 policy that Blacks "could be only temporary residents in white towns."[72] In 1953, the Bantu Education Act was passed as part of the policy of removing "the control of African education from the hands mainly of missionaries to the control of central government," essentially an institutionalized inferior educational system for Blacks.[73] The Bantu Self-Government Act of 1953 created the Bantustan program, which divided the African population by ethnic background and forced Africans to move out of allocated "white" areas and live in barren and arid reserves akin to the American Indian reservations. It is precisely because of the terror and totalitarianism of apartheid articulated here that it has been likened to Nazism.[74]

In 1952, the ANC launched its Program of Action, sparking the Defiance Campaign in tandem with the South African Indian Congress and the Colored People's Congress on June 26. The action involved deliberately breaking numerous apartheid laws such as those on passbooks, the Group Areas Act, the Suppression of Communism Act, the Separate Representation of Voters Act, and the Bantu Authorities Act by trained volunteers. These actions resulted in over 8,577 being arrested during 1952 and 1953.[75] The campaign was strongest in Port Elizabeth in the Black township of New Brighton and in parts of Kwazulu-Natal. The membership of the ANC swelled from 7,000 to 100,000 nationally. The Defiance Campaign drew the attention of the international community, and the United Nations established a formal commission of inquiry into apartheid.[76]

In 1955, the ANC attempted to present a united front in its resistance to the apartheid system, coalescing with organizations like the South African Colored People's Congress, the South African Indian Congress, and the Congress of White Democrats, forming the Congress Alliance. The result was the adoption of the Freedom Charter at Kliptown, Johannesburg, in 1955 by these various groups. Although the Freedom Charter espoused essential democratic principles, the preamble tacitly accepted the occupation of South Africa by European colonizers when it stated that "South African belongs to all who live in it, Black and white." It is the Freedom Charter upon which the current ANC majority government predicates its vision of a post-apartheid South Africa. The historical issue of land dispossession of South Africa by European colonial settlers that has now been elevated to the fore in the litany of Black grievances places the ANC in a politically embarrassing position. The organization has apparently accepted the immorality of the European theft of African lands in South Africa as a fait accompli and such historical occurrences cannot be fundamentally undone owing to the adherence of the ANC to the Kliptown Freedom Charter. The result has been the growing frustration among the Black masses, many of whom are even members of the ANC and who essentially put the new government into office, yet find themselves being overlooked in the crucial area of land dispossession and redress. The new government has already committed itself to maintaining the structures of land dispossession that accrued after 300 years of colonial conquest, with minor modifications.

Several people who helped organize the Kliptown meeting were arrested a few months later, and 156 of them were tried in the long trial famously known as the "Treason Trial." Throughout 1955 and 1956, thousands of people marched in the streets, vehemently denouncing passbooks and other racist statutes.

In 1959, the Pan-Africanist Congress of Azania was formed by Robert Mangaliso Sobukwe and P. K. Leballo, erstwhile members of the Congress Youth League. The formation of the PAC was the culmination of the cumulative frustrations among more radical-minded members of the Youth League who felt that the ANC had progressively compromised its objective of independence for the Africans and had capitulated to appeasing white liberals as well as catering to the needs of other oppressed apartheid groups such as the "Indians" and "Coloreds." The 1955 interracial Kliptown gathering where the Freedom Charter was adopted was viewed as a detraction from the goal of African self-determination. In April 1959, following ongoing tensions between the Transvaal branches of the ANC and the rest of the organization over the philosophy of broad-based, multiracial political struggle, the secessionist tendency won the day and the PAC was formally inaugurated at a meeting in Orlando Communal Hall, Johannesburg, with about 400 people in attendance. The clarion cry was "Africa for the Africans, Cape to Cairo, Morocco to Madagascar!" and "Imperialists Quit Africa!"[77] The slogan "Izwe Lethu!" (The land is ours!) was coined at this meeting and became the official identification slogan of the PAC. Robert Mangaliso Sobukwe was the most important thinker and articulator of the PAC tendency. At the Orlando meeting, he expressed the essential basis and goals of the PAC and its position on nationalism, Pan-Africanism, and race. Given the nexus of this discourse to the Pan-Africanist struggle, it is necessary to quote Sobukwe extensively here:

> In conclusion, I wish to state that the Africanists do not at all subscribe to the doctrine of South African exceptionalism. Our contention is that South Africa is an integral part of the indivisible whole that is Africa. She cannot solve her problems in isolation from and with utter disregard of the rest of the continent.
>
> Against multi-racialism, we have this objection, that the history of South Africa has fostered group prejudices and antagonisms, and if we have to maintain the same group exclusiveness, parading under the term of multi-racialism, we shall be transporting to the new Africa these very antagonisms and conflicts. Further, multi-racialism is in fact pandering to European bigotry and arrogance. It is a method of safeguarding white interests irrespective of population figures. In that sense it is a complete negation of democracy. To use the term "multi-racialism" implies that there are such basic differences between the various national groups here that the best course is to keep them permanently distinctive in a kind of democratic apartheid. That to us is racialism multiplied, which is what the term truly connotes.

We aim, politically, at government of the Africans by the Africans for Africans, with everybody who owes his [or her] only loyalty to Africa and who is prepared to accept the democratic rule of an African majority being regarded as an African. We guarantee no minority rights, because we think in terms of individuals, not groups.

Economically, we aim at the rapid extension of industrial development in order to alleviate pressure on the land which is what progress means in terms of modern society. We stand committed to a policy guaranteeing the most equitable distribution of wealth.

Socially, we aim at the full development of the human personality and a ruthless uprooting and outlawing of all forms or manifestations of the racial myth. To sum it up, we stand for an Africanist Socialist Democracy.

Here is a tree rooted in African soil, nourished with waters from the rivers of Africa. Come and sit under its shade and become, with us, leaves of the same branch and branches of the same tree.

Then Sons and Daughters of Africa, I declare this inaugural convention of the Africanists open! *Izwe Lethu!!*[78]

The PAC subscribed to the philosophy of African nationalism, analyzing the South African situation as one of colonialism, whereby Europeans had stolen the land from the African people by force and conquest, and the solution was to repossess this stolen land through armed struggle.[79]

Although the PAC later contradicted its position of thinking in terms of "individuals not groups" by excluding nonindigenous Africans from its membership (today it has members from all Black groups in South Africa, and some whites), the movement nevertheless represented a significant development in the history of Pan-African organization on the African continent. It came shortly after the independence of Ghana from British colonialism in 1957 and forged links with the freedom struggles waged by the likes of Kwame Nkrumah of Ghana and Sekou Toure of Guinea. Sobukwe dedicated his life to the intensification of the Azanian independence struggle, through leading the Sharpeville protest of 1960 and various other subsequent protests and marches during the 1960s. He was imprisoned on Robben Island for a few years and died a short time after being released in 1979.[80]

It is worth noting that when the Colored People's Congress dissolved in the mid-1960s, it was invited to join the Pan-Africanist Congress. A communiqué released at the time of unification of these organizations resolved to intensify the struggle against colonization and imperialism, declaring that

the liberation struggle is not merely aimed at the removal of white supremacy in order to attain a so-called liberal democracy which leaves untouched the super-exploitation of the people by monopoly capitalism. To be completely free and independent the South African revolution aims at the complete elimination of all imperial monopolies and preferences in the country and transfer of all imperial enterprises to the people as a whole.[81]

The question of monopoly capitalism, together with the centrality of white supremacy, loomed at the fore of Pan-African political organizing during that formative period of the PAC.   .

A History of Black Working-Class Women's Resistance

Black women have been involved in the struggle for liberation since the early part of the twentieth century, when they protested against passbooks in 1913.[82] In the 1920s, women were organized in the laundry, clothing, mattress, furniture, and baking industries. In the 1940s, they became prominent in organizing workers within the manufacturing sector. For example, Christine Okolo, a worker in the clothing industry in the Transvaal, launched a movement that went to the courts, where she insisted that she and other Black women be defined as "employees" under the Industrial Conciliation Act of 1924 (a law that only recognized white workers). In December 1944, the courts ruled in Christine Okolo's favor. The result was that wages were increased significantly, work hours reduced, and many other worker benefits allocated.[83] When apartheid officially came into being in 1948, however, these limited rights were stripped away since racism became institutionalized. Despite repressive conditions, women continued to resist.

The ANC Women's League was formed in 1943, and the Federation of South African Women was organized in 1954. The former essentially represented Black women, whereas the latter consisted of women from both Black and white communities. On April 17, 1954, over 150 women attended the first National Conference of Women in Johannesburg, where a Women's Charter was adopted, highlighting the situations of gender discrimination, disparities between rich and poor, conditions of women's labor, the need for education against sexism, and the cause of national liberation of the oppressed.[84] A strong call for women's rights, elimination of repressive laws, and justice for all was made at the conference.

In 1950, Elizabeth Mafeking, active in union organizing since 1941, led a strike of the African Food and Canning Workers Union. She was also involved in activities of the Federation of South African Women (FSAW) and the ANC. She was banned in 1959 because the Food and Canning Workers' strike was considered illegal, forcing her to flee to Lesotho and leave her husband and five children behind, except for the youngest. Elizabeth Mafeking's example of courage inspired many other Black women to resist, and by the 1960s, half of the secretary generals of the unions were headed by women. On the political front, FSAW urged a boycott of apartheid schools in 1954, but under threat of expulsion of children by the apartheid state, the boycott was abandoned.

On March 11, 1956, thousands of women gathered outside the white government's parliamentary buildings, demanding the revocation of the dreaded passbook law. Many women were beaten by police, yet the women remained defiant and sang in unison outside white prime minister J. G. Strijdom's office:

Strydom, wathint'abafazi   *Strijdom, You have*
Wathint'imbokodo   *touched the women*
Uzakufa   *You have struck a rock*
                                  You have dislodged a boulder
                                  You will die.[85]

On August 7, 1956, more than 20,000 women protested against the dreaded passbook law for African women and filed petitions to Strijdom. In November 1956, in Lichtenburg in the Western Transvaal, more than 1,000 women protested. When the police charged the gathering, women stoned the police in response. The police callously opened fire, killing two women. Similar protests involving significant numbers of women were organized in Sophiatown in May 1957. In January 1957, a bus boycott was organized principally by women to protest fare increases, with workers walking nine miles to and from work. Apartheid police stormed the boycotters, arresting thousands. However, the boycott worked in that the fare increases were rescinded. In Nelspruit, in the Eastern Transvaal, women attacked the car of the magistrate as a way of registering their opposition to carrying passes. Five women were arrested, prompting 300 women to march and demand their colleagues' release. Again, the police responded with brute force, wounding four with gunfire. The following day, another eight protesters were wounded. In Standerton in the Southeast Transvaal, women protested against passes and 914 were arrested. In Gopane Village in the Baphurutse area, many women burned their passbooks. Thirty-five were arrested. In June that year, 2,000 women stoned apartheid officials, and in July, more than 3,000 women confronted apartheid passbook authorities, resulting in the white officials retreating. In October, thousands of women protested against passbook laws and 2,000 were arrested. In Evaton, near Vereeniging, 2,000 women marched seven miles and handed in 10,000 protest forms with the Native commissioner.[86] Notwithstanding all this consistent and vigorous protest, over 3 million Black women were forced to accept carrying passbooks.

In 1957, uprisings emerged around various rural and semiurban areas of South Africa—Groot Marico in the Transvaal, Sekhukuneland, Witzieshoek, Kwazulu-Natal, and Pondoland. Women and the chieftains were involved in most instances; collaborators with the apartheid authorities were killed and forced to leave their positions, while these communities established their own democratic form of leadership and representation. Many Black leaders who refused to cooperate with white authorities in the relocation of their communities were exiled to isolated areas. The essential thrust of this phase of the liberation movement was the intensification of the struggle and the dissolution of all internal colonial structures imposed by the apartheid system. Black leaders handpicked by the regime were denounced and eliminated as a way of delegitimating the oppressor's institutions. The Bantu Authorities Act and the Promotion of Self-Government Act, so named by apartheid's rulers and specifying that Africans would be divided according to language grouping and forcibly relocated to reservation lands, precipitated this wave of rural peasant resistance.

In 1959, in Cato Manor outside Durban, over 2,000 women entered the state beer halls, destroyed apartheid-brewed beer, and chided men for not supporting women's brewed beer, resulting in several arrests by police. In that year, over 20,000 women participated in mass protests, with 1,000 arrested. In 1974, 374 strikes occurred, with 57,665 workers participating. As many as 841 workers were arrested, including many women.[87]

Women in the more radical Pan-Africanist Congress and Black Consciousness Movement also participated decisively in the Black liberation movement, with both organizations espousing Pan-Africanist revolutionary consciousness and the latter promoting socialist principles. At the historic Sharpeville massacre organized by the PAC on March 21, 1960, 40 of the 186 people wounded were women. Former ANC president and Nobel Peace Prize–winner Albert Luthuli asserted that women were the cornerstone of the Black struggle: "Among us Africans, the weight of the resistance has been greatly increased in the last few years by the emergence of our women. It may even be true that, had the women hung back, resistance still [might] have been faltering and uncertain.[88]

Several women emerged as formidable leaders in the Black Consciousness Movement, such as Thenjiwe Mtintso, a formerly banned journalist who is now serving in the new government; Asha Moodley, a leader of the Black People's Convention (BPC) who worked closely with founder Steve Biko; Nkosazana Dlamini, a former vice president of the South African Students Organization; and Sikose Mji, a leader of the Soweto student uprising of 1976, among numerous others. The first leader of the Black People's Convention, a major Black Consciousness grouping, was Winnie Kgware, a woman.[89] The formation of the Black Women's Federation (BWF) in 1975 was another extension of the Black Consciousness Movement, teaching gender awareness to young Black women and conducting literacy, nutrition, and health classes in both urban and rural areas. When it started to organize Black women in cottage industries and unions and to agitate for Black women's rights in areas of housing, the apartheid regime clamped down with draconian force, banning the organization on October 19, 1977, together with several other Black Consciousness formations.

Perhaps one of the most dynamic areas of women's resistance has been the activism of women in some of the most impoverished sectors of Black life, in the shanty slums of Crossroads outside Cape Town and in Alexandra Township outside Johannesburg. Crossroads has a population of over 700,000 people and was established as a result of Black women's refusal to be separated from their husbands, who had come to work in Cape Town from the rural parts of the Eastern Cape. Under apartheid laws, most urban areas of South Africa were decreed "white" areas, and Black men were permitted to enter the industrial cities solely as laborers for the white-owned economy. They were prohibited from bringing their wives with them. The spouses of many of these men defied apartheid authorities' demands to be returned to the Transkei in the Eastern Cape and formed the Crossroads Women's Movement. The women rallied around the slogan "We shall not move." In June 1978, over 200 women demonstrated at the Bantu Affairs Administration Board, airing their grievances at the flagrant injustices imposed against them. The authorities refused to listen and stormed Crossroads with bulldozers and armed cars, arresting 800 people in September and shooting three persons, one fatally. Still, the women persisted in their demand to be left alone. Finally, with protracted and determined resistance offered by the women and mounting international pressure, the rulers of apartheid agreed to leave the people of Crossroads alone.

Armed Struggle of the 1960s, 1970s, and 1980s

March 21, 1960, marked the watershed of South African liberation politics, when 69 Black people were killed and 188 wounded in Sharpeville, most shot in the back in cold blood after protesting in an anti-pass campaign organized by the PAC. There was an instantaneous international outcry at the massacre, and foreign governments protested against the horrific act, but to no avail. It was at that juncture that it dawned upon the Black population that "nonviolent" passive resistance was inapplicable in the South African context because of the savagery of brute force employed by the repressive apartheid regime in responding to peaceful protest. The level of intransigence on the side of apartheid compelled the ANC and the PAC to rethink their approach to resolving the South African problematic of oppression and colonial occupation.

The ANC, which had not endorsed the Sharpeville protest, called for a national strike in support of the mourning families. Strikes and protests grew widespread following the Sharpeville incident. On March 24, 1960, 2,000 men gave themselves up for arrest at Langa police station in Cape Town, and police refused to arrest them. On March 28, 50,000 people gathered in Langa to bury the dead, with speakers at the funeral demanding the abrogation of passbook laws, a £35 minimum wage, and the right to strike. The white authorities suspended passbook arrests temporarily. On March 30, the police entered Langa and Nyanga Townships and attacked and beat up PAC members. The result was a massive march to Cape Town, led by Philip Kgosana, the regional secretary of the PAC in the Western Cape region. Kgosana was subsequently arrested, together with other organizers of the PAC.[90] April 1960 saw outbursts of protests, marches, and strikes throughout Durban, Port Elizabeth, and Bloemfontein. On April 8, the apartheid parliament promulgated the Unlawful Organizations Act, declaring the ANC and the PAC and the activities of both organizations illegal.[91] The ANC initially refused to accept proscription and vowed to continue the struggle outside the country until freedom had been realized since it was ill-prepared for an underground operation.

The PAC decided to move to Maseru in the neighboring country of Lesotho and to execute its campaign of mass resistance from there. The apartheid regime did not hesitate to respond with the full weight of its draconian laws, pursuing ANC and

PAC activists outside the country, arresting those who were attempting to cross the border into South Africa from neighboring countries. Potlako Leballo, who endeavored at administering the underground operation from outside South Africa, was unsuccessful primarily because the governmental authorities in the adjacent states of Lesotho and Swaziland were essentially subordinate to the political and military might of white South Africa. The Basotho Army of Lesotho clamped down on PAC operatives shortly after the latter's move into Maseru and submitted documentation on PAC activists to the South African police.

Following the arrests of prominent ANC leaders like Walter Sisulu, Govan Mbeki, Raymond Mhlaba, Ahmed Kathrada, and five others at Rivonia, Johannesburg, in 1962, and the conviction of these men to Robben Island prison terms (and in the aftermath of the banishment of Nelson Mandela to Robben Island), the ANC realized that it had no option but to engage in a campaign of armed struggle from organizational nerve centers in the neighboring independent countries. At the Rivonia trial, Mandela explained the rationale for the underground level of armed struggle, essentially against property and not targeting civilians:

1. We believed that, as a result of government policy, violence by the African people had become inevitable and that unless a responsible leadership was given to control the feelings of our people there would be an outbreak of terrorism which would cause bitterness between the various races of the country.
2. We felt that without sabotage there would be no way open to the African people to succeed in their struggle against the principle of white supremacy. All other means of opposing this principle were closed by legislation.[92]

In 1962, a conference was held in Lobatse, Botswana, where ANC leaders from inside the country were reunited with those from outside the country, such as Oliver Tambo. The conference was a reflective analytical session on the domestic situation of repression, the affirmation of rejection of neocolonial strategies of sham independence planned for the Transkei by the white regime, and the echoing of the need for massive organization of the rural peasantry and the urban proletariat. The ANC convened a conference of its external mission in Tanzania, where it established a formal mission in 1962. These meetings resolved to initiate a sophisticated and covert exchange of strategies of leaders inside and outside the country so that the struggle against apartheid could be successfully waged from internal and external quarters.[93]

An important meeting of the ANC was held in Morogoro, Tanzania, in April 1969, where formal strategies of guerrilla warfare combined with internal mass worker resistance were proposed, with emphasis on a carefully thought out critical theory that was able to circumvent the designs of imperialism and reaction as manifest in the apartheid military regime. It was at the Morogoro meeting that the ANC began to plan for a protracted infiltration of ANC cadres into the country, trained and equipped in neighboring African countries and such socialist-bloc nations as the Soviet Union, East Germany, and Cuba. The objectives of attacking apartheid military and industrial targets were to demonstrate that the apartheid regime was indeed militarily vulnerable and to frustrate the tactical "reformist" strategies of the system that sought to legitimate its existence among Black people.

Both the ANC and PAC initiated a small-scale guerrilla war against the apartheid regime, the ANC with its military wing, Umkhonto We Sizwe, and the PAC with its armed wing, POQO. The 1970s and 1980s saw sporadic military assaults by Umkhonto We Sizwe involving bomb attacks on apartheid military installations, some industrial centers, and metropolitan streets. Generally, the guerrilla attacks endeavored at preserving human life wherever that was possible, while directing the thrust of its tactical warfare against the symbols of apartheid power. These attacks were relatively frequent, to the point that a blast occurred once every seventy-two hours.[94] In 1981, over fifty-five major guerrilla attacks against railway lines, military installations, apartheid institution offices, power stations, police stations, and commercial centers occurred.

During 1982, there were twenty-three incidents of bombings reported in various parts of the country, the most significant being the attacks on the President's Council Building in Cape Town on June 4 and the quadruple consecutive explosions at the Koeberg Nuclear Reactor on December 18, 1982, causing extensive damage to the facility and precipitating a moratorium on work at the plant. In 1983, the armed attacks continued, the major event being on May 19, when an ANC car bomb destroyed the air force headquarters in Pretoria, killing twenty people, including many white air force personnel. The bombardment was the deadliest and boldest in the history of the ANC's strategy of sporadic guerrilla warfare. Oliver Tambo, then acting president of the ANC, responded glumly after the attack: "Never again, we will never shed our blood again. We are tired of turning the other cheek. We have run out of cheeks."[95]

The apartheid regime retaliated through air force attacks on the neighboring states of Southern Africa, including Mozambique, Lesotho, Swaziland, Botswana, and Zimbabwe. At the most, though, the guerrilla campaigns waged by the ANC were symbolic in nature and hardly formidable enough to topple the military might of the apartheid machine.

The Black Consciousness Movement

In the early 1970s, the Black working-class movement in South Africa regained momentum after the brutal suppression of the Black liberation movement during the 1960s. Wildcat strikes by Black workers in Durban, Johannesburg, Cape Town, and Port Elizabeth in 1973 were the result of Black working-class cultural resistance and solidarity, aided by organizations of political-cultural resistance like the Black People's Convention, the South African Students Organization (SASO), and Black Community Programs.[96]

It was organizations such as these that came to constitute the Black Consciousness Movement. The Black Consciousness Movement represented a political formation that sought to unite Black people across ethnic and political lines, rejecting the dehumanization of the term "non-white" for all Blacks as defined by apartheid (since being white was equated with full humanity) and supplanting this sense of nonpersonhood with the nomenclature "Black." Steve Biko, the founder of the Black Consciousness Movement, defined the racial colonization of Blacks by whites as the principal problematic of South African life and proposed the Black Consciousness philosophy as the corrective. At a meeting of the South African Students Organization in 1971, a grouping that Biko inspired, he explicated the principles of Black Consciousness in elaborate and succinct terms, historically redefining what "Black" meant:

> We have in our policy manifesto defined Blacks as those who are by law and tradition, politically, economically and socially discriminated against as a group in the South African society and identifying themselves as a unit in the struggle towards the realization of their aspirations. This definition illustrates to us a number of things:
>
> 1. Being Black is not a matter of pigmentation—being Black is a reflection of a mental attitude.
> 2. Merely by describing yourself as Black you have started on a road to emancipation, you have committed yourself to fight against all forces that seek to use your Blackness as a stamp that marks you out as a subservient being. From the above observations therefore, we can see that the term Black is not necessarily all-inclusive, i.e., the fact that we are all not white does not necessarily mean that we are all Black. Non-whites do exist and will continue to exist for quite a long time. If one's aspiration is whiteness but his pigmentation makes attainment of this impossible, then that person is a non-white. Any man who calls a white man "Baas," any man who serves in the police force or Security Branch is ipso facto a non-white. Black people—real Black people—are those who can manage to hold their heads high in defiance rather than willingly surrender their souls to the white man.
>
> Briefly defined therefore, Black Consciousness is in essence the realization by the Black man [and woman] to rally together with his [or her] brothers [and sisters] around the cause of their operation—the Blackness of their skin—and to operate as a group to rid themselves of the shackles that bind them to perpetual servitude. It seeks to demonstrate the lie that Black is an aberration from the "normal" which is white.[97]

Biko underscored the preponderant fact of the totality of white power established in South Africa as a result of the colonization of Azania by Europeans, even though liberal whites attempted to obscure the material reality of the racial divide and obfuscate the issue of white power by calling for "color-free" and "race-free politics." His critique of white liberalism was scathing:

> So while we progressively lose ourselves in a world of colorlessness and amorphous common humanity, whites are deriving pleasure and security in entrenching white racism and further exploiting the minds and bodies of the unsuspecting Black masses. Their

agents are ever present amongst us, telling us that it is immoral to withdraw into a co-coon, that dialogue is the answer to our problem and that it is unfortunate that there is white racism in some quarters but you must understand that things are changing. These in fact are the greatest racists for they refuse to credit us with any intelligence to know what we want. Their intentions are obvious; they want to be barometers by which the rest of the white society can measure feelings in the Black world. This then is *what makes us* believe that white power presents itself as a totality not only provoking us but also controlling our response to the provocation. This is an important point to note because it is often missed by those who believe that there are a few good whites. Sure there are a few good whites just as much as there are a few bad Blacks. However what we are con-cerned here with is group attitudes and group policies. The exception does not make a lie of the rule—it merely substantiates it.[98]

Biko was cognizant of the dimensions of power that were being progressively and deliberately concealed by liberal whites. Contrary to many who simply viewed (and continue to view) Black Consciousness as a tactical deployment, Biko clearly articu-lated the objectives of Black Consciousness both as a philosophy and a political program:

One must immediately dispel the thought that Black Consciousness is merely a methodology or a means towards an end. What Black Consciousness seeks to do is pro-duce at the output end of the process real Black people who do not regard themselves as appendages to white society. This truth cannot be reversed. We do not need to apologize for this because it is true that the white systems have produced throughout the world a number of people who are not aware that they too are people. Our adherence to values that we set for ourselves can also not be reversed because it will always be a lie to accept white values as necessarily the best. The fact that a synthesis may be attained only relates to adherence to power politics. Someone somewhere along the line will be forced to ac-cept the truth and here we believe that ours is the truth.[99]

The Black Consciousness Movement was instrumental in advancing the levels of cultural resistance and defiance of the Black working class. It sought to bridge exist-ing chasms between Black workers and the Black student intelligentsia. It con-fronted the policy of ethnic fragmentation of Black people legislated by the colonial state and forcefully worked to override ethnic differences so that a powerfully new rejuvenated and dignified Black identity and culture could be forged. It organized workers, students, and community people in diverse geographic and ethnically dis-parate Black communities, teaching Black Consciousness and instructing the rejec-tion of oppressor-defined identities such as "Bantu," "Zulu," and "Colored."

The basis and potency of Black Consciousness resided in its emphasis on Black solidarity, in a response to the colonial process that viewed all Europeans regardless of ethnic background as "white" and all Blacks as distinctive and disparate "non-white" units. It was this sense of "whiteness" that transcended ethnic divisions

among Europeans that became associated with power and synonymous with privilege, establishing a firm bulwark of "white unity" that implied and institutionalized
"non-white disunity." This historical prerogative of definition of the oppressed by
whites in South Africa substantiates the assertion that whites played and continue to
play the roles of essential colonizers and oppressor in whatever capacity they functioned and function. Whites found and still find it extremely difficult to break with
their traditional identities of colonizer and oppressor, given that there was such privilege and power associated with both roles, making it impossible for them to engage
in untainted moral action that would involve a surrender of their oppressor identities and in so doing, an abdication of their preferential position in society. Most
whites viewed, and still view, the decision to refute white privilege as racial and class
suicide, compelling Biko to come to the painful realization that the task of Black liberation was not going to be executed by white liberals. Black cohesiveness was thus
an antidote to the pathology of indivisible white racist unity. Biko cogently explained the intrinsic exigency of Black solidarity as a prerequisite for liberation praxis
on the part of the Black oppressed:

> The importance of Black solidarity to the various segments of the Black community
> must not be understated. There have been in the past a lot of suggestions that there can
> be no viable unity amongst Blacks because they hold each other in contempt. Coloreds
> despise Africans because they [the former], by their proximity to the Africans, may lose
> the chances of assimilation into the white world. Africans despise Coloreds and Indians
> for a variety of reasons. Indians not only despise Africans but in many situations also ex-
> ploit the Africans in job and shop situations. All these stereotype attitudes have led to
> mountainous inter-group suspicions amongst the Blacks.
>
> What we should at all times look at is the fact that:
>
> 1. We are all oppressed by the same system.
> 2. That we are oppressed to varying degrees is a deliberate design to stratify us not
>   only socially but also in terms of aspirations.
> 3. Therefore it is to be expected that in terms of the enemy's plan there must be this
>   suspicion and that if we are committed to the problem of emancipation to the
>   same degree it is part of our duty to bring to our attention of the Black people
>   the deliberateness of the enemy's subjugation scheme.
> 4. That we should go on with our programme, attracting to it only committed peo-
>   ple and not just those eager to see an equitable distribution of groups amongst
>   our ranks. This is a game common amongst liberals. The one criterion that must
>   govern all our action is *commitment*. (italics mine)[100]

The Black Consciousness Movement fostered a new sense of cultural pride and
power, replacing the European colonial definitions of ethnicity among the various
Black communities in South Africa with a cohesive unified Black identity.[101] It mobilized Black working-class people by utilizing various elements of historical cultures
as contemporary communal edifices against the alienation and atomization wrought

by apartheid ethnicization and distortions of indigenous Black culture.[102] It infused Black working-class culture with an outlook that was unapologetically and uncompromisingly Black, African in its core, and revolutionary socialist in orientation.

On the first and final days of his testimony at the trial of nine leaders of the SASO-BPC in 1976, what came to be known as the "Trial of Black Consciousness," Steve Biko emphasized that the BPC was committed to bringing about a total change in South Africa "which is opposed to the Western Capitalist System."[103] The Black Consciousness philosophy was perspicacious in that not only did it elevate the need for Black solidarity and independence to a paramount principle but it also critiqued the superficiality of capitalist culture introduced to Azania by the colonialists. It was for this reason that integration was repudiated by the Black Consciousness Movement, eloquently articulated by Biko:

> The concept of integration, whose virtues are often extolled in white liberal circles, is full of unquestioned assumptions that embrace white values. It is a concept long defined by whites and never examined by Blacks. It is based on the assumption that all is well with the system apart from some degree of mismanagement by irrational conservatives at the top. Even the people who argue for integration often forget to veil it in its supposedly beautiful covering. They tell each other that, were it not for job reservation, there would be a beautiful market to exploit. They forget they are talking about people. They see Blacks as additional levers to some complicated industrial machines. This is white man's integration—an integration based on exploitative values. It is an integration in which Black will compete with Black, using each other as rungs up a step ladder leading them to white values. It is an integration in which the Black man and woman will have to prove himself and herself in terms of these values before meriting acceptance and ultimate assimilation, and in which the poor will grow poorer and the rich richer in a country where the poor have always been Black. We do not want to be reminded that it is we, the indigenous people, who are poor and exploited in the land of our birth. These are concepts which the Black Consciousness approach wishes to eradicate from the Black person's mind before our society is driven to chaos by irresponsible people from Coca-Cola and hamburger cultural backgrounds.[104]

It was for the advocacy of Black working-class political and cultural resistance that Steve Biko, the founder of the Black Consciousness Movement, paid the ultimate price on September 12, 1977, when he was brutally beaten to death by police. Black Consciousness was viewed as such a formidable liberation movement that it had to be crushed by the colonial authorities, primarily because it began to inculcate a new sense of *Africanity* and *Blackness* that refused to accept European definitions of African identity and personality and bestowed a profound spirit of fearlessness and cohesiveness among members of the Black working class that had last been witnessed during the Pan-Africanist Movement in the early 1920s and late 1950s. The effectiveness of the Black Consciousness Movement was that it began with a program of mental decolonization, emphasizing the psychic liberation of the oppressed Black mind in order to engage in transforming the materiality of Black oppression.[105]

The Black Consciousness Movement was the principal catalyst in propelling the South African revolutionary movement to its present phase, predicated on the principles of radical Black working-class culture.[106] White industrial bosses had sought to forge ethnic and social divisions by hiring nonstriking workers, a pervasive strategy of capitalism in South Africa. It even set up its own union, the Trade Union Council of South Africa, as a way of attempting to confuse Black workers about political objectives.

Between 1973 and 1975, thousands of miners went on strike, resulting in over 200 killed and 1,000 injured in clashes with police. In 1975, members of the Metal and Allied Workers Union organized a strike at Heinemann. The police were summoned and beat up stewards and union members. What is clear is that there has never been fair play between Black labor and white capital in South Africa, particularly in times of articulating worker grievances. Black Consciousness has functioned to philosophically remind Black workers of their allegiance to the Black working class and the fraudulence of capitalism.

In five months of 1981, fifty strikes occurred, involving 50,000 workers. Concomitantly, the Council of Unions of South Africa, a Black Consciousness grouping, grew by leaps and bounds from nine affiliates and 30,000 members in 1980 to twelve affiliates and 147,000 workers five years later. These were the Broom and Brush Workers Union with 1,000 members; the Building, Construction, and Allied Workers Union with 27,264 members; the Food Beverage Workers Union with 16,124 members; the National Union of Wine, Spirits, and Allied Worker Union with 5,000 members; the South African Chemical Workers Union with 30,000 members; the South African Laundry, Dry Cleaning, and Dyeing Workers Union with 4,771 members; the Steel, Engineering, and Allied Workers Union with 28, 927 members; the Transport and Allied Workers Union with 23,327 members; the United African Motor and Allied Workers Union with 10,873 members; the Vukani Black Guards and Allied Workers Union with 514 members; and the Textile Workers Union of the Transvaal with close to 1,000 members.[107]

In 1981 and 1982, strikes organized by Black mining unions were quite effective, with over 13,000 miners going on strike in July 1981, resulting in seven miners being killed and over R1 million (South African rands) lost in mining revenues (over $1 million in 1981 terms). In 1984, mine workers from the National Union of Miners forced the Chamber of Mines to listen to their demands. 50,000 mine workers went on strike against low wage increases and protested against unsafe working conditions, since 46,000 miners had died in mining accidents since 1900 and sixty continue to die each year. The result was the scrapping of job-restriction legislation that protected white miners passed in 1922. So, too, in the Cape in the 1980s, the Food and Canning Workers Union organized "Colored" and African workers in a demonstration of Black solidarity, resulting in prolonged strikes at the Fattis and Monis food company and the demand that fired striking workers be reinstated. When these tactics failed, the union organized a successful boycott of all Fattis and Monis products, and after seven months, all fired workers were reinstated. In another instance, the Western Province General Workers Union refused to handle meat from the Karoo Meat Exchange until fifty-nine striking workers were reinstated. The union called for a boycott of red meat that resulted in plummeting sales, forcing the management to rehire the fired workers. These successful strikes showed that Black unions working closely with the broader Black community could expedite achievement of objectives, an important principle for all worker organizations to practice.

In 1984, the Azanian Congress of Trade Unions was formed, consisting of the African Allied Workers Union, the Amalgamated Black Workers Union, the Black Electronic and Electrical Workers Union, the Black General Workers Union, the Hotel, Liquor, Catering, and Allied Workers Union, the National Union of Workers of South Africa, the Black Allied Mining and Construction Workers Union (BAMCWU), and the Insurance and Assurance Workers Union of South Africa (IAWUSA), the latter two each claiming a membership of 30,000 out of a total membership of 75,000. These unions waged continuous strikes through 1983 and 1984. In 1988, IAWUSA went on strike at Liberty Life, one of South Africa's largest commercial conglomerates, demanding union recognition. Similarly, BAMCWU struck at the Penge Asbestos mine in Northern Gauteng, involving 1,700 workers. BAMCWU demanded recognition of their union by industry and that workers be removed from working in asbestos-hazardous environments.

Today, the growing Azanian Workers Union (AZAWU), which predicates the struggle for revolutionary socialism on independent Black praxis, emphasizes that it is Black workers who will free all Black people, as opposed to the working class being freed by liberal or radical whites.[108] It has enfleshed Black Consciousness in its most crystallized form, resonating with the Black cry for resistance, "Black man [woman], you are on your own!"[109]

The Black Consciousness Movement thus assisted in the intensification of the politics of resistance on the part of the Black working class, by reclaiming and reappropriating the dynamism of indigenous African culture so that this historical culture could become mobilized in a manner that dauntlessly confronted the machinery of European colonialism and capitalism. Black working-class people were no longer embracing a colonially imposed identity that denigrated Blackness and emulated whiteness. Instead, Black people elevated their Blackness in raised clenched fists, became immersed in organic social organization and resistance, and proudly adorned a recreated Black identity that boldly embodied a resuscitated Black nation that fearlessly resolved that nothing would stand in the way of Black decolonization and liberation. The Black Consciousness Movement cannot be dismissed as a marginal antiwhite organization; rather, it must be credited for elevating the question of class in its assertion of the racial politics of liberation. Gary van Straden, a researcher with the South African Institute of International Affairs, confirms this view:

Black Consciousness is a highly sophisticated ideology and to dismiss it as anti-white, a fringe element, irrelevant or elitist, is reductionist and a grave injustice to some of the most brilliant minds produced by South Africa such as that of Bantu Steve Biko, Mapetla Mohapi, and Mthuli ka Shezi . . .

Black Consciousness leans far more toward a class analysis of South African society than does Africanism. Thus demands for the return of the land, while sharing the Africanist belief that this was "stolen" during the colonial period, contain new more sophisticated elements of scientific socialism.[110]

The historic Soweto student insurrection of 1976, in which over 1,000 Black students were killed in demonstrations against the dreaded Bantu educational system, signified a watershed in Black Consciousness politics and the politics of national resistance and liberation struggle. Even though the Black Consciousness Movement was not the formal organizer of the 1976 uprising, the bulk of the leadership of the South African Students Movement that led the protest, such as Tsietsi Mashinini, Khotso Seathlolo, Barney Mokgatle, and Selby Semela, were directly influenced and empowered by the philosophy of Black Consciousness to challenge the apartheid system.[111] SASM was viewed as a youth contingent of SASO, the Black Consciousness student organization on college and university campuses.[112]

Catalyzed by a limited campaign of rejecting Afrikaans as the premium language of instruction in Black schools, the Soweto movement was transformed into a revolutionary upsurge against the apartheid system. Black workers downed tools in major metropolitan centers in solidarity with the student movement. Thousands of students were arrested and fled the country to Botswana, Zambia, Mozambique, and Tanzania to join the ranks of the guerrilla war against white rule. A report by the Police Ministry revealed that 2,430 people had been detained following the Soweto uprising, of whom 817 were tried and convicted.[113] Students as young as ten and twelve years old were interrogated and severely tortured by security police. The eruption of Soweto marked a new aura of defiance and spirit of resistance on the part of Black youth that was determined to liberate South Africa from white minority rule through all avenues possible, with these youth fearlessly declaring their resolve to defend the Black cause, including the use of armed struggle and guerrilla praxis.

Several Black Consciousness organizations were banned by the colonial state in October 1977. Black working-class culture regained its momentum in early 1979, when the Federation of South African Trade Unions was formed, coordinating the mass mobilization of Black workers nationally. Textile workers in New Germany in Kwazulu-Natal, railway employees in Johannesburg and Durban, furniture workers in Bophuthatswana, gold miners in Germiston and Carletonville, stevedores in Cape Town, and thousands of workers in other industrial sectors walked off their jobs in the demand for justice at the workplace and the liberation of their country.[114]

An outstanding example of worker mobilization was the Port Elizabeth Civic Organization (PEBCO) under the skillful leadership of Thozamile Botha, which utilized the precepts of Black Consciousness in mediating unified worker resistance in the Port Elizabeth area. It claimed a membership of 10,000 workers by June 1980, representing Black workers' interests particularly in the motor assemblies and tire manufacturing industries. The United Automobile, Rubber, and Allied Workers Union worked closely with PEBCO, escalating worker strikes and stoppages and forcing the commercial establishment to deal directly with Black working-class demands. Other unions that also mobilized worker resistance were the Transport and General Workers Union; the African Food and Canning Workers Union; the South African Allied Workers Union; the Black Municipal Workers Union; the Motor Assemblies and Component Workers Union; the National Union of Metalworkers of South Africa; the National Union of Miners; the Paper, Pulp, Wood, and Allied Workers Union; the Glass and Allied Workers Union; and the Union of Black Miners.[115] Much of the courage for this level of open protest and defiance derived from the ingredients of Black working-class culture, specifically the cultural potency and political foresight of the Black Consciousness Movement formed in the late 1960s.

When Allan Boesak, a church leader and anti-apartheid activist, made the call for a national united front against the planned tricameral parliament that would include "Coloreds" and "Indians," resulting in the formation of the United Democratic Front (UDF), few people realized that this was originally an idea proposed by the National Forum Committee, a Black Consciousness consortium organized in 1983. The National Forum represented a coalition of major political, student, and worker organizations subscribing to Biko's Black Consciousness philosophy, whereas the UDF adopted a nonracial line in its program of resistance. Essentially, Boesak and other UDF organizers "hijacked" the broad front strategy, without acknowledging the role of the Black Consciousness Movement at its inception.[116]

Mass Mobilization of the 1980s: United Fronts and Trade Unions

By philosophical contrast to the Black Consciousness tendency in worker organizing, the Federation of South African Trade Unions espoused nonracial politics and in 1984 consisted of the Chemical Workers Industrial Union; the Jewellers and Goldsmiths Union; the Metal and Allied Workers Union; the National Automobile and Allied Workers Union; the National Union of Textile Workers; the Paper, Pulp, Wood, and Allied Workers Union; the Sweet, Food, and Allied Workers Union; and the Transport and General Workers Union. These unions were involved in various strikes throughout the 1980s, such as automobile workers at a Uitenhage assembly plant in 1980, and fifty strikes involving 50,000 workers in the East Rand metal industry in 1981 and 1982. In the 1980s and 1990s, workers generally demanded:

> the right to strike; the right to safe and healthy working conditions; the right to have a say over retrenchments; the right to have access to company information; the right to protection from hazards to pregnancy; the right to attend clinics before and after pregnancy; the right to control pension contributions; an end to sexual harassment; maternity rights; child-care facilities; equal pay for equal work; May Day, June 16 and March 21 as paid public holidays; an end to apartheid; troops out of the townships; and one person, one vote.[117]

Although the last three have been achieved, the remainder of worker demands have not been generally met in their entirety. Part of the problem is the fact that many of these unions still had reformist demands (toward gaining more egalitarian rights at the company workplace) rather than radically insisting on controlling the very industries that employed Black workers.

It is important to note that it was Black working-class solidarity that was able to maintain South Africa's longest mining workers strike in 1987, notwithstanding the fierce level of repression and intimidation by the white state and mining establishment, because Black workers collectively shared responsibilities and resources in the historical African communal sense and consequently intensified the culture of resistance.

During this strike organized by the National Union of Miners, which became known as the "Great Miners' Strike," over 340,000 Black miners were involved for three weeks, crippling mining production and resulting in over R250 million (valued at $60 million in 1987) in lost production revenues.[118] Eleven workers were killed, 600 were injured, and over 500 were arrested. Over 50,000 miners were dismissed by Anglo American, Gencor, Johannesburg Consolidated Investments (JCI), and Goldfields, all white capitalist–owned mining conglomerates that constitute the cornerstone of South Africa's gold-and-diamond-mining driven economy. This act of worker resistance was described as "Twenty one days that shook the Chamber [of Mines]."[119] Although the mining strike was successful in that it forced the management to address miners' grievances and demonstrated the muscle of the Black unions, the weakness of the strike resided in its failure to coalesce with other major industrial unions and union coalitions such as the Congress of South African Trade Unions (COSATU), the National Council of Trade Unions (NACTU), and the Azanian Congress of Trade Unions (AZACTU) so that the country would become economically paralyzed. It also neglected to organize workers in the rural areas and was unprepared for the harsh and brutal response of Anglo American Corporation, which the mining union viewed as "liberal," "amenable," and "reasonable." Following the strike, the National Union of Miners critically assessed the operation of the strike. An official of the union pointed out both its strengths and weaknesses:

> It revealed the depth of our organization. We surprised ourselves at the organizational capacity we had despite the deaths, killings, lack of facilities and adequate records . . . we underestimated the power of its opponents and had not anticipated the level of repression and violence which the mineowners unleashed . . . .We had misjudged the power of capital . . . .We learned that there is nothing like a liberal employer. . . . We were expecting the harsh action less from Anglo than from Gencor and others. But it came the other way around.[120]

Although this longest-enduring strike in South Africa's history could be considered unsuccessful in that it failed to achieve the desired 30-percent wage increase, lost 50,000 workers, and only resulted in a 17-percent wage increase coupled with added leave and death benefits, it did instill new levels of confidence among Black miners in particular and workers in general, reaffirming this notion: Undeterred resolve and organized action can mobilize Black workers for justice. Bobby Godsell, a spokesperson for Anglo American, the largest conglomerate in South Africa, which owned 57 percent of shares on the Johannesburg Stock Exchange in 1987, did acknowledge that since mid-1986: "[T]here was a pattern of wildcat strikes, go-slows, half-shifts, sit-ins. All of this became prevalent. When the 1987 strike happened it was as much a battle for physical control of the workplace as it was about wages."[121] Black workers had clearly demonstrated their political muscle.

In 1983, two sectors within the Black community that the white minority regime had attempted to co-opt into a tricameral white-run parliament, the so-called Coloreds and the Indians, overwhelmingly rejected this effort to further entrench apartheid, with the "Colored" turnout about 18 percent and the "Indian" response about 16 percent.[122] Similarly, too, with the formation of the regional services councils among the urban African community as part of the 1982 Black Local Authorities Act, a policy whereby local Black officials were given the responsibility of managing apartheid by local Black officials, 80 percent of Black township residents across the country, including 90 percent of the residents of Soweto, the country's largest township, boycotted these bogus structures.[123] Turnout in the Vaal area was 12 percent.[124] Numerous people who opted to participate in this political charade were either assassinated or forced to resign after realizing that the Black community did not view collaborators with apartheid amicably. Instead, young Black activists established independent political structures and educational cells within the Black townships, to serve the function of conscientization of the Black masses and to dispel the political obfuscation wrought by the regime's imposed regional service councils and related municipal bodies.

Protest actions intensified in 1983. Acts of sabotage and political violence increased from 59 in 1980 to 395 in 1983. The number of persons killed in these incidents reached 214 in the same year, up from 39 in 1980. From July through September 1983, over 90 people were killed in the Ciskei alone, owing to activism in a bus boycott.[125] Apartheid authorities engaged in cold-blooded murder and reprisals against apartheid opponents.

The "Koornhof Bills," which were drafted by Piet Koornhof, a government minister, for the purpose of reinforcing apartheid policies in Black townships, precipitated the 1984 revolts.[126] For months, thousands of Black residents living in townships of the Transvaal (Gauteng) and the Cape refused to pay rents on their homes to the white authorities and resisted their foisting rent increases on them. Sowetan residents refused to pay rents partially in response to the increase of the electricity service charge when municipal authorities doubled it. After June 1986, the levy was repealed.[127] On July 15, 1984, over 1,000 residents of Tumahole, in the Vaal region, protested against rent increases; 51 people were arrested, of whom one, Johannes Ngale, died in police custody. The anger of the people of Tumahole erupted to the point that some of the town's councillors appointed by apartheid authorities resigned.[128]

On March 21, 1985, 19 people were massacred in Langa, Cape Province, at an event commemorating the 1960 Sharpeville protest in which 69 people were killed and 188 wounded. The authorities claimed that they were not aware that the gathering represented a commemoration observance. Waves of anger and shock rippled through the Black communities of South Africa, strengthening the spirit of defiance and resolve toward making white colonial rule unmanageable. The Black townships, where police repression was the order of the day, soon became transformed into training areas for political revolution and organized mass resistance. Local businesses unsympathetic to the revolutionary Black cause were burned. Councillors who refused to join the demonstrations had their houses and cars burned. Four councillors in Sharpeville and neighboring townships were killed in 1984, and seven resigned. Security forces were unable to restrain Black anger against Black collaboration with the apartheid system.[129]

In the "Colored" township of Athlone in Cape Town on October 15, 1985, there occurred a tragic shooting by police that ended with three people being killed and seventeen wounded. The police had baited militant youths by entering the township in a disguised police vehicle, with 8 policemen hidden in crates. When the youths threw the first stones, the policemen crept from the crates and opened fire with bird shot, killing and wounding the youths. Thirty-nine rounds of ammunition had been fired in a space of 14.5 seconds.[130] Black communities around Cape Town were outraged and intensified their protests and demonstrations.

On July 20th, 1985, the white president, Botha, declared a national state of emergency in response to the escalating Black revolt, affecting 36 magisterial districts in the country. This announcement basically licensed the military sector of South Africa to execute necessary action to contain the flames of wide-scale dissent engulfing the Black townships. More than 50,000 police and security personnel were involved in the military clampdown on the Black community. Military occupation took the form of Hippo carriers and Saracen armored cars manned by soldiers being stationed permanently in several Black townships. In tandem with the policy of military occupation of townships such as Soweto, Sebokeng, Kattlehong, and Boipatong in the Vaal Triangle and in Zwide and Langa in the Cape, there were vigorous actions of intimidation, arrest, torture, and imprisonment by the military authorities of thousands of Black women, men, and children thought to be responsible for the swelling agitation.[131] In one incident, over 800 children between the ages of nine and fifteen were arrested and held by the military. By the end of 1985, white government figures cited that 763 people had been killed by the police and that "only three were white, and 201 were children and 2,571 had been shot and wounded."[132] Over 20,000 people had either been arrested or detained by white authorities. Scores of Black activists disappeared without trace. Hundreds of children were detained and held incommunicado in indoctrination detention camps. Between September 1984 and August 1, 1986, over 2,000 people had been killed and 19,000 arrested or detained.[133] The situation indicated a graduation into a political crisis of proportions not witnessed since the Soweto insurrection of 1976.

Boycotts of classes in Black schools heightened in the mid-1980s, with militant youth coining the slogan "Liberation before education!" The defiant tone of student resistance marked a foundational rejection of the entire apartheid educational and sociopolitical system by the younger generation. For many parents, however, this marked a crisis that threatened to erase any possibility of intellectual and technical training that they viewed as desperately needed in the struggle against apartheid. It was out of this attempt to come to grips with the rapidly spreading educational degeneration that the National Educational Crisis Committee (NECC) was formed in

March 1986 and popularized a new slogan that articulated the organization's independent sociopolitical position: "Education for liberation." The NECC viewed its responsibility as fashioning an alternative educational curriculum and set-up that would address the angry rebuttals of apartheid education by protesting Black youth. The NECC, notwithstanding its diverse and even conflicting ideological assertions, viewed itself as an instrument of "people's power" in the face of an increasingly harsh and draconian white-ruling state.[134]

As part of the strategy of resistance and noncooperation with apartheid capitalists, the Port Elizabeth area in the Cape Province became enveloped in a series of boycotts of white stores by Black consumers, precipitating the closure of many white stores, the owners of which complained directly to P. W. Botha. PEBCO was instrumental in this active movement of boycotts and isolation of white-owned commercial enterprise.

Of critical import in November 1985 was the formation of the Congress of South African Trade Unions, representing some 500,000 Black workers and currently estimated to have a membership of 1,500,000 people. There are over 300 unions affiliated to this coalition grouping. The other two are NACTU and AZACTU. These worker organizations, together with the United Democratic Front, a broad-based resistance organization and the National Forum, a Black Consciousness group, mobilized 1,500,000 Black workers on May 1, 1986, in one of the largest worker stayaways in South African history. Similar action occurred on the occasion of the tenth anniversary of the Soweto uprising on June 16.

By the end of 1986, an estimated 8,800 children were still being held in detention, according to reports of the Detainees Parents Support Committee, an organization functioning to highlight the condition of children in detention and campaign for their release.[135] From September 1, 1985, through August 1, 1987, 2,000 persons were killed and 19,000 had been arrested or detained.[136] Over 33 Black schools were closed in the Eastern Cape and many more in the Transvaal area. The State of Emergency was extended for another year by the white minority regime in June 1987. The situation around the Pietermaritzburg area of Kwazulu-Natal Province deteriorated into one of open political conflict as tensions between the Inkatha organization, led by apartheid regime appointee Chief Gatsha Buthelezi, and the United Democratic Front surfaced. By the end of 1987, 180 people had been killed.[137] Buthelezi's status as Kwazulu "Bantustan" leader made him a consistent target of the Black resistance, as most Black people viewed him as a collaborator with apartheid and one who legitimated the regime's "tribalistic" divisions of Black people, facilitating the ancient colonial policy of "divide and rule." Some measure of rapprochement between the UDF and COSATU, on the one hand, and Inkatha, on the other, was mediated by a peace pact signed in August 1988.

Meanwhile, the white authorities had proceeded with their policy of Bantustanization, forcibly dislocating 3.5 million Black people from their ancestral lands and relocating them into arid reserves that are akin to those of the Native reservations in the United States. In May 1986, more than 30,000 Black people were forced to flee their homes from Crossroads, the slum settlement outside metropolitan Cape Town that was established by women who insisted on living with their migrant-worker husbands. In August 1988, over 5,000 people had their informal dwellings razed to the ground at Emzomusha (Love Zone) outside Durban in Kwazulu-Natal Province, the area earmarked for "property development."[138]

The resistance of the Black community against oppression continued throughout the 1980s, notwithstanding the fact that eighteen major Black trade union, political, educational, and civil organizations had been banned since February 1988. As part of the ongoing defiance campaign waged by the new broad-based Mass Democratic Movement, all major banned organizations such as COSATU, UDF, the Azanian People's Organization (AZAPO), and others declared themselves unbanned on the weekend of August 12, 1989, while affirming the principle of unity among the varying tendencies within the resistance. Massive protests and demonstrations against segregated hospital facilities and beach resorts occurred over the month of August, in a spirit of noncooperation with the structures of apartheid planned by the Mass Democratic Movement, a major component of which was COSATU. The strategy of escalating civil protest in August served as a buildup for the racist white minority elections on September 6. The day of the elections was marked by bloodshed, with twenty-three Black people killed and 200 wounded by police gunfire as Black people took to the streets in a show of anger at the whites-only election.

The underground movement for liberation persisted, despite a posture of unmitigated repression and unrestrained violence by the apartheid regime. Organizations of the working-class and underclass communities prevailed amid a tightening security noose foisted on the Black urban townships and rural areas. The armed struggle waged by the African National Congress via its military wing Umkhonto We Sizwe and the Pan-Africanist Congress intensified, to the point that an average of one major explosion or act of sabotage occurred daily in South Africa's towns and cities. At the same time, the white backlash of conservative colonialism and ultra-right-wing racism had steadily grown, as evidenced by the gains that the white Conservative Party had made in the white Parliament in the 1989 elections, holding thirty-nine seats in that body. In this regard, the history of the white election process in South Africa demonstrated two patent facts:

1. that the white minority was fundamentally incapable of solving the South African political and economic crisis precisely because the whites had been the very architects of the apartheid problem and were consequently essentially irrelevant to Black liberation experience.
2. that the white community in general was still entrenched in retaining the apartheid superstructure and generally affirming a unanimity of the policy of white domination, while varying on the manner of implementation. Outside the sphere of white parliamentary politics, the Nazi-styled Afrikaner Weerbestandigheid group, which openly espoused racist violence against Black people, continued to be a growing force among Afrikaner conservatives.

These waves of political dissidence led by both Black workers and students that were solidly organized during the 1960s and protracted in the 1970s culminated in a situation of political ungovernability over Blacks and economic instability for whites in the 1980s, compelling the white commercial and industrial establishment to realize that apartheid was cracking in certain quarters. In 1985, a group of liberal white businesspeople and political leaders traveled to Lusaka, Zambia, to discuss issues of change in South Africa with members of the senior leadership of the exiled ANC.[139] The result was a bifurcated approach by the ANC: Although it claimed to be directed toward overthrowing the apartheid regime by force, it was also engaged in a series of discussions with elements of the South African government through the latter part of the 1980s. This was problematic for the principal reason that the groundswell of the liberation forces *inside* the country—represented by organizations within the Black working-class movement, Black student organizations like the South African Students Convention and the Azanian Student Movement, the United Democratic Front, the Azanian People's Organization, the Workers Organization of South Africa (WOSA), and the Unity Movement—were involved in an intense struggle against the apartheid system. These cadres were committed to a protracted movement that would wear down the apartheid forces from within, through organized and systematic events of protest and rebellion.

However, the ANC forces *outside* the country were entering a phase of discussions with elements of the white South African establishment that would lay the groundwork for the opening of negotiations that eventuated in the historic climax of 1990. Some of the tensions in the dynamics between the "internal" sector of the ANC, that is, those members of the organization who did not leave the country, and the "external" segment of activists, those who were forced to go into exile since the 1960s, were evident in the elections process in the 1990s when nominations for national governmental and regional leadership positions were being solicited within the ANC. It needs to be pointed out that Nelson Mandela had made overtures toward earnest negotiations between the ANC and the apartheid regime already in March 1989, declared in a secret letter sent to then ruling president P. W. Botha. Part of the letter read:

> At the outset, I must point out that I make this move without consultation with the ANC. I am a loyal and disciplined member of the ANC . . . The step I am taking should, therefore, not be seen as the beginning of actual negotiations between the government and the ANC. My task is a limited one, and that is to bring the country's two major political bodies to the negotiation table. The renunciation of violence by either the government or the ANC should not be a pre-condition but the result of negotiation.
>
> The key to the whole situation is a negotiated settlement and a meeting between the government and the ANC will be the first major step toward lasting peace in the country . . . Two political issues will have to be addressed at such a meeting: first, the demand for majority rule in a unitary state; secondly, the concern of white South Africa over this demand, as well as the insistence of whites on structural guarantees that majority rule will not mean domination of the white minority by Blacks. . . . The move I have taken provides you with the opportunity to overcome the current deadlock and to normalize the country's political situation. I hope you will seize it without delay.[140]

This communication may sound somewhat disturbing to radical participants in the liberation struggle because it suggests that Nelson Mandela was preempting negotiations and *appealing* to the apartheid regime for peace. However, if one peruses the discussions transpiring between the ANC and the white corporate community since 1985, this letter represents the logical result of a series of events that began much earlier. Mandela was not acting entirely on his own prerogative; the ANC, of which he was head, had already participated in exploratory prenegotiation talks in Lusaka. Concomitantly, white capital in South Africa realized that the writing was on the wall: Black majority rule was on the horizon. All that the representatives of capitalism were concerned with was protecting their vital and lucrative economic interests in South Africa, under an intensely reformed apartheid system or Black rule. One realizes with hindsight, then, that the move toward a negotiated settlement to the apartheid conflict was mutually and reciprocally reached between the regime and the ANC: the regime seeking assurances of white economic security, and the ANC unable to defeat the regime militarily through its limited guerrilla warfare strategy.

The 1990s Counterrevolution by Capitalist and Neocolonialist Forces

It was not too surprising when Gavin Reilly, then chairperson of Anglo American, the largest capitalist corporation in South Africa and one of the largest in the world, made a public address in 1987, in which he called, inter alia, for the dismantling of various statutes of apartheid such as the Group Areas Act and the Communism Act, the freeing of Nelson Mandela and other political prisoners, the unbanning of the ANC and other liberation organizations, and the beginning of a dialogue between the apartheid regime and the ANC that would lead to a negotiated and peaceful settlement of the South African conflict.[141] In late 1989, P. W. Botha, the ailing white president of the Republic of South Africa, was replaced by F. W. De Klerk, who was then hailed as the new South African leader with the role of principal architect responsible for crafting the process of transition of the country from a situation of apartheid to that of postapartheid. All of the reforms called for by Anglo American were subsequently executed by the De Klerk regime. The rest is now history.

Nelson Mandela was released on February 21, 1990. The apartheid regime legalized the ANC, PAC, and the South African Communist Party, and launched a détente with the ANC, culminating with the formation of the Convention for a Democratic South Africa (CODESA), in which the ANC, the ruling white Nationalist Party, and other apartheid political formations, including representatives from the Bantustans such as the Inkatha Freedom Party, actively participated. In December 1991, 228 delegates from nineteen political organizations met at the World Trade Center in Johannesburg to discuss the process of the transition from apartheid and the constitutional process for change.[142] By June 1992, CODESA had agreed on basic principles for a constitutional document, and then it collapsed over the question of the proportion needed for a majority adoption of the new constitution. The National Party insisted on a three-fourths majority and the ANC, a two-thirds majority. In the same year, there was a tragic eruption of violence by workers from a migrant workers' hostel who attacked residents of a slum community in Boipatong outside Johannesburg. Forty-six people were killed, and the ANC broke off negotiations, suspecting police involvement. However, it was agreed in February 1993, that an election would be held in April 1994. Cyril Ramaphosa was the chief negotiator for the ANC and Roelf Meyer was the National Party representative.

COSATU, meanwhile, continued to play an important role in the transition process, reminding all of the major political players that workers' interests were paramount. It organized a major anti-VAT (valued-added tax) strike in October 1991, drafted a workers' charter, and led national demonstrations by workers, challenging the intransigence of the apartheid regime in the negotiations process. It planned a very successful march to the World Trade Center in October 1993 to protest against the clause that granted employers the right to lock out striking workers. It was responsible for advancing the Reconstruction and Development Accord in tandem with the ANC to elevate workers' interests.

In April 1993, Chris Hani, the general secretary of the South African Communist Party, was gunned down by a white assassin, causing angry eruptions among the Black community throughout South Africa. Hani was believed to be the most popular leader in South Africa after Nelson Mandela, and his death was deeply mourned by all freedom-loving people in South Africa.

The result of the negotiations was an election based on the discussions and agreements reached within CODESA, held on April 27, 1994, with the ANC winning over 60 percent of the vote. The ANC agreed to a coalition government, describing itself as a Government of National Unity and casting itself as the majority, with the minority composed of the National Party and other smaller parties from the "right," such as the Freedom Front and the Conservative Party, and the left, such as the PAC and the SACP. Nelson Mandela was elected president of this new coalition government, with the ANC holding a 60 percent majority in government, the National Party holding about 25 percent, and the remainder split among parties like the PAC, SACP, Freedom Front, and other minor parties. The ANC began its rule with a very fragile coalition involving the SACP and COSATU, both of which have been openly critical of the government's latest GEAR (Growth, Employment, and Redistribution Program).

Several major events have occurred since 1994. Black workers have organized several stay-away strikes and boycotts to ventilate their frustrations over wages incongruent with inflation and over massive privatization initiatives in areas like transportation, telecommunications, and energy utilization. The country's new and democratic constitution was formally adopted in May 1996, and the Truth and Reconciliation Commission was formed in the same year to investigate crimes and atrocities perpetrated during the apartheid era, with the power to offer clemency to those who have committed crimes.

On May, 8, 1996, the National Party announced that it was withdrawing from the coalition government to form the official opposition party in the South African Parliament. It was apparent that the "honeymoon" between the ANC and the National Party might have reached an abrupt end. Upon receipt of this news, the South African rand plunged to its lowest rate in relation to the U.S. dollar, with $1 worth R4.50. Gold and reserves plummeted by R2.3 billion, falling to R11.3 billion.[143] The markets were thrown into disarray because it was believed that the country's economy was at the precipice of collapse. Why? The principal reason was that there would be no major white party in government to assure foreign capitalist investors that their investments would be protected under a "Black majority" government that appeared to be ruling without the collaboration of South Africa's leading white party! Capitalist "stability," or what most economists call "economic stability," is still associated with European ownership and control. The markets clearly felt far more secure when F. W. De Klerk and the Nationalist Party were in government, because white politicians are still regarded as more trustworthy than Black ones, even though the white ones may have derived their legitimacy from apartheid and the Black ones may be legitimately democratic from the era of "post-apartheid."

The conditions of life for working-class and impoverished Black people have generally worsened in every area. Unemployment has reached 45–50 percent in major urban metropoles and about 60–70 percent in rural areas, with the former facing increased disintegration of infrastructural supports and the latter remaining essentially underdeveloped. Housing remains a critical shortage, with organizations like the South African National Civic Organization struggling to assert the rights of the poor to decent housing in the townships.[144] Commercial banks continue to deny loans to many Blacks buying houses in the townships on the grounds that property values in the townships are steadily declining. Government housing subsidies have barely scratched the surface, with a 3–4 million-unit housing shortage, with only about 680,000 houses being built or already built. However, most residents of these new houses are disgusted with the substandard quality and minuscule size of these family houses (in Alexandra, 90 percent had cracks or serious structural defects), built by private developers contracted by the government, and many are attempting to sell them and use the raised money for other needs.[145]

On the political front, the Azanian People's Organization has split, due to ideological and personality differences between post-1990 exiles returning to the country and AZAPO members active inside the country. The Socialist Party of Azania was formally launched on March 21, 1998, headed by Lybon Mabasa, a former leader of AZAPO and a historic Black Consciousness activist. Both groups participated in the 1999 elections yet garnered less than 1 percent of the national vote. The New National Party, basically a carryover of the old Nationalist Party, continues to be the leading opposition party in Parliament, with seriously waning support from the "Colored" population in the 1999 elections, particularly in the Western Cape.[146] The PAC has begun to see its support climb, especially in the Eastern and Western Cape, even though it received less than 1 percent of the vote, and the Inkatha Freedom Party saw its support dive in Kwazulu-Natal, a historical enclave for the party, though it still received 41.9 percent of the vote. The ANC received 66.4 percent of the overall 1999 vote, with landslide victories in seven of the nine provinces, just 0.2 percent short of a two-thirds majority.[147] Other smaller parties such as the African

Christian Democratic Party, the Democratic Party, the Freedom Front, and the United Democratic Movement all received tiny proportions of votes in the election.

In the area of education, the post-1994 situation has been quite turbulent and evinces policy confusion.[148] Although the post-apartheid government has built 100,000 new classrooms and brought in 1.5 million new learners and 100,000 higher education students into the educational system, it is clear that a giant educational crisis looms.[149] Most Black public schools are still overcrowded and understaffed by teachers, with student-teacher ratios hovering at 40:1. In 1995, over 150,000 primary school teachers were underqualified to teach, and new books had to be devised for the post-apartheid curriculum, most of which were not available and are still hardly available presently.[150] Due to financial constraints in the arrangement of the South African political economy, as it is a continuation of the capitalist economic framework from apartheid and has no governmental restrictions or mandatory demands on large private enterprise for educational support, there is a significant shortage of teachers in Black schools in both urban and rural areas and lack of monies for educational budgets. The government's policy of retrenching teachers and instituting early retirement packages worth millions of rands has resulted in educational bureaucracies becoming bankrupt. Even Nelson Mandela was shocked at the high cost of these retirement packages.[151] In January 1995, two teachers' unions went on strike for higher salaries, crippling schools for days. Teacher "chalk-downs" have occurred sporadically over the past few years, often disrupting classes and in some instances, delaying examinations. Civil service workers, including teachers, struck for higher salaries again in August 1999. In such instances, Black students at public schools suffer the most, given the historically inferior and distorted educational system under apartheid.

At Black universities, acute financial shortfalls have taken a serious toll on enrollments, with such places as the University of Venda seeing its student population drop from 15,000 to 5,000 in 1999. Other historically Black universities such as the University of Fort Hare, the University of Transkei, and the University of the Western Cape have seen massive student strikes and protests over the inability of low-income students to pay tuition fees and over administrative ineptitude and inefficiency. Vice chancellors at the University of Fort Hare and at the University of Transkei and the acting vice chancellor at the University of the Northwest were all suspended on grounds of administrative shortcomings and irregularities in 1999.[152]

It is evident that much of what has happened since 1990 has functioned to undermine the revolutionary potential of the Black working class and its allies. It is palpable that although the post-apartheid government gives the impression that it is committed to Black liberation and empowerment, in actuality, more Black working-class and underclass people have become poorer, and the tiny Black elite has grown wealthier. The resistance movement has been hijacked by neocolonial politics bent on adhering to the ideology and practice of the global capitalist economy. The result is frustration, anger, and despondency on the part of the Black oppressed.

The Black working-class struggle against neocolonialism and capitalism, for the return of dispossessed lands and for national independence, like the historic indigenous struggle against European colonial penetration into Azania in the 1500s and 1600s, continues.

Summary and Conclusion

As of 2000, the new government has been in office for about six years. Although there have been some changes within the political and social spheres—such as the 680,000 houses constructed or under construction, electrification now reaching 63 percent of the population, 3 million more people having access to clean drinking water, 5 million school children now receiving free lunch meals—much of this change has been *cosmetic* in character and more *symbolic* than substantial.[153] The new government has constructed houses that are defective and woefully small to address Black family needs in 90 percent of the cases. The electrification of townships has not addressed the fundamental structures of overcrowded urban townships, vitiated by pollution, catastrophic unemployment, chronic crime, and poorly equipped schools and the fact that most people continue to remain poor. Over 5 million people either still live in shacks or are landless. Unemployment hovers at 42 percent, and the private sector has seen a loss of 500,000 jobs since 1994, with the working-age population increasing by 5 million people.[154] The Bank of England's recently auctioning 25 tons of gold with plans to sell off another 390 tons has resulted in the loss of 5,000 mining jobs in August 1999, with another 80,000 jobs threatened to be lost as the gold price rapidly falls.[155] Ironically, South Africa had agreed to the planned sale of 300 tons of gold by the International Monetary Fund (IMF).[156] In mining, 28,000 retrenchments were announced, causing a ripple effect of 20,000 jobs lost in related industries.[157] Spoornet, the transportation agency, will lay off 27,000 people.

Social violence continues unabated, particularly against girls and women, with at least two children being raped every minute of the day and night, on average.[158] Child prostitution and street children have become normative phenomena in post-apartheid South Africa, issuing from the influences of the global economy and the high unemployment and poverty rates in the country. Rural areas, where the bulk of Black people live, are essentially untouched by the symbolic changes of the post-apartheid dispensation.

The essential structures of the white-owned capitalist economy, monopolized landownership, and racial residential separation with white suburbanization and Black ghettoization remain intact. White capitalist power remains firmly in charge of changes in post-apartheid South Africa, or "new" South Africa, as many in South Africa describe it. It is obvious that though Nelson Mandela was elected the first president of a post-apartheid South African republic within an ANC majority government and has now been succeeded by Thabo Mbeki, neither of these leaders nor the ANC possesses the *power* to effect radical structural changes toward justice and equality for the oppressed. Many are watching the South African situation rather closely, wondering how different the next five years will be, given the promises Mbeki and the ANC have made to deliver houses, jobs, and economic empowerment for *poor* Black people.

The apartheid regime was acutely adroit and deceptive to the point that it was willing to concede political changes in which white domination would be transformed, while being fully conscious that economic power would reside firmly in the hands of the country's white capitalists, who are intimately connected to the centers of global capital. This is one of the reasons I would argue that the election process in April 1994 was fundamentally flawed in that the apartheid regime controlled the entire process, a subject that will be explored in greater detail in the next chapter. The result, in essence, was *change* in South Africa with *no real change*.

Notes

1. William Beinart, *Twentieth Century South Africa* (Oxford: Oxford University Press, 1994), 4. Even though Beinart does make some valuable contribution to the discourse on South African history, he nevertheless must be criticized for his Eurocentric oversight, as he states that "any author writing a general book on South Africa must decide whether to start with whites or blacks" (p. 9). Such consideration depends solely on whether the historian views South Africa as an intrinsic part of Africa or an extension of Western Europe, with most Africans inclined toward the former and most white historians pondering the latter. Similarly, Beinart suggests that the "romance of Mpondo history—or that of the Sotho or Xhosa or Zulu was that they survived and retained some of their culture and land," responding to C. W. de Kiewiet's suggestion that South Africa was not "a romantic frontier like the American west"(in Kiewiet's *History of South Africa, Social and Economic* (London: Oxford, 1941). European historians still reflect their own ethnic biases when such remarks are made. For the indigenous people of North America, white colonialism meant the genocide of the Native people. For the Africans of Southern Africa, European colonialism meant genocide of the Abathwa, Khoi Khoi, and Herero people, as well as the continued subjugation, enslavement, and dispossession of the Zulu, Xhosa, and Sotho nations by terroristic European powers. There is nothing romantic about physical or cultural genocide!

2. Notwithstanding classical Marxist theoreticians' attempts to perceive a class analysis without seeing the confluence of the categories of race and class, it must be stated that racism is the prevailing and dominant ideology by which capitalism survives. See George Frederickson, *White Supremacy* (New York: Oxford University Press, 1981), and Stephen Cell, *The Highest Stage of White Supremacy* (Cambridge: Cambridge University Press, 1984).

3. Bernard Magubane, *The Political Economy of Race and Class in South Africa* (New York: Monthly Review Press, 1979), 3.

4. The most incisive work that dramatically illustrates the European colonial conquest and plunder of Africa, the Americas, Asia, and the Pacific is Ihechukwu Chinweizu's *The West and the Rest of Us: White Predators, Black Slavers, and the African Elite* (New York: Random House, 1975).

5. W. M. Tsotsi, *From Chattel to Wage Slavery: A New Approach to South African History* (Maseru: Lesotho Printing and Publishing Company, 1981), 11.

6. Ibid.

7. Richard Elphick's book *Khoikhoi and the Founding of White South Africa* (Johannesburg: Ravan Press, 1985) describes the history of the Khoi Khoi nation prior to European colonial intervention. Another work that furnishes concrete vignettes on Khoi Khoi society and the enslavement of its people by the Dutch is found in Robert Ross's *Beyond the Pale: Esssays on the History of Colonial South Africa* (Hanover: Wesleyan University Press, 1993).

8. Alfred Molleah furnishes a sound critique of Eurocentric histories of Southern Africa by many white historians in *South Africa: Colonialism, Apartheid, and African Dispossession* (Wilmington, Del.: Disa Press, 1993) and describes the detailed history of plunder of east and central Africa by the Portuguese, as well as the resistance by the Khoi Khoi on p. 107.

9. Motsoko Pheko, *Apartheid: The Story of a Dispossessed People* (London: Marram Books, 1984), 25.

10. Paul Maylam, *A History of the African People of South Africa: From the Early Iron Age to the 1970s* (New York: St. Martin's Press, 1986), 3. Cited also in Jordan Ngubane, *Conflict of Minds* (New York: Books in Focus, 1979), 234.

11. Tsotsi, *From Chattel to Wage Slavery*, 14.

12. Cited in Stephen Taylor, *Shaka's Children: A History of the Zulu People* (London: HarperCollins, 1994), 9. See also Eric Walker, *A History of Southern Africa* (London: Longmans, 1957), 32. For an illumination of Columbus's resentful attitudes and violent actions toward the indigenous people of the Americas, see James Loewen's *Lies My Teacher Told Me* (New York: Harper, 1994), and Howard Zinn's A *People's History of the United States* (New York: Harper and Row, 1980), chapter 1.

13. Freda Troupe, *South Africa: An Historical Introduction* (Harmondsworth, Eng.: Penguin, 1975), 53.

14. See, for example, H. J. and R. E. Simons, *Class and Colour in South Africa: 1850–1950* (Baltimore: Penguin Books, 1969), 14–29.

15. Ibid., 30.

16. Tsotsi, *From Chattel to Wage Slavery*, 19.

17. Troupe, *South Africa*, 51.

18. Ibid., 52.

19. See James Gump, Ph.D. diss., "Revitalization Through Expansion in Southern Africa, c. 1750–1840: A Reappraisal of the Mfecane," University of Nebraska, 1980, especially 83–100.

20. Timothy Stapleton, *Maqoma: Xhosa Resistance to Colonial Advance 1798–1873* (Johannesburg: Jonathan Ball, 1994), 15.

21. Ibid., 10.

22. J. D. Omer-Cooper, *The Zulu Aftermath* (Evanston, Ill.: Northwestern University Press, 1966), 111. See also Jeff Guy's *The Destruction of the Zulu Kingdom: The Civil War in Zululand, 1879–1884* (Pietermaritzburg: University of Natal Press, 1994), 39.

23. John Jackson makes this point in his classic *Introduction to African Civilizations* (New York: Citadel Press, 1995), 188.

24. K. A. Jordaan, "The Land Question in South Africa," *Points of View*, Vol. 1, No. 1, October 1959, cited in Allison Drew, ed., *South Africa's Radical Tradition: A Documentary History, Vol. 2, 1943–1964* (Cape Town: Buchu Books, Mayibuye Books, and UCT Press, 1997), 325–341, provides a foundational knowledge of land struggles, dispossession, and the alienation of indigenous labor for capital accumulation.

25. Frank Molteno, "The Historical Significance of the Bantustan Strategy," in Martin Murray, ed., *South African Capitalism and Black Political Opposition* (Cambridge: Schenkman, 1982), 553.

26. In 1994, Transvaal was changed to "Gauteng Province" and the name of the province in the southeast formerly called Natal was changed to "Kwazulu-Natal."

27. Duncan Innes, *Anglo American and the Rise of Modern South Africa* (New York: Monthly Review Press, 1984), 25.

28. Molteno, "The Historical Significance of the Bantustan Strategy," 553.

29. Innes, *Anglo American and the Rise of Modern South Africa*, 27.

30. Stanley Greenberg, *Race and State in Capitalist Development* (New Haven: Yale University Press, 1980), 37.

31. Walter Rodney, *How Europe Underdeveloped Africa* (Washington D.C.: Howard University Press, 1980), 152.

32. Ruth Milkman, "Apartheid and Economic Growth and U.S. Foreign Policy in South Africa," in *South African Capitalism and Black Political Opposition*, 406.

33. Innes, *Anglo American and the Rise of Modern South Africa*, 27.

34. Ibid., 33.

35. I. B. Tabata, *Imperialist Conspiracy in Africa* (Lusaka: Prometheus Publishing Company, 1974), 99. On p. 99, Tabata furnishes the shocking statistic that in 1972, the bill for wages in the gold mining industry was R95 million, and pre-tax profits were R548 million! The increase in profits alone over the previous year was more than twice the wages paid to Black miners. Such facts substantiate the foundational role that mining played in profit accumulation in South Africa and the subsequent emergence of industrial capitalism. It is this legacy that needs to be destroyed for any authentic socioeconomic change to occur in post-apartheid South Africa.

36. Bengt Sundkler's *Bantu Prophets in South Africa* (London: Oxford University Press, 1975), 38–43, provides a rudimentary reading on the subject of independence asserted by the indigenous African churches in South Africa.

37. Thomas Karis and Gwendolyn Carter, eds., *From Protest to Challenge: A Documentary History of African Politics in South Africa, 1882–1964*, Vol. 1 (Stanford: Hoover Institution Press, 1972), 9–10.

38. Nelson Mandela, *No Easy Walk to Freedom* (London: Heinemann Educational Books, 1965).

39. For an understanding of the manner in which whites crafted land policies and subsequent measures for dispossessing and disenfranchising Africans, C. M. Tatz's work, *Shadow and Substance in South Africa: A Study in Land and Franchise Policies Affecting Africans, 1910–1960* (Pietermaritzburg: University of Natal Press, 1962), is helpful.

40. Albert Luthuli, *Let My People Go* (New York: McGraw Hill Book Company, 1962), 88.

41. The effects of the "Color Bar Act" are discussed in Laurence Solomon's "Socio-Economic Aspects of South African History, 1870–1962," Ph.D. diss., Boston University, 1962, 86–95.

42. American Friends Service Committee, *South Africa: Challenge and Hope* (New York: Hill and Wang, 1982), 50.

43. Alex Hepple, *South Africa: A Political and Economic History* (New York and London: Frederick Praeger, 1966), 93.

44. Solomon Plaatje, *Native Life in South Africa* (Johannesburg: Ravan Press, 1982), 17.

45. Ralph Horwitz, *The Political Economy of South Africa* (London: Weidenfeld and Nicolson, 1967), 135.

46. C.F.J. Muller, ed., *Five Hundred Years: A History of South Africa* (Pretoria and Cape Town: Academica, 1969), 383.

47. Rob Davies, Dan O' Meara, and Sipho Dlamini, *The Struggle for South Africa: A Reference Guide*, Vol. 2 (London and Atlantic Highlands, N.J.: Zed Press, 1984), 285.

48. Quoted in Francis Meli's *A History of the ANC: South Africa Belongs to Us* (Harare: Zimbabwe Publishing House, Bloomington: Indiana University Press, and London: James Currey, 1988), 36.

49. Peter Walshe, *Black Nationalism in South Africa: A Short History* (Johannesburg: SPRO-CAS Publications, Ravan Press, 1973), 33.

50. Meli, *A History of the ANC*, 44.

51. It was the passbook law that made South African law so distinctively heinous and inhuman, under which all Africans were required to carry little documents chronicling their life history and requiring submission upon demand, similar to the slave codes adopted during the slave era in the U.S. South.

52. Beinart, *Twentieth Century South Africa*, 99.

53. Karis and Carter, *From Protest to Challenge*, 145.

54. T.R.H. Davenport, *South Africa: A Modern History*, 3d ed. (Toronto and Buffalo: University of Toronto Press, 1987), 299–300.

55. Karis and Carter, *From Protest to Challenge*, 145. This incident came to be known as the Bulhoek Massacre, which had its roots in indigenous religio-political and nationalistic sentiments for the reclamation of Africa from the clutches of European colonialism.

56. The account of the 1920 strike is documented in *Freedom from Below: The Struggle for Trade Unions in South Africa* (Johannesburg: LACOM, a project of SACHED, 1987), 26. This book has a good chronology of the development of Black worker organizations under the yoke of apartheid.

57. Tsotsi, *From Chattel to Wage Slavery*, 58.

58. Solomon, "Socio-Economic Aspects of South African History," 89.

59. Davenport, *South Africa*, 300.

60. D. Hobart Houghton and Jenifer Daught, *Source Material on the South African Economy: 1960–1970, Vol. 3, 1920–1970* (Cape Town: Oxford University Press, 1973), 83.

61. I. B. Tabata, *The Awakening of a People* (Nottingham, Eng.: Spokesman Books), 1974, 2.

62. Muriel Horrell, *South Africa's Workers: Their Organizations and the Patterns of Employment* (Johannesburg: South African Institute of Race Relations, 1969), 7.

63. Ibid., 8.

64. Neil Parsons, *A New History of South Africa* (London and Basingstoke: Macmillan, 1982), 274.

65. Drew, *South Africa's Radical Tradition*, 21.

66. *Freedom from Below*, 89.

67. Parsons, *A New History of South Africa*, 277.

68. Alf Stadler, *The Political Economy of Modern South Africa* (London and Sydney: Croom Helm, 1987), 148.

69. Gail Gerhart, *Black Power in South Africa: The Evolution of an Ideology* (Berkeley and Los Angeles: University of California Press, 1979), 60–61.

70. Horrell, *South Africa's Workers*, 13.

71. Parsons, *A New History of South Africa*, 271.

72. Ibid., 272.

73. M. Ncholo, *Equality and Affirmative Action* (Cape Town: Community Peace Foundation, Sig Publications, 1994), 181.

74. See Sipho Mzimela's *Apartheid: South African Naziism* (New York: Vantage Press, 1993) for an illumination of the striking similarities between the rule of Nazi Germany and that of apartheid South Africa. Mzimela makes the point that the entire Western European world is culpable for the horrors of apartheid since it is the European capitalist powers that financed and maintained this inhumane system, though fully cognizant that its effects were akin to those inflicted on the victims of Nazism.

75. Walshe, *The Rise of African Nationalism in South Africa*.

76. Parsons, *A New History of South Africa*, 280.

77. Gerhart, *Black Power in South Africa*, 181.

78. Sobukwe, Robert Mangaliso, *Speeches of Mangaliso Sobukwe 1949–1959* (Johannesburg: Pan Africanist Congress of Azania, 1993), 20–21.

79. *Time for Azania* (Toronto: Norman Bethune Institute, PAC in Perspective Series, 1976), 36.

80. For a detailed treatment on the life of Mangaliso Sobukwe, see Benjamin Pogrund's *How Can Men Die Better: Sobukwe and Apartheid* (London: P. Halban), 1990.

81. *New Phase in Azanian Struggle* (Lusaka: Department of Information, Pan Africanist Congress of Azania, 1965), 18.

82. Ernest Harsch, *South Africa: White Rule, Black Revolt* (New York: Monad Press, 1980), 193.

83. *The Plight of Black Women in Apartheid South Africa* (New York: UN Department of Public Information, 1981), 21.

84. Drew, *South Africa's Radical Tradition*, 115–121.

85. Meli, *A History of the ANC*, 133.

86. Mary Benson, *South Africa: The Struggle for a Birthright* (London: International Defence and Aid Fund, 1985), 183.

87. *The Plight of Black Women in Apartheid South Africa*, 22.

88. Luthuli, *Let My People Go*, 191.

89. Asha Moodley, "Black Woman, You Are on Your Own," in *Agenda: A Journal about Women and Gender*, No. 16, 1993 (Durban, South Africa: Agenda Collective).

90. In his autobiography, *Lest We Forget* (Johannesburg: Skotaville Publishers, 1988), Philip Ata Kgosana provides a detailed account of his role as PAC leader in the Cape and the organized resistance against the apartheid system during the 1960s.

91. Gerhart, *Black Power in South Africa*, 246.

92. Cited in Pierre van den Berghe, *South Africa: A Study in Conflict* (Berkeley and Los Angeles: University of California Press, 1967), 163–164.

93. Meli, *A History of the ANC*, 151.

94. *Africa News*, January 17, 1983.

95. *Africa News*, October 3, 1983, 12.

96. For an elucidation of the interconnection between Black Consciousness organizations and Black worker resistance praxis, see Gerhart's *Black Power in South Africa*, 290–294.

97. Steve Biko, "The Definition of Black Consciousness," in *I Write What I Like* (London: Bowerdean Publishing Company, 1996), 48–49.

98. Ibid., 50–51.

99. Ibid., 52.

100. Ibid.

101. Steve Biko, "The Quest for a True Humanity," in *I Write What I Like* (San Francisco: Harper and Row, 1986), 97.

102. Steve Biko, founder of the Black Consciousness Movement of Azania, addressed a conference of the Interdenominational Association of African Ministers of Religion (IDAMASA) and the Association for the Educational and Cultural Development of African People (ASSECA) at Edendale in 1971, in a speech entitled "Some African Cultural Concepts." The speech contains a classical indictment of the working of European colonial-capitalist culture and identifies some cornerstone African notions that are innately antithetical to

the former. See his seminal address, which later proved to be the stepping-stone for the formation of the historic Black People's Convention in *Frank Talk*, Vol. 1, No. 4, September–October 1984, Secretary for Publicity and Publications of the Azanian People's Organization.

103. Steve Biko, *Black Consciousness in South Africa*, ed. Millard Arnold (New York: Random House, 1978), 318.

104. Biko, "The Quest for a True Humanity," 91.

105. See for instance, the chapter on Black Consciousness in Thomas Ranuga's *The New South Africa and the Socialist Vision: Positions and Perspectives Toward a Post-Apartheid Society* (Atlantic Highlands, N.J.: Humanities Press, 1996). Ranuga argues that the philosophy of Black Consciousness is still very relevant as a medium for articulating transformation paradigms in post-apartheid South Africa.

106. Robert Fatton Jr., *Black Consciousness in South Africa: The Dialectics of Ideological Resistance to White Supremacy* (Albany: State University of New York Press, 1986), 40.

107. *Freedom from Below*, 177.

108. A version of the program of liberation and decolonization that AZAPO and AZAWU have adopted is found in the appendix of Motsoko Pheko's *Apartheid: The Story of a Dispossessed People* (London: Marram Books, 1984), where the platform of the Azanian People's Manifesto was adopted at a meeting of the National Forum Committee, a Black Consciousness/Pan Africanist coalition held in Hammanskraal in 1983.

109. Biko, "The Quest for a True Humanity," 97.

110. Gary van Straden, "Outside the MDM: An A-Z of Azanian Politics," *Indicator South Africa*, Vol. 7, No. 2, Autumn 1990, 9.

111. Interview with Selby Semela, September 1981, and with Barney Mokgatle, July 1995. See also George M. Frederickson, *The Comparative Imagination: On the History of Racism, Nationalism, and Social Movements* (Berkeley and Los Angeles: University of California Press, 1997), 206.

112. Ranuga, *The New South Africa and the Socialist Vision*, 103.

113. Mokgethi Mothlabi, *The Theory and Practice of Black Resistance to Apartheid* (Johannesburg: Skotaville, 1984), 34.

114. Harsch, *South Africa: White Rule, Black Revolt*, 317.

115. The political and cultural resistance of these Black worker organizations is described in Ken Luckhardt and Brenda Wall's *Working for Freedom* (Geneva: Program to Combat Racism, World Council of Churches), 1981.

116. Alex Callinicos, *South Africa Between Reform and Revolution* (London: Bookmarks, 1988), 62.

117. *Freedom from Below*, 192.

118. Baskin, *Striking Back: A History of COSATU* (Johannesburg: Ravan Press, 1991), 235.

119. Ibid., 224.

120. Ibid., 238.

121. Ibid., 225.

122. Peter Parker and Joyce Mokhesi-Parker, *In the Shadow of Sharpeville: Apartheid and Criminal Justice* (London: Macmillan, 1998), 21.

123. Stanley Greenberg, "Resistance and Hegemony," in W. G. James, ed., *The State of Apartheid* (Boulder: Lynne Rienner, 1987), 54.

124. Parker and Mokhesi-Parker, *In the Shadow of Sharpeville*, 21.

125. Lyle Tatum, ed., *South Africa: Challenge and Hope* (New York: American Friends Service Committee, Hill and Wang, 1987), 68.

126. Thozamile Botha, "Civic Associations as Autonomous Organs of Grassroots' Participation," *Theoria: A Journal of Studies in the Arts, Humanities and Social Sciences*, No. 79, May 1992 (Pietermaritzburg: University of Natal Press), 63.

127. Mark Swilling and Khehla Shubane, *The Soweto Experience: Urban Popular Movements in "Post Apartheid" South Africa* (Montreal: Centre d'Information et de Documentation sur le Mozambique et l'Afrique Australe [CIDMAA], 1992), 5. These increased charges were part of the exploitative machinery around electricity and the manner in which electricity is used as a political and economic weapon against the Black community in Soweto. Swilling and Shubane note that Johannesburg purchases R600 million (about $100 million) worth of electricity each year and sells this at a 25 percent profit. Most of the profits are used to subsidize the cost of services to other white rate payers, keeping down white payer rates for electricity and other services. Only 36 percent of Johannesburg's electricity supplies domestic customers with power; the remainder is sold to commercial and industrial enterprises. It is evident from this information that electricity costs for Black people are elevated and used to make profits, which are then used to subsidize white electricity and civic needs! Considering that 23 million Black people had no access to domestic electricity in 1992, this is yet another damning economic disgrace and another shame of apartheid.

128. Parker and Mokhesi-Parker, *In the Shadow of Sharpeville*, 21.

129. Ibid., 12.

130. Ibid.

131. Martin Murray, *South Africa: Time of Agony, Time of Destiny: The Upsurge of Popular Protest* (London: Verso, 1987), 253–256.

132. Anthony Sampson, *Black and Gold* (New York: Pantheon Books, 1987), 182.

133. Tatum, *South Africa*, 73.

134. Robert Price, *The Apartheid State in Crisis: Political Transformation in South Africa, 1975–1990* (Berkeley and Los Angeles: University of California Press, 1991), 214.

135. *Africa News*, June 11, 1988.

136. Sampson, *Black and Gold*, 73.

137. *Africa News*, November 16, 1987.

138. *City Press*, August 28, 1988.

139. Price, *The Apartheid State in Crisis*, 238–241.

140. Cited in the *Christian Science Monitor*, January 26, 1990.

141. Quoted in David Pallister, Sarah Stewart, and Ian Lepper, *South Africa Inc.: The Oppenheimer Empire* (New Haven and London: Yale University Press, 1987), 294. This point will be illuminated in detail in the following chapter.

142. Leonard Thompson, *A History of South Africa* (New York and New Haven: Yale University Press, 1995), 247.

143. *Reuters News Service,* Johannesburg, May 8, 1996.

144. Mzwanele Mayekiso's *Township Politics: Civic Struggles for a New South Africa* (New York: Monthly Review Press, 1996), provides an excellent insight into the struggles of the South African National Civic Organization on behalf of poor people in urban areas like Alexandra waging serious battles against white commercial banks that refused to offer lower rates of interest for low-income housing buyers (pp. 174–189), so that only the top 10 percent of the working class could qualify for R35,000 homes, the standard price for new homes in Alexandra Township. It is evident from Mayekiso's book that the question of a capitalist financial and commercial system, which empowers and privileges large white banks and disadvantages the poor section of the Black working class, is at the core of the problem around

housing. Mayekiso also expresses his frustrations with the new government in its hostile posture toward grassroots struggles for housing and land (p. 277).

145. *Mail and Guardian,* Johannesburg, May 21–27, 1999.

146. *New York Times,* June 7, 1999.

147. Carl Brecker, "The ANC After Mandela," *International Viewpoint,* June 1999.

148. See the author's article "Indigenous African Philosophies and Socio-Educational Transformation in Post Apartheid South Africa," in Philip Higgs et al., eds., *African Voices in Education* (Cape Town: Jutas, 2000), for a comprehensive illumination and assessment of the current educational crisis facing South Africa.

149. *Mail and Guardian,* Johannesburg, May 21–27, 1999.

150. Thompson, *A History of South Africa,* 264.

151. *Financial Mail,* May 30, 1996.

152. *Chronicle of Higher Education,* April 17, 1999.

153. Jordi Martorell, "ANC Victory: Masses Expect Action," Black Radical Congress Distribution on the Internet, June 16, 1999. See also *New African,* June 1999.

154. Martorell, "ANC Victory: Masses Expect Action."

155. *Sunday Independent* editorial, cited in *South Africa: Highlights from the South African Media* (Chicago: South African Consulate General), August 1999.

156. Ibid.

157. *In Defence,* published by the Socialist Party of Azania, Vol. 1, No. 1, September 1999.

158. *New African,* June 1999.

2

Why Apartheid Changed
Its Character in 1990

When F. W. De Klerk, the newly elected leader of the then-ruling white National
Party declared in February 1990 that all previously banned organizations would be
legalized and that Nelson Mandela and other imprisoned leaders of the African National Congress and other liberation movement groups would be released from
prison, many uncritical observers outside South Africa viewed these measures as
"historic" and an indication that the apartheid regime had finally come to its moral
senses and realized that apartheid was wrong. When one comprehensively considers
the sequence of events that led to the unusually conciliatory steps on the part of
apartheid's rulers, however, one realizes that there is much more to these "reformist"
acts than meets the eye. It was certainly no "road to Damascus" experience that led
to the apartheid system conceding these political changes within the apartheid landscape. As Martin Murray begs the question:

> Can De Klerk's actions be understood as a magnanimous gesture, symbolizing, like Paul
> on the road to Damascus, a genuine "change of heart" or were they cynical ploys de-
> signed to "buy time" for the propertied white oligarchy? . . . Is apartheid really dead, or
> will it continue to survive, tragically metamorphosed in a ghastly new guise?[1]

Although many will argue that the compromise agreement that was reached between the National Party and the ANC was "better than nothing" and that the negotiated settlement was a necessary step in the right direction to avoid further bloodshed, it is crucial to underline the point that for most of the Black masses in South
Africa, life is not getting better. For most Black people who generally belong to the
working or sub-working class, life has become much harder through the escalating
rise in the cost of living, with wages either decreasing or remaining stagnant in terms
of real value. The unemployment rate is still incredibly high, and with the devaluation of the South African currency, Black workers are required to work much harder

now than in 1995 solely to make ends meet at the level of basic subsistence. The obscene disparities of wealth between the wealthy, and the lower classes continue to
widen, and the white rich and their surrogate Black elites become richer, while the
Black masses grow poorer.[2]

Palpably, the line "something is better than nothing" cannot be asserted when
"something" refers to the compounding of economic exploitation and impoverishment for a people who have had "nothing" since the entrenchment of colonization
on South African soil. The chains of colonialist domination and capitalist exploitation in South Africa have neither been weakened nor removed; they have simply
changed their coloration and assumed a modernistic character.

Capitalism Promotes Post-Apartheid

In the South African context, race is a foundational and decisive socioeconomic indicator, with whites generally wielding economic affluence and social influence and
Blacks basically remaining economically disadvantaged. The class stratifications that
exist in the Black community, which will be discussed in detail in the following
chapter, are essentially superficial creations of the white capitalist and colonialist
power structure. The white capitalist ruling class, which is the legacy of Western European colonial penetration of South Africa in the mid- to late 1800s, controls economic power while delivering material benefits and according a bourgeois lifestyle to
the bulk of the white community. This class coheres with the political-colonial elite,
and together they share political, economic, and social power, while concurring in a
united movement to suppress and exploit the overwhelming majority (up to 90 percent) of Black people who are working class or sub-working class, essentially a colonized population. When discussing the South African economy in Western capitalist
terms, the economy is "white," in that it is essentially controlled by and basically
benefits white people, to the peripheralization of the Black majority. In 1993, "all
Black owned businesses combined account[ed] for less than 1 percent of South
Africa's total economic output."[3] In 1998, this figure increased by a few percentage
points so that Blacks owned between 6 percent and 10 percent of shares on the Johannesburg Stock Exchange, with a 4 percent beneficial ownership.[4] The totality of
white economic benefit from Black income is illustrated, for instance, in the fact
that in 1987, though Soweto's workforce earned about R2 billion (about $1 billion
then), over R1 billion (about $500 million in 1987) was spent back in Johannes-
burg's central business district, an essentially white populated area, according to
PLANACT, a progressive organization working with civic groups in Johannesburg.[5]
Johannesburg derives large portions of nondomestic sources of income that accrues
from Black labor and consumer spending.

Capitalism has historically defined and continues to define the quintessential contours of South Africa's political economy. As the apartheid edifice began to crumble
because of sociopolitical pressures building through wide-scale Black discontent and
global repudiation of apartheid from many socialist-bloc and underdeveloped na-

tions, capital was perspicacious. Capital realized that the only way to ensure the inevitability of capitalist entrenchment in a future Black-governed state was through carefully planned strategizing that conveyed the impression of "Black economic advancement." This strategy would necessarily involve the creation and nurturing of a vibrant and visible yet minuscule Black elite that would be co-opted into supporting white capital through monetary incentives such as corporate board memberships and stock ownership options. Martin Murray reinforces this contention:

> In maneuvers designed to pre-empt radical restructuring of big corporations through anti-trust legislation and to avoid compulsory affirmative-action programmes meant to increase employment and training opportunities for black employees, several conglomerates moved fast to bring black people into management and to appoint them to corporate boards of directors, to transfer some assets to black-owned or -managed consortia, and to "unbundle" huge holding companies.[6]

Capital hoped and continues to hope that this class, once fully cultivated, would champion the capitalist cause, envisaging that the Black elite would eventually become so thoroughly enamored with capitalism as to believe that capitalism is the only route to Black economic power. This elite would be urged and indoctrinated to oppose Black worker interests and even identify poor Black people as being problematic and obstacles to the "engine of economic growth" underway in the post-apartheid South Africa. In essence, this Black elite would be drawn into a politics of accommodation with capital to the point that it would be anesthetized to Black working-class and underclass poverty and exploitation.

The class stratification within the Black community is an integral product of capitalist strategy in South Africa and is designed to exacerbate socioeconomic tensions and foster internecine conflict, basically maintaining an environment of group instability and preventing collective Black solidarity. It is geared toward undermining any potential for revolutionary socialistic aspirations among Black working-class organizations through institutionalized marginalization of such tendencies. Ultimately, it is a systematic attempt by capital to legitimate monopolistic European and non-African capitalist ownership of South Africa's resources. This class division is spearheaded by white capital, and its plan is to ensure the maximum cooperation of the Black rulers of South Africa through promises of lucrative economic gain.

Within the white community, the divisions between the white political elite and the white owners of capital and between the so-called liberal white establishment and the ultra-right wing in South Africa are more *tactical* than *real*. This is not to diminish by any extent the contributions of dedicated individuals like Helen Joseph, Beyers Naude, the late Joe Slovo, the late Ruth First, the late Eddie Webster, Alby Sachs, Braam Fischer, Trevor Huddleston, and several other white activists who have fought tirelessly against the apartheid system, with some sacrificing their lives.[7] The unmitigated fact is that the overwhelming majority of the white population staunchly adheres to the ideology of white supremacy and demands retention of the historic white privilege foisted upon Black people from the early colonial period and entrenched through the era of apartheid. White privilege and protection, particularly Afrikaner privilege, was so heavily enshrined under apartheid because the system functioned to protect unskilled white Afrikaners from the demands of the private-sector labor market.[8] The result was a bloated apartheid civil service that employed hundreds of thousands of Afrikaners to enforce the historical white-beneficiary system.

Although there has been an increase of low-income whites due to changing economic dynamics of capital, the rates for white poverty are negligible. Most whites, even those in the working class, still earn enough to employ poor Black women as maids and cooks. *Generally*, there is still a consensus among the various white political tendencies (with the Afrikaner ultra-right wing expressing disgruntlement with the negotiation process so as to create further Black discord and confusion in the country) that the most genial way needs to be found to convey the impression to the Black majority that freedom is at hand, while preserving the essential pillars of white economic control and power. Under these circumstances, whites would have continued monopolistic ownership of land and material resources. In essence, the variables and nuances of "change" that are proposed by the white community more often than not function within a settler-colonial ethos whose very existence is predicated on the dispossession and conquest of the indigenous Black population. It may sound simplistic, yet it is true, borne out by the history of Europeans in South Africa since the seventeenth century: All of the initiatives that emanate from the white population are generally designed to legitimize and normalize the colonial occupation of South Africa by whites.

Consider, for instance, the fracas that resulted when Malegapuru Makgoba, the first Black deputy vice chancellor to be appointed at the "liberal" white-administered University of the Witwatersrand, was compelled to resign after he proposed the "Africanization" of the university.[9] In the eyes of many of the Black colonized, these actions erode the moral legitimacy of whites, substantiating the view that whites are determined to tenaciously adhere to the philosophy of white supremacy and dominance and viscerally oppose the Africanization of an African country that was Europeanized by colonialism.

To demand that Blacks accept the historical immorality of the colonial occupation of South Africa as moral is to insist that indigenous peoples deny their fundamental historical and inalienable right to ancestral lands that they have lived on and with for millennia. The moral dimension of the political struggle cannot be overlooked in assessing the South African political equation. Just because many theorists utilize a Eurocentric version of history to rationalize existing civilizations does not imply that nations that have been victims of these civilizations need to accept their (the colonizer's) standards of morality and their assumptions as definitive.[10] Because the land was pilfered from the indigenous African people by Europeans does not necessarily imply that Africans today have to accept this reality as a given. For indigenous peoples of color who have suffered colonial dispossession in the world, not all processes of history are irreversible. Although past history is of course unchangeable, the *effects* of history can certainly be changed through measures of redress, reparation, repossession, and reclamation.

The view promulgated by some theorists who contend that apartheid was no longer viable as it existed in 1989 because it stymied industrial capitalist growth owing to its rigid racial codification may have some plausibility. Ruth Milkman describes this position as "the fundamental incompatibility of the industrial economy and apartheid,"[11] necessitating the ultimate modification of apartheid through a comprehensive reform process. It is becoming abundantly clear from the unfolding of events in South Africa today that the colonialists who ruled South Africa were also capitalists. The neoclassical analysis of the South African apartheid state that disassociates apartheid from capitalist development is implausible in view of the fact that the mineral discoveries in South Africa in the mid- to late 1800s were intrinsic to the emergence of industrial capitalism conceived in Western Europe. Neo-Marxist criticism analyzes the South African situation as one of "superexploitation" of Black workers, yet it is weak in not adequately acknowledging the colonial trajectory that established white supremacy as a ruling ideology as codified in apartheid in 1948, marking the convergence of race and class.[12] Apartheid signified a culmination of European domination, following three centuries of indigenous African dispossession by racist colonialism and the violent hegemonic ascent of capital.

Anthony Lowenberg and William Kaempfer are incorrect when they argue that "South African capitalists . . . could not have possibly have been inherently racist, because if they were . . . why was it necessary to enforce discrimination by law?" and that white capitalists were generally inclined to support Black workers rather than white workers because of the low cost involved in Black labor.[13] What these authors fail to acknowledge in their dependence on such works as W. H. Hutt's *The Economics of the Color Bar,* Mats Lundahl's *Apartheid in Theory and Practice: An Economic Analysis,* and Walter Williams's *South Africa's War Against Capitalism*[14] is the pervasive historical character of racism that developed as a global ideology following the establishment of the institutions of slavery and colonialism that conditioned all Europeans, including capitalists who were bent on accruing optimal profits by every means possible. Although these capitalists were profit driven and inclined to reduce costs wherever possible, they were nevertheless willing to forgo profit maximization when it came to white workers, following the 1922 strike by white miners, for instance. It was a verification that both industrial owners and workers belonged to the same racial group: white colonizers, albeit from different economic positions.

It is imperative to bear in mind that capitalism developed as a logical extension of slavery and colonialism, in which Black and all indigenous non-European life was essentially devalued and dehumanized by *all* sectors of European society. Believing that white capitalists disowned or distanced themselves from white workers owing to the capitalists' possibility of making larger profit margins on cheap Black labor is to downplay the very deep and penetrating foundations of white racism, which, notwithstanding its economic variegations and distinctions, is still determined to protect all members of the white race.[15] White supremacy as a result of slavery and colonialism may have manifested itself in differentiated forms during the early colonial era and in the industrial apartheid era, but it nevertheless remained a racist cornerstone in that all whites were viewed as human and all Blacks were perceived to be subhuman beings.[16]

At a critical juncture of their history of oppression in the mid-1980s, these oppressors realized that apartheid was incongruent with global capitalist evolution and stability. They recognized that "naked" apartheid or "grand" apartheid as we knew it was inimical to the acquisition of maximal profits and ironically grew "bad for big business." They decided that they would discard the label "apartheid" and the more repulsive elements of the colonial system. To demonstrate their political goodwill, they went even further, and in October 1992, abolished the racist legislation that had informed the myriad statutes of apartheid for over four decades. They wanted the world to know that apartheid was a part of the past and was *dead* (as one ruling Nationalist Party member responded to a question that was raised at a meeting of various political groups held in Durban in 1992) and that as in the United States, racial discrimination of any kind would be absolutely intolerable in a "new South Africa." Economic and political considerations were foremost in the minds of apartheid's rulers as they began dismantling components of the apartheid machine.

It needs to be borne in mind that already in the 1970s, the vicissitudes of capitalist evolution and political volatility in South Africa had much to do with the state of crisis that loomed in the late 1980s and early 1990s.[17] Between 1973 and 1979, the South African capitalist economy underwent severe strains and entered a recession, marked by deep deficits in its balance of payments.[18] The waves of massive Black resistance exacerbated an already pulverized and volatile capitalist climate, particularly with the historic Soweto student insurrection of 1976. However, as is customary with many other instances of capitalist behavior, the microcosmic situation of a vulnerable South African economy was rescued by the intervention of global capital via the ascent in the price of gold, which reached close to $800 per ounce in 1980, rising from a low figure of $350 per ounce. The abundant revenues from an inflated gold price helped alleviate the crisis within the deficits of balance of payments in government financial circles. The regime earned over $3.8 billion from gold sales in 1980 alone. The euphoria from the sharp increases in the gold price was short-lived, however. In 1981, the gold price dropped to about $350 per ounce. The stability of South Africa's economy was further shaken with the Islamic revolution in 1979 in Iran that overthrew Shah Pahlavi, a longtime ally of the apartheid junta. Pahlavi had continued supplying oil to apartheid South Africa, despite an international oil embargo, since Iran was a serious U.S. imperialist surrogate at that stage of the Islamic nation's history.[19]

The global recession of the early 1980s certainly had an impact on the further weakening of the apartheid economy, as it became increasingly difficult to justify the escalation of corporate involvement in South African apartheid. From 1975 through 1989, over 50 percent of net savings were sent abroad, and existing debt repayments cost the country almost $8 billion until 1993.[20] The impact of limited trade and commercial sanctions by many of the Western capitalist nations, notably, the United States, Canada, Australia, and the European Community, the divestment of capital, and the demand of repayment on outstanding international bank loans had all taken their toll on the South African economy since the mid-1980s, costing "South Africa $32 billion to $40 billion between 1985 and 1989, including $11 billion in net capital outflows and $4 billion in lost export earnings."[21] Between 1981 and 1986, South Africa lost $3.7 billion in capital flight.[22] From 1980 to 1985, indebtedness rose to 41 percent of the country's gross domestic product, forcing the regime in the post-1985 era to dip into its domestic savings to fund domestic development and pay off external creditors.[23]

Scores of transnational corporations and their subsidiaries based in North America, Europe, and Japan withdrew from South Africa during that period, even though many like IBM deceptively sold their franchises to local white entrepreneurs with the assurance of the continued supply of components and technology from parent corporations. Some, such as Coca-Cola, moved across the border to places like Swaziland. A few companies, such as Kodak, which was involved in developing film for the nefarious passbook system, withdrew totally. Some of these companies have returned since 1994 or are considering returning to South Africa, although the general mood of U.S. companies is cautious.[24] The consequences of the divestment and disinvestment campaign in the United States had a deleterious effect on South Africa's economic capacity and drained the country of much-needed capital at a time when the global recession in the early 1980s was being felt even among the powerful capitalist nations. The Pretoria regime experienced serious economic isolation and lack of economic self-confidence since so much of the apartheid economy was predicated on global trade and foreign investment. Limited economic sanctions, though grudgingly imposed by many Western capitalist governments such as the United States via the Comprehensive Anti-Apartheid Act of 1986, which banned new corporate investments and bank loans, and the European Community, which prohibited the imports of gold, steel, gold coins, and new investments by European firms, began to have an impact on South Africa's domestic economy:

> A 1989 econometric study by the economics division of South Africa's Trust Bank found that since 1985, sanctions and disinvestment had led to a cumulative foreign exchange loss of about $15.21 billion. The indirect impact was far greater than those directly measurable and more visible consequences on specifically targeted industries. Production losses (because of the macro-multiplier effect of $15 billion on GDP) may be about $30.42 billion, and the total spending or "standard of living" loss approaches $38 billion. And as consumer spending is estimated to have decreased by approximately 15 percent because of sanctions, there were 500,000 fewer jobs than would have been the case without sanctions.[25]

Between 1985 and 1990, the gross domestic product growth rate was under 2 percent.[26] The white and Black middle class was thus being affected by sanctions, particularly in terms of material prosperity and financial mobility. Apartheid, as it stood, had benefited white capitalists and the white bourgeoisie for decades through exploitation of cheap Black labor and draconian laws of repression; but the time had come when it was clearly too costly, especially for big business.[27] In the minds of apartheid's architects and capital's power brokers, a resolution had to be found in which whites would retain economic power while allowing some Black people to assume the reins of political leadership.

Black Resistance: Pressure for Post-Apartheid Rhetoric

Nevertheless, the hand of history was shaking the corridors of the apartheid machine. Domestically, the regime found that it faced a Black majority population that was no longer tractable. The level of resistance by the youth, especially in the townships, reached new proportions.[28] The security forces were subject to physical attack by angry cadres of liberation organizations. People refused to pay rents for miserable housing in many townships, particularly in townships around Vereeniging in Gauteng Province (formerly the Transvaal) and in Soweto, where the rent boycott was efficacious and lasted from 1986 through 1990.[29] Local municipal authorities and state-appointed councillors were targeted by many in the Black community because in the eyes of Black people, it was unpardonable for Blacks to willfully collaborate with the structures of apartheid and perform the dirty job of apartheid management for the white colonial master.

An unrelenting spirit both of defiance and fearlessness swept through many Black townships, creating a situation of ungovernability and forcing apartheid's rulers to declare a State of Emergency in 1985 when it appeared that rule by a white minority over a Black majority was becoming unmanageable. The days of apartheid rule, as the system was defined in the pre-1990s, were clearly numbered. It just did not make too much political and economic sense for a group of 4.5 million people of European descent to rule a 30-million Black majority with an iron fist and to expect that the response from the subjugated population would be amicable. Most important, the financial cost of maintaining apartheid grew unbearable, and the weakening of the South African economy due to sanctions and other economic pressures was penetrating enough to render apartheid "cost ineffective" and bad for big business. For the first time, the defenders of the apartheid system began to seriously examine their policies under "grand apartheid." They subsequently spent the next five years evolving a calculated strategy aimed at containing the dissent within the ranks of the oppressed Black majority, as well as devising schemes that would extricate South Africa from an unprecedented economic mess in order to attract new foreign capital investment.

It was becoming increasingly evident that apartheid as it existed then in its crude form of overt and legalized racial domination and discrimination would not be able to survive under such conditions of external pressure and internal resistance. Some new initiative had to be promulgated to enable the beleaguered white minority regime to rule an African majority more effectively. This new initiative was conceived with the principal objective of primarily protecting the interests of the white capitalist ruling class and secondarily ensuring that the white citizenry of South Africa would be shielded from further economic isolation, social insecurity, and political uncertainty. It was not, as Heribert Adam and Kogila Moodley argue, because whites "wished to stay and cooperate with the black majority."[30] The white minority was primarily concerned with ensuring its own survival "by any means necessary," including accepting a form of Black rule, so long as their economic and social interests were protected. Their willingness to accept change was motivated by cynical group self-interest as opposed to desiring genuine interracial cooperation and democracy.

This is another important point that needs to be underscored in the equation of South African politics: In addition to South Africa representing a colonial entity on the African continent, it also functions as a pivotal capitalist enclave within the domain of global geopolitics. The fact that capital investments in South Africa by the European capitalist world, Canada, the United States, and Japan exceed those nations' investments in all other African nations combined illustrates the precedence of economic considerations when analyzing the South African situation. This assertion is further substantiated by the fact that there were still over 135 U.S. corporations functioning in South Africa even in the aftermath of economic sanctions legislation that curtailed investments in South Africa with the passage of the Comprehensive Anti-Apartheid Act by the U.S. Congress in 1986.[31]

Another decisive factor that changed the dynamics and tide of global geopolitics and that had a bearing on the South African struggle was the collapse of the Soviet Union in 1989. As the leading organizational protagonist against the apartheid regime, the African National Congress was heavily dependent on the military, technical, and financial assistance that it received from the Soviet Union, a world power, and other Eastern-bloc socialist countries. When the events of 1989 unfolded, leading to the disintegration of the Soviet Union, it was obvious the assistance the ANC had counted upon would no longer be forthcoming. In turn, Gorbachev's Soviet reform program also emboldened De Klerk and functioned as a reassurance to his white constituency that Moscow was no longer a "Communist threat" pressuring negotiations toward Black majority rule.[32] There was subsequent pressure exerted on the Southern African states that had provided refuge and shelter to the ANC by Western governments to desist from continuing to extend such aid and to urge the ANC to pursue a negotiated settlement to the South African conflict, abandoning armed struggle as a realistic option.[33]

There is no question that the dismantling of the Soviet Union precipitated the wave of political maneuvers that occurred within South Africa, seriously effecting both South Africa and Namibia, which had been previously colonized by South Africa, and culminating with the institutionalization of the democratic process as extant in South Africa today. As diplomatic relations between the United States and the Soviet Union warmed in the late 1980s, so too both superpowers sought to find a negotiated settlement to white minority rule in Southern Africa.[34] In fact, both the United States and the Soviet Union insisted that the protection of white "minority rights" be assured in any new post-apartheid dispensation.[35]

The heroic involvement of Cuban military, technical, and medical personnel in Angola resulted in the decisive defeat of South African troops in Cuito Cuinavale in

July 1988, through the combined tactical strength of the Angolan and Cuban forces. For the first time, the apartheid military forces realized that they were not as invincible as they had projected to the Southern African nations. The change in military power balance between South Africa and Angola led to a process where a peaceful withdrawal of Cuban troops from Angola was brokered by the United States and the Soviet Union in early 1989, laying the groundwork for the independence of Namibia from South African colonial occupation in 1990 as part of the mediated agreement. Following a relatively amicable resolution to the Namibian situation, South Africa remained the last bastion of white minority rule in the region and could not viably survive in its then apartheid proportions. There was a unanimous resolution among the various parties involved, namely, the ANC, the apartheid regime, the United States, and the Soviet Union, that an equally congenial process needed to be initiated to pave the way for a post-apartheid South Africa.

The ANC had little choice, given its position of vulnerability within the sphere of global politics and its location within the two Frontline States of Zambia and Tanzania. Since its military forces were based outside in neighboring states such as Angola and Zambia, where pressures of the guerrilla war against South Africa and economic destabilization by South Africa had begun to take their toll, the ANC found that it had no allies upon which it could depend to pursue an intensification of the armed struggle. Such events as the demise of the Soviet Union may have also impinged on the change in economic policy within the ANC once it became immersed in negotiations, moving from a position of "nationalization" to one of supposed pragmatism: coexistence with capitalist instrumentalities and collaboration with the market system.

Post-Apartheid: The Politics and Economics of Survival for the White Capitalist Class

The historical and contemporary dynamics of the South African situation as described above point to one unequivocal truth: The white ruling regime in South Africa sought to find creative ways to ensure its very survival, particularly in the interest of preserving the domination of capital "by any means necessary." The visible result was the dramatic unfolding of the South African political drama as seen at the World Trade Center in Kempton Park, close to Johannesburg International Airport. The stage for this "historic change" in South Africa, as the media hailed the measures of the De Klerk junta, had been set. "Historic" was the adjective used to describe the process of what had transpired in South Africa, because for the very first time in colonial South Africa, Black people were being granted the opportunity to cast their votes for a "new South Africa." Since Black people had never voted in national elections in colonial South Africa before, this "historic" moment supposedly signified that Blacks would now be able to count in the same proportions as whites did, since both Blacks and whites would be considered on an equal footing at the ballot box. For the first time, it was believed that Black life in South Africa would be viewed as equal in value to white life. At least this is what people the world over were led to believe.

From the vantage point of the conscious Black oppressed segment that realized what was really occurring, however, such as the Azanian People's Organization, a political grouping based on the Black Consciousness philosophy espoused by Steve Biko, the question of the evasion of the truth of neocolonialism was still an issue that needed to be spelled out and understood by the Black majority of South Africa and the rest of the world.

There are thus several subquestions that emerge from the euphoria that swept parts of South Africa after 1990, a sanguineness often promoted by the Western mainstream media in newspapers such as the *New York Times* and the London *Times*. It is imperative that many of the distorted perceptions created by such media be corrected by raising some further clear-cut questions.

Negotiations and Post-Apartheid: A Black-Consciousness Critique

First, the question of the intention of the ruling white regime in South Africa in dismantling apartheid after decades of its rigid and violent enforcement needs to be raised. There was certainly no overnight conversion on the part of the historical capitalist oppressors in South Africa. It was not a case where whites underwent a "road to Damascus" experience in 1990 and remorsefully believed that Black people were human after all and therefore needed to be brought to eat at the table of economic prosperity, a position that whites in the country had always taken for granted and enjoyed. F. W. De Klerk and his associates did not signify an exception in the historical accounts of relations between oppressors and oppressed, by volitionally abdicating power to the oppressed at the cost of losing the regime's position of power, privilege, and advantage. F. W. De Klerk was not going to vote himself and his party out of political power, particularly while he was in power. No ruling class in history has given up power voluntarily, and there are no tangible data to suggest that the events of 1990 in South Africa were any different.

It needs to be borne in mind that De Klerk's primary constituency was the white electorate of South Africa. He was elected by whites for the sole purpose of effectively adhering to the mandate of preserving white power and privilege. For him to have abruptly reversed his stance and become beholden to the dispossessed and disenfranchised Black majority would have been tantamount to committing political and economic suicide. In actuality, F. W. De Klerk was not radically different from his predecessor, P. W. Botha. He was considered the "most conservative of the four contenders for the National Party's leadership," following Botha's resignation as leader of the National Party in February 1989.[36] De Klerk in fact challenged the position that apartheid was declared "a crime against humanity," and he refused to acknowledge this reality even after his "reform" program was underway, in his second submission to the Truth and Reconciliation Commission.[37] He was initially adamantly opposed to any form of negotiations with the ANC, as his brother Willem De Klerk declared.[38] If De Klerk had felt such revulsion for apartheid, he would in all probability have resigned from the white-controlled Nationalist Party and joined one of the country's liberation movements. De Klerk was not acting totally unilaterally and independently when he announced the decision to unban previously banned parties like the ANC, PAC, and the South African Communist Party, as T.R.H. Davenport argues.[39] Neither was he embarking on a venture that he feared he would lose. Former police operative Dirk Coetzee, a perpetrator of apartheid crimes against opponents, claimed, "They [the National Party] did not unban the ANC without deciding that they would steer the course."[40] Thus, to view De Klerk as a new political "time bomb" who had emerged suddenly is politically naive. De Klerk's ascent to power was at the instigation of capital and the "enlightened" sector of the Afrikaner community, both of which viewed the intransigent and defiant Botha as an obstacle to the course of "reform" needed for capitalist stability and prolonged white survival in South Africa, a situation no longer tenable under grand apartheid.[41]

In 1993, during the height of the climate of negotiations between the ANC and the then-ruling National Party, there were public announcements made claiming that Blacks and whites were going to "share power" in South Africa.[42] Practically speaking, power in capitalist societies cannot be equally shared politically and economically. Ultimately, in most instances someone or some group or organization has power while other groups and organizations do not. "Power sharing" was and is mythic, and when it comes to oppressed people sharing power with their oppressors, the oppressed almost always end up being more powerless than they were when they first entered the negotiation process in the course of liberation struggle. Archie Mafeje, the well-known Azanian sociologist, balked at the petit-bourgeois character of the negotiations between the ANC and the De Klerk regime, arguing that the "programmes of the South African liberation movements which were conceived 30–45 years ago, historically belong to the era of petit-bourgeois nationalism of the independence movement."[43] He explains that the absurdity of "power sharing" had to do with securing bourgeois rights and not the rights of workers and peasants, which the revolutionary movement ought to champion. He asserts:

> It is obvious that in the so-called power sharing that is being touted in South Africa, this category of people will be shunted off, as has happened elsewhere in Africa. The self-seeking Black petit-bourgeois elite is not unaware of the fact. This is why it is important that the new democracy must include the appropriation of value. In a society in which differentials in income between Black and White are palpable, it is meaningless to talk about "power-sharing" without raising the question of equity or redistribution of wealth.[44]

One of the serious drawbacks in the entire negotiation settlement process in South Africa was that the two organizations of the liberation movement, namely, the African National Congress and the Pan-Africanist Congress, were negotiating from a point of weakness rather than strength. These organizations were unable to inflict a military defeat of the apartheid regime, especially in the aftermath of the collapse of the Soviet Union in 1989, which had previously given economic and military support to the ANC in its armed offensive against Pretoria. Consequently, they opted for settling the South African conflict peacefully, a process urged by Nelson Mandela while in prison in 1989, as a tactical way of saving lives and preventing further bloodshed. The mandate from the Black majority prior to these decisions is still unclear. Peaceful solutions are desirable but not always efficacious, particularly for oppressed and colonized peoples when such peoples negotiate their freedom from a point of weakness. Under such terms of negotiations, there was bound to be significant compromise of the original objectives as declared by the liberation movement because the liberation organizations had no solid ground to stand on or defense to fall back on should the series of negotiations have failed.

The ANC found itself in the invidious position of negotiating from a weaker position, and as a result, it compromised on major principles of freedom in its negotiations with the apartheid oppressors.[45] The De Klerk regime portrayed itself as a weak negotiating partner so as to secure the optimal concessions, and the ANC falsely believed that it was entering the negotiations from a point of superior strength. The end result was that a watered-down resolution to the South African conflict was reached in which the Black oppressed were generally given a raw deal, with freedom described more in vague and rhetorical terms acceptable to the white minority than in concrete and tangible ways that would have empowered the impoverished and dispossessed Black people. Strini Moodley, a historical leader of the Black Consciousness Movement and a former publicity secretary of AZAPO, issued a perspicacious caveat in this regard:

> In today's euphoric climate of negotiations the BCM believes that the black working class is still too fractured and too weak to rush to the negotiating table. We believe that the negotiations that can take place now are those between and amongst the components within the broad liberation movement. It is pointless rushing to the negotiating table when the might of the ruling class has not been sufficiently dented by the collective pressure of the working class.[46]

Pointedly, the ANC capitulated on the question of its use of armed struggle prior to the negotiation and agreed to demobilize its military forces too easily. The ANC abandoned armed struggle to meet this prerequisite set by the apartheid regime. It surrendered knowledge of its armed wing to the white government. Although the ANC had prided itself in the past on being a liberation organization based on revolutionary principles, it did not realize that a fundamental objective of any liberation movement is to protect and maintain its military wing and resources, so that it has a backbone for support should the course of negotiations fail. It ought to have realized that an initial step in any course of radical struggle for liberation is to neutralize the military establishment of the oppressor, since the oppressor thrives on spreading fear among the oppressed via its military machinery.

Archie Mafeje faults the ANC for its lack of honesty in openly acknowledging that armed struggle was impractical following the onset of negotiations, while giving the impression that the Soviet Union was willing to "give it as much arms as before," a specious claim.[47] The ANC continued the propagandistic deception during the course of early negotiations that it would continue the armed struggle but would meticulously assure that "nobody gets hurt," and as the late Chris Hani conveyed in a BBC interview, "nobody" referred to white South Africans.[48] In reality, the armed struggle by the ANC was totally abandoned in 1990, as the Soviet Union withdrew its erstwhile level of tactical support in arms to the organization following its own demise in 1989.

Even in Zimbabwe, former President Canaan Banana, who was a freedom fighter with the Zimbabwe African National Union (ZANU), pointed out that ZANU was able to assume control of the Rhodesian military with immediate effect upon taking power in 1980 and that a serious weakness in the South African negotiations scenario was allowing the apartheid South African Defense Force (SADF) to remain intact even after elections, with peripheral or incidental changes in personnel and structure.[49] Should the apartheid forces opt for a military coup at some stage in the future, the SADF was and is certainly in a position to execute subversive action against a radical state. The leadership of the SADF continues to remain in the hands of the old apartheid generals, with a sprinkling of military cadres from the liberation movements in some senior positions.

The ANC made a tragic mistake in prematurely exposing its military cadres above ground and failing to arrange a complete takeover of the SADF. The reason for the ANC's weakened posture was its inability to inflict a military defeat of the apartheid regime. The ANC lacked the strategic capability to engage in sufficiently intensified and protracted armed struggle or to organize mass revolutionary resistance strong enough to overpower the military might of the apartheid regime. It was thus compelled to work within the existing armed forces structure of the previous regime as a face-saving measure in the wake of its poorly organized guerrilla war effort.

Second, even if De Klerk and his regime had reflected a modicum of integrity (and Black people in South Africa have had too much experience of racism and deception to believe this of any of the stalwarts of apartheid), the land question, which is of the greatest consequence in any discourse on change in South Africa, remains unanswered and was never raised as a primordial issue during the negotiations.[50] Richard Levin and Daniel Weiner write:

> Despite the central role which forced removals and land dispossession played in the historical development of colonialism and apartheid, land has not featured very highly on the ANC's agenda. This is partly a result of particular conceptions of development in South Africa in which the role of industrialization and the creation of a working class are the major priority, with little recognition that agrarian questions include the relationship between industry and agriculture, and town and country.[51]

As was the case in Zimbabwe with the Zimbabwe African National Union Patriotic Front government, urban industrial strategies were viewed as principal in economic transformation, discarding the critical role of land in rural peasant economies that functions to sustain people in these areas.[52] Themba Sono, from the Institute of African Studies at the University of Bophuthatswana, notes that during negotiations, unlike the PAC and AZAPO, which were concerned with repossession of rural lands, the ANC viewed the "urban landless and homeless group" as the focus of land reform policy.[53]

Today, as was the case under formal apartheid, 87 percent of South Africa's land area belongs to the 13 percent white minority population as a result of forced dispossession, relocation, and disinheritance of Black people during three and a half centuries of colonial rule. There will never be any peace or justice while 35 million Black people must still contend with living on the 13 percent remainder of the most barren and unproductive land. In actuality, fifty years after the enactment of the 1937 Land Act, which extended the area earmarked for Black occupation from 7 percent to 13 percent, 1,259,000 hectares of land earmarked for African occupation are still held by the South African Development Trust.[54] It is only after this land has been occupied by Blacks that one can accurately state that Black people reside on 13 percent of South Africa's land area. This is the only land that was being considered for redistribution under the apartheid government's White Paper on Land Reform.[55]

The new South African government has declared that 3.5 million people who have been forcibly removed under apartheid between 1913 and 1990 would be able to claim restitution for dispossessed land on condition that they can verify their claims. Should they be successful, the state would buy such land with some form of compensation to the white landowner.[56] Yet the question of the majority of the 40 million Black people whose ancestors were forced off their lands by the European colonizers since the latter's arrival in South Africa three and a half centuries ago still remains. European colonialism cannot be accepted as a de facto characteristic of the South African landscape, as most white critics and academics, including those on the "left" do. With the exception of a few white writers such as Richard Levin and Daniel Weiner, most are silent on the issue of whites owning the bulk of the land that was acquired through illegitimate means during the colonial occupation of South Africa beginning in 1652. The fact of this quiescence in discussions on postapartheid change by the white academic community, business community, government agencies, and nongovernment extraparliamentary organizations on such a decisive issue as land is amazing, even though everyone is fully aware that landlessness is the root cause of Black poverty.[57] So long as most Black people are landless in South Africa and forced to live in either overcrowded urban ghettos or arid and barren rural reserves, Black people, particularly the rural peasantry, will be poor and remain economically depressed, exploited, and powerless, forced to work for wealthy white farmers and industrialists for substandard wages.

Many white critics argue that for "pragmatic reasons" and for the cause of minimizing "white fears of Black retribution" and "white flight" such a policy of compromise by remaining diffident on the land question is necessary. They argue that land distribution is "unrealistic and even romantic."[58] They also insist on land compensation for white farmers whose land may be needed for redistribution among the 50 percent of people in the Bantustans who do not have access to arable land.[59] Since

Blacks were dispossessed of their ancestral lands, they do not need to be forced to buy back the *same* land that was stolen by European colonizers.

The prospect of compensation to wealthy white farmers who are beneficiaries of the European colonial legacy of theft of African lands is.problematic, particularly given that the average white farmer in South Africa still owns 988 hectares (2,470 acres) of farmland (even more than the farmer in the United States and Canada, who owns an average of 161 hectares), representing 25 percent of the country's 600,000 commercial farmers.[60] This wealthy farming elite produces 80 percent of the nation's food and earns the bulk of income from agricultural production.[61] The lands that make up the bulk of the South African agricultural land area must be redistributed among impoverished Black farmers, most of whom own an average of 22 hectares in the Bantustans.[62] White farmers owning farmland must count in the same proportions that Black farmers would, in ratio to the respective numbers of white and Black farmers, in any "free" South Africa. Whites cannot expect to be accorded preferential treatment in a new dispensation if post-apartheid society is geared toward *justice and equality.* If whites or any non-Africans desire to live in Africa, they must expect to live on the same terms that indigenous Africans do. Otherwise, they will continue to be perceived as colonizers and predators in Africa.

The fact of the matter was that under the negotiation agreements, justice was not an underlying principle. Concern for "minority rights," translated as "white privilege and power," has been the substitute for justice. The issues of land redistribution and colonization were never hammered out at the negotiation table at Kempton Park, although they held the key to liberation, justice, and peace in South Africa. The ANC was forced to compromise in negotiations with the National Party on this issue by agreeing to subsume land issues and land restoration under the Property Clause, which eventually became amended in the new constitution as the Restitution of Land Rights Act No. 78 of November 1996. Yet the damage was done, since land is not merely private property but a public possession, but it became a private possession under the violence of colonialism.[63] Khehla Shubane, a researcher at the University of Natal, emphasizes the point that colonialism was a given assumption from the inception of negotiations and, from the vantage point of the colonized, was never really and adequately dealt with. He argues:

> South Africa, although not dominated by a colonial master, is in very important respects yet to transcend its colonial origins. There is no disagreement about the historical origins of the South African state in colonialism. Many analysts assume, however, that colonialism ceased with the advent of the union of South Africa in 1910 which removed the nexus between the country and the then colonial power. This explanation neglects the issue of economic inequalities, and hence decolonisation was incomplete. As far as explaining the colonial problem from the point of view of the vast majority of the rest of the population, it has been most unpersuasive.
>
> The decolonization of South Africa involved the transference of political power from the colonialists to the white population. Institutions of self-rule subsequently designed for the Black people under apartheid resembled those under colonialism . . . Thus, for

black South Africans, colonial domination never ceased but merely changed form. Full decolonisation should therefore be sought.[64]

The skirting of such a decisive material issue implied a blatant disregard for the fundamental causes of internecine conflict in South Africa: poverty, hunger, homelessness, police violence, fragmentation of Black communities, and the intense frustration that has enveloped Black communities around the country.

Inanda, outside Durban in Kwazulu-Natal, for instance, is only one area where close to half a million Black people eke out an existence in shacks and subhuman dwellings within an area of twenty-five square miles. Overcrowding of Black residential areas and confining of Black people to small, scattered, and arid pieces of land produces anger and animosity within the Black community. The inevitable outcome is premature death and self-destruction by the impoverished over the inability to gain access to economic and material resources, especially land. There can be no genuine freedom *for* or power exercised *by* Black people who are landless in the country of their birth. Any discussion of transformation must entail the redistribution of land *among* and the return of the land *to* the dispossessed Black population, painful and traumatic as such a policy may be for wealthy white industrialists and large farmers and landowners.

Third, it is ironic that the string of reforms that F. W. De Klerk legislated in 1990 sounded precisely like the measures that were contained in the annual statement issued by Gavin Reilly, chairman of South Africa's largest corporation, Anglo American Corporation in 1986: unbanning political parties, freeing political prisoners, repealing key statutes of apartheid, and initiating the negotiation process.[65] It is no coincidence that these steps were called for by the owners and managers of the capitalist enterprise in South Africa so as to make capitalism look respectable in a "free and fair South Africa." Given capitalism's tarnished image due to its historical shameless and unabashed alliance with the unmitigated racism enshrined in apartheid, such "enlightened thinking" on the part of capital was clearly understandable and even logical.[66] It is worth noting that the South African regime that had previously controlled up to 50 percent of the country's gross domestic production had launched a massive program of privatization, selling off statal and parastatal sectors of the economy to private corporations such as in road and transport, airlines, telephone, telecommunications, health care, and so on since 1987.

In areas of wealth possession, such as on the Johannesburg Stock Exchange, in 1990, Anglo American Corporation owned close to half of all shares traded on the exchange, controlled sixteen of the top fifty companies, whose combined worth in assets was R98.8 billion (about $27 billion) and owned over 600 subsidiary companies.[67] Anglo's wealth exceeded that held by South Africa's statal corporations, some of which have become privatized, including the Reserve Bank; SASOL Oil from Coal; ISCOR, the steel corporation; ESKOM, the electricity commission; and Armscor, the regime's arms manufacturer.[68] In early 1995, the wealth of Central Selling Organization, the conglomerate that incorporates Anglo American and De Beers, was worth an estimated £21 billion (about R200 billion).[69] The mammoth tentacles of Anglo American have permeated virtually every sector of leading industries, making it

> The largest and most powerful South African-based conglomerate and multinational . . . Anglo owns 69 percent of the total capital invested in the South African mining industry. It has vast industrial holdings and is a primary force in agriculture. The Corporation also has a major presence in the South African insurance, finance, property, press and service sectors. It has extensive interests in all SADCC [Southern African Development Coordination Conference] countries. In 1981, Anglo became the single largest foreign investor in the United States, and it has substantial interests in many European and "Third World" countries.
>
> Anglo American and its immediate past chairman, Harry Oppenheimer, have long been regarded as the leading liberal monopoly force in South African politics. The group plays a key role in "reform" initiatives of monopoly capital, embodied in the Total Strategy.[70]

Together with three other corporations, namely Sanlam, Rembrandt, and S. A. Mutual, Anglo American controlled 80 percent of the values of stocks traded at the Johannesburg Stock Exchange in 1993, and it owns about 60 percent of total stocks today.[71] In 1994, the total capitalization value of the Johannesburg Stock Exchange was $264 billion, worth more than three times the stock exchanges of the rest of Africa put together and one of the largest of its kind in the world, largely on account of South Africa being a haven for foreign-investor capitalism.[72]

One would be hard-pressed to convince conscious Black people in South Africa that the wave of negotiations over apartheid's abolition was not being spearheaded and controlled from behind the scenes by transnational and local capitalist conglomerates. The desperation on the part of capital to perpetuate its hegemony in South Africa and create the impression of genuine commitment to democratic ideals and the eradication of racism has induced some of these large corporations like Sanlam to sell portions of their consortiums to Black financial organizations for the first time, since Black businesses constituted a mere 1 percent of South Africa's economic output in 1990.[73] These measures were part of the plan to legitimate essentially morally illegitimate institutions in the country, in view of the fact that these companies have historically exploited cheap Black labor and always had whites at the helm of ownership. It becomes more apparent with the evolution of events in South Africa that the interests of capital were foremost in the minds of the political negotiators representing the apartheid regime prior to 1990 and through the early 1990s.

Between the first post-apartheid elections on April 27, 1994, and the second on June 2, 1999, little has changed within the *fundamental* economic and social apartheid structures of the country. The changes that have occurred are highly cosmetic and profoundly superficial, essentially irrelevant to emancipating the Black poor, particularly in rural communities. White settlers still own the bulk of the land, either as farmers or corporate tycoons, while the vast majority of Black people still languish in overcrowded townships and rural reservations and are illiterate, often homeless, impoverished, and exploited by the continuance of the economic system under apartheid. There are still some 7 million Black people either homeless or living in subhuman housing in South Africa, Africa's wealthiest country. Over 50 percent of children born still die before the age of five in some rural areas. Forty percent of the Black urban population is unemployed, and most Black people still earn one-fifth of what whites earn. In 1991, 45 percent of the population earned below R600 per month—some 16 million people—and of this number, over 2.3 million suffered from acute malnutrition.[74] According to recent studies, reports reveal:

> Inequality is among the highest in the world. One-fifth to one-quarter of children under the age of six (in urban areas) are malnourished; only about one-fifth of Black families have incomes above the minimum effective level. The whites, although they only contribute 15 percent of the population of South Africa, lay claim to well more than half of the national income, and more than 90 percent of the wealth of the country is in their hands. More than one-fifth of all Black South Africans between the ages of 15 and 60 are unemployed, and in some areas unemployment is already as high as 50 percent. Every year the number of unemployed grows by nearly a quarter million.[75]

The Human Sciences Research Council provides contemporary statistics that are even more disturbing:

> [T]he poorest 40 percent of households in South Africa earn less than 6 percent of total income, while the richest 10 percent earn more than half the income. While Africans make up 76 percent of the population, the African share of income amounts to only 29 percent of total income. Whites who make up less than 13 percent of the population, take away 58.5 percent of total income. [The total Black population, which makes up 87 percent of the population, thus receives only 41.5 percent of the national income][76]
> . . . Of the poorest part of the population, a third live in shacks or "traditional dwellings." Of the poorest 53 percent of the population, about 80 percent have no access to piped water to their premises, and more than 80 percent no access to modern sanitation. Inequalities in education and health care are as striking. Every indicator, from attendance at university, to the incidence of TB or stunting rates among children, tell the same story: the wealthy have a great quality of life, and our community pays the price for their lifestyle . . . For those who become parents, the maternal mortality rate is 70 times higher among Africans than among whites. The cumulative effect of such inequity carries through life. Per capita, whites earn 9.5 times the income of Blacks and live, on average, 11.5 years longer. In sum, South Africa exhibits that most bitter of social incomes: destitution amid plenty.[77]

It is ironic that the statistics in the second paragraph are cited from a World Bank report, given that the World Bank promotes the kind of blatant injustices and disparities between the wealthy and the poor all over the underdeveloped world.

The obscenely skewed ownership of South Africa's resources is further substantiated by the following facts from 1994:

> Much of the economy is owned by a minority of a white minority; only 5 percent of all South Africans own 88 percent of the country's wealth; four large corporations—Anglo American, Rembrandt, S. A. Mutual, and Sanlam—control 81 percent of share capital; twenty families hold shares worth R10.7 billion plus other undisclosed personal assets; 87 percent of the land is owned by whites; 50,000 white farmers own 85 percent of all agricultural land . . . [78]

Seven hundred thousand Black farmers owned an average of 22 hectares in the arid Bantustans in 1990, a figure that has not radically changed since then.[79] In terms of commerce and business, the totality of white power is still unequivocated:

> Ninety-five percent of managerial jobs are held by whites; Blacks hold a mere 2 percent of the total of 2,550 directorships in the top 100 companies (*Star*, March 30, 1993); despite claims of overtaxation and declining living standards, in 1993, nearly half the top companies gave their board of directors pay rises which far outstripped company performance and the rate of inflation, and this was on top of incentive and pension schemes which top-up basic fees (for instance, Gencor gave its board a 100 percent salary increase despite a drop of 18 percent in earnings per share and dividend growth of less than 5 percent).[80]

It is thus not too difficult to comprehend why De Klerk and his regime demonstrated an unbelievable willingness and even enthusiasm in the negotiation process. The objective was the co-optation of legitimate Black leaders into the inner circle of capital's "free-market" system, with the promise of perquisites for all participants.

The apartheid regime was highly successful in this effort of co-optation, owing to its skillful, diplomatic, and adept maneuvering, in which leaders of the liberation movements were so overawed with the prospect of sitting in the same corporate boardrooms and parliamentary chambers as their previous apartheid oppressors that they were willing to overlook critical issues directed toward the transfer of political and economic power to the Black majority. In essence, these figures, who were originally committed to the elimination of the apartheid system because they categorically concluded that apartheid was incorrigible and therefore had to be destroyed, sad to say, abandoned their goal of being freed from the apartheid scourge and opted for *inclusion* within it. The leaders who historically viewed freedom from apartheid rule as their vocation have even hailed some of apartheid's former rulers as "heroic" and "courageous." The regime that was historically seen as oppressive and essentially responsible for the problem was viewed in the early 1990s as liberating and indispensable in being the key to the solution. Quite an ironic twist of history! Archie Mafeje contends that anything short of liberation that signifies "freedom of Blacks from White oppression and exploitation" is tantamount to betrayal by the liberation movements, principally the ANC.[81]

Clearly, the tantalizing tentacles of capitalism's promise of wealth and prosperity have been overpowering, to the extent that they were successful in seducing the historic leaders of the South African struggle into accepting an agreement in which capital would be assured of protection under the new regime, this time a seemingly legitimate one since it would reflect a Black-majority participation. The entrenchment of the monopoly of economic power by the same white elitist capitalist clique was ensured by expanding the ranks of this group to include *some* members of the disenfranchised Black majority.

The first "free and fair" democratic national election held in South Africa in April 1994 provided the much-needed international stamp of recognition for this façade of freedom. The world would no longer need to hear about the inhumanity of apartheid because everyone would be led to believe that Black people had been set free from the world's most hideous system since Nazi Germany. One must concede that it was indeed a masterful plan by oppressors to continue oppression, especially because the oppressors made the oppressed feel appreciative, as they were led to believe that their oppressors had been actively involved in setting them free.

The elections of 1994 were flawed primarily because the negotiations that led to the elections were problematic in the first place, in that they principally consisted of two major groups, the apartheid National Party and the ANC, the latter of which had already agreed to compromise and had been conducting secret negotiations even before the actual negotiations began under the auspices of the Convention for a Democratic South Africa (CODESA). Even external observers involved in promoting urban development in post-apartheid cities commented on the manner in which the ANC capitulated and accepted an accommodationist position during the negotiations, the legacy of which is evident in the weakness of the post-apartheid order today:

> The current trend indicates that the ANC will be forced to function within a "coalition" government with a long-term commitment to protect the fundamental features of the status quo. This imposed coalition is the result of the "war of exhaustion" that the government organized against the democratic movement. As a result, there is the danger that the ANC will achieve power in a situation where it is unlikely to effect substantial changes. The consequence might be uncertainty and economic instability, under a very weak government.[82]

The ANC recognized an illegitimate regime as legitimate as part of this compromise, because it was unable to defeat the apartheid machine decisively. Instead of seeking common ground with other liberation groups that represented the disenfranchised Black majority such as AZAPO, PAC, WOSA, and other Black civil organizations that were involved in radical struggles to eradicate apartheid, thus representing a unified front of the oppressed and colonized vis-à-vis the oppressor, the ANC sought to find unity with the forces of apartheid.[83] In fact, the ANC and the PAC were even willing to expunge AZAPO at the behest of Democratic Party leader Zac De Beer, when AZAPO called for the unity of the liberation forces and the expulsion from the negotiations of apartheid representatives such as the Bantustan leaders.[84] Yet the ANC itself had been engaged in a fierce war of resistance in the

Bantustans against these apartheid surrogate authorities in areas like the Ciskei and Transkei shortly before and after 1990, as in the Bisho confrontation in 1992, which resulted in several ANC members being massacred.

The ANC was little concerned about the principle of unity of the oppressed groups of the country prior to any negotiation with the oppressor. In essence, it was willing to forgo the establishment of consensus of the oppressed in its haste to sit at the negotiation table with the powerful oppressors. There was no referendum held among the oppressed Black population that called for unity and sought to harmonize philosophical, political, and strategic differences among the liberation movements so that a Black "patriotic front" could be created to stave off the devious and divisive tactics of the forces of apartheid. The absence of this solid and necessary political principle and the uncaring posture of the ANC on the issue of principle led to a distorted political platform representing the interests of the Black oppressed: CODESA. Given the presence of a preponderance of apartheid forces at the negotiations table, the foundation laid for the election was inevitably impuissant and toothless. The logical fruit of such hasty and ill-conceived negotiations was skewed elections in which a charade of democracy was evident, with the vying of such ridiculous parties as the KISS Party, the Football Party, and so forth. What could have been a genuine and earnest election in which the Black working class and its interests were well represented turned out to be a façade. Khehla Shubane refers to this phenomenon as a rejection of civil society and its supplantation by a liberation movement claiming to represent all tendencies, which was precisely the case in the negotiations scenario. Shubane contends:

> In the pursuit of liberation, the various interests of the colonially dominated people were collapsed into a single, overriding endeavour. Thus arose the notion of movements as authentic representatives of the people. This notion is not consonant with the plurality which is inherent in the notion of civil society. All interests had been reduced into an overriding one which was expressed by the liberation movement. Again, the autonomy which must be a characteristic of groups in civil society was done away with.[85]

The question of which groups were considered representative of the "liberation movements" is another critical issue that was never fundamentally addressed, as it was always assumed that the ANC was the most important liberation movement and all other groups were subsumed under this "umbrella movement," a fatal political flaw, as South African Blacks are progressively discovering.

The organizers of the elections were more interested in demonstrating the "new South Africa's" commitment to diversity and free expression by having a plethora of frivolous and highly irrelevant newly formed political parties for Western consumption and approval than they were in articulating the voices and aspirations of the millions of dispossessed and colonized Black people. The fundamental question of colonization was never raised by the principal organization ostensibly representing the colonized, since the objective of the ANC was never to *decolonize* South Africa. The colonized status of South Africa, along with its deformed structural inequalities and injustices, particularly regarding land and the natural resources, was accepted as a de facto situation by all parties involved throughout the course of the negotiations.

Fourth, even while negotiations continued through the early 1990s, there were many mass killings of Black people, and the police and security forces were generally unable to apprehend the killers, even though the security apparatus was highly effective during the days of formal apartheid. In a space of five weeks in July and the first week of August 1993, close to 600 people were killed, either by the security forces or by vigilante groups. Between July 1990 and June 1993, 9,325 deaths occurred as a result of "political violence," marking an average of 259 deaths per month and 8.5 deaths per day.[86] The media and white ruling class blamed the violence on Black "tribal conflict" or ideological differences between the ANC and the regime's surrogate party, the Inkatha Freedom Party. Both reasons were racist because they did not get to the heart of the cause of the real violence in South Africa: Brutal colonization of a Black majority by a racist white minority. Coupled with this fact was the indisputable evidence that a so-called third force orchestrated by the ruling National Party was created to cause havoc and perpetrate violence in the townships so as to maintain an aura of social instability.[87] The Goldstone Commission, or the Commission of Inquiry Regarding the Prevention of Public Violence and Intimidation, formed in 1991 by F. W. De Klerk in response to Black community pressure, revealed that the apartheid state and its security apparatuses were involved in township violence and that senior police officers were engaged in the trafficking of illegal arms and ammunition in Kwazulu-Natal in support of the Kwazulu police and elements within the Inkatha Freedom Party.[88]

The shocking statistics regarding the annihilation of Black life in South Africa between 1985 and 1994 substantiated this conviction: Over 10,000 Black people were killed in "political violence." Most commentators never raised the point that the overwhelming majority of those killed were desperately poor and were the most socially vulnerable within South African society. These "experts," often ensconced in the comfort of their plush white suburbs, never cited the underlying reasons for such violence being the poverty, squalor, and insecurity of Black people being forced to live in the overcrowded townships of Iphumalanga, Kwa Mashu, and Kattlehong or the dilapidated slums of Kwazakhele, Crossroads, Joza, and Canaan.

In South Africa in 1992, 23 million Black people had no access to electricity; there were approximately 604,000 hostel beds (with COSATU estimating that there were as many as six residents per hostel bed), 20 percent of the population (some 4 million people) had inadequate water supplies, and some 7 million people had poor sanitation.[89] In 1996, millions of Black people still lacked decent or any housing and were without electricity and running water; 400,000 Black people lived and continue to live in huge slums consisting of shacks and hovels in Crossroads, in Cape Town, which was South Africa's most scenic and beautiful city in 1999; and over 300,000 people continue to live in poverty in Alexandra, outside Johannesburg, South Africa's wealthiest city. In 1993, only 30 percent of Black South Africans had electricity, with only 10 percent in the urban areas receiving electric power.[90] Thirty-five percent of the population is still illiterate, and 60 percent of the Black population lives in absolute poverty. According to a World Bank report, "South Africa's health status relative to income was one of the worst in the world," and the July 15, 1993, issue of the *Star* newspaper of Johannesburg reported, "53 percent of South African children between two and five suffered from stunting as a result of malnutrition" compared with "an average of 39 percent in the rest of Africa." Further, in 1990–1991, "63 percent of children in South Africa under one received immunization against diphtheria, whooping cough and tetanus," compared with 89 percent of the children in Zimbabwe and 79 percent of Zambian children who also received equivalent shots. One report states: "There are over two million South Africans who are nutritionally deficient, with childhood nutrition being as high as 31 percent among pre-school African children because of rural poverty."[91]

These conditions must be described as *structurally violent* because they continue to decimate Black people in South Africa. They represent "primary or institutionalized violence" that leads to "secondary factional violence."

Regarding the issue of "Black on Black violence," as the media still unthinkingly use the phrase, two incisive points need to be made: First, it is the apartheid regime that stood to benefit maximally from the situation of internecine conflict in the Black community in South Africa, and second, the police and security forces were and are always, almost without exception, unable to apprehend the killers of Black people primarily because Black life is considered expendable. During the 1960s, 1970s, and 1980s, the police force with its massive internal security apparatus had little difficulty in locating the opponents of apartheid, many of whom were in hiding, and subsequently arrested people by the thousands. However, in the case of violent acts against white citizens, the police have generally been successful in finding the culprits, through infusion of more personnel and the offer of rewards for information. Many people in the Black community are convinced that the security forces have had a major hand in secretly funding and organizing these assassinations of Black people, dating from the era of apartheid through to the present. The evidence collected from newspaper reports and eyewitness accounts at various crime scenes point clearly in this direction.[92] Further, Black people unable to speak any South African languages were apprehended by local activists at massacre sites, an indication that foreign mercenaries were being brought into South Africa to destabilize Black communities and terrorize Black residents, all while the "peace" of negotiations was being pursued.

These incidents lead to a deeper questioning of the integrity of the previous De Klerk regime, widely believed by the Black community to have been responsible for fomenting violence in Black communities so as to create sharper cleavages and devolve the South African situation into a climate of uncontrollable and random violence. The objective was to overwhelm Black people with such terror and fear by these atrocities that they would be coerced, manipulated, and lulled into accepting any watered-down solution in a desperate bid to end the bloodshed. In the words of a mother who lost her relatives in Daveyton Township, "[A]nything is better than this violence . . . and when are we going to stop killing each other?"[93] On the one hand, the regime mouthed the niceties of "peace." On the other hand, it either killed

Black people or allowed Black deaths randomly, corroborating the point that in the eyes of the apartheid regime, Black people were totally dispensable.[94]

Fifth, the twenty-four parties that were part of the negotiation process, with the exception of the ANC and the PAC, were all essentially apartheid parties, representing the Bantustans, the tricameral groups, and other pro-apartheid entities. It was a flagrant violation of fundamental democratic principle that parties that have never been and were totally unrepresentative of the vast Black majority were involved in representing the aspirations of the same Black majority. There was no referendum taken among members of the Black community to gauge which groups or organizations Black people would choose to represent their interests at the negotiating table.[95] Barring the ANC and the PAC, the grouping of Kempton Park participants had no earnest mandate to represent Black people in South Africa, making it possible to challenge whether the group had a legitimate mandate to draw up an interim constitution.

Finally, even if one were to take the South African regime at its word (which, given its history of duplicity, was worth little or nothing) that the elections planned for 1994 were authentic, there was no equal proportionate representation of whites and Blacks in the national assembly. It took four or five Black votes to elect a Black person to the assembly but only one white vote to elect a white representative, as the result of a compromise reached between the National Party and the ANC. Minority parties were represented in Parliament in excess of their showing in the April 1994 elections. Unlike most other democratic nations with proportionate national representation, whites in South Africa continued to have overrepresentation in the Black majority Parliament.

The agreement to draw up a new constitution was signed between De Klerk's National Party and the ANC on November 17, 1993, and clearly evinced that whites would "still dominate the economy and civil service."[96] Whites still do control South Africa's economy and civil service, although the constitutional agreement was ostensibly about a transition to democracy and majority rule. The precise reason for this situation is that the agreement was not about *the transfer of power* from the colonial white minority to the Black majority but instead was about "power sharing," as De Klerk had insisted upon and many in the ANC had subsequently denied.

The elements of the federalism model of dividing South Africa into nine provinces as proposed by De Klerk's colonial order, to which the ANC tragically consented, represented another sophisticated version of the entrenchment of white domination of South Africa. These changes were predicated on a racist perception of humanity that accords the white person a value four or five times that of a Black person even in an exercise of the fundamental civil right of voting. It is clear that the division of the country into these different regions ultimately reproduced similar enclaves of white domination, since whites and Blacks have lived in separate areas under apartheid for generations (the April 27, 1994, election results in the Western Cape, the most scenic and beautiful province of South Africa, for instance, confirmed this view, in that the National Party was able to win a majority because of white and "Colored" support, even though the 1999 election results putatively changed this).[97] It was unnecessary to further divide the country into nine provinces because such divisions were inevitably going to result in a further fragmentation of the Black vote, giving minority whites larger majorities in these newly carved provinces. Whites were thus assured of receiving representation within the South African legislature that exceeded their composition within the country's overall population. This rezoning was akin to the racist move in the United States in which predominantly Black districts were ruled unconstitutional by the high courts in 1995 because the courts claimed that these districts supposedly entrenched racial divisions.[98] The actual effect of redistricting was fragmentation of the Black vote. Given the obduracy of white racist attitudes and proclivities in South Africa (an issue that was not fundamentally addressed in the wave of negotiations of the early 1990s except for a superficial bill of rights stating that discrimination on the grounds of race, gender, sexual orientation, physical ability, and age are prohibited), such divisions cannot be viewed tritely.

On May 8, 1996, many Western newspapers reported that the South African Parliament had finally ratified the country's 200-page Constitution, ushering in a new era that would lead South Africa into the "democratic circle of *Western* nations (italics mine)."[99] The Constitution would make it possible for the completion of Western parliamentary democracy, with the winning party of the 1999 elections being empowered to form the government based on its majority victory without needing to constitute a coalition government with opposition parties, as is the case currently. The Constitution is reflective of broad democratic provisions found in other major democracies:

> Its bill of rights contains everything in the U.S. constitution and more. It specifically prohibits discrimination on the basis of race, gender, sex, pregnancy, marital status, ethnic or social origin, color, sexual orientation, age, disability, religion, conscience, belief, culture, language, and birth . . .
>
> . . . The constitution also enshrines socio-economic rights, making access to housing, health care, food, water, and social security inalienable rights, "*within [the state's] available resources.*" (italics mine)[100]

Inasmuch as these constitutional provisos and protections are important in defining principles of fairness and equity, they are woefully *inadequate* in eradicating the legacy of colonial injustices such as land dispossession and economic disparities because these principles (noble as they are) function within a capitalist "democracy." Capitalist societies are inherently unequal, though the nations of which they are part may legislate equality through constitutional measures, because capitalism favors the rich ruling class that owns the bulk of land and property, a class of individuals that ultimately wields dominant social, political, and cultural power by virtue of the vast economic resources at its disposal.

The United States is a classic case in point. Although there are constitutional provisions that legislate social equality among races, such groups—for instance, Blacks, Native people, and other people of color—are never full recipients of social justice because of the capitalist superstructure in which the laws of the land favor the rich and propertied capitalist class. The Native people in the United States are not able to acquire social justice or retain ownership of indigenous land bases primarily because the courts generally rule on the side of large companies that prey on Native lands with lucrative mineral and energy resources, as with the Black Hills of South Dakota or Black Mesa in Northern Arizona.[101]

Another instance would be that of the phenomenal African American prison population in the United States, which is approximately half of the overall prison population of about 2 million people.[102] Despite constitutional provisions that supposedly guarantee protection of individual civil rights regardless of race and class, the National Justice Commission reported that "African Americans, 12 percent of the population and 13 percent of all monthly drug users, represented 74 percent of those sentenced to prison for drug possession."[103] Clearly, the fact that most of those who are arrested and imprisoned for illicit drug use in the United States are Black and poor is one indication of the class- and race-biased system of justice and law enforcement, reinforcing the contention that capitalist societies, regardless of constitutional enactments against social inequality, are unable to practice egalitarianism because it is the rich who enjoy the bulk of societal and political protection. If Black people in the United States generally had significant economic wherewithal, they would be able to attain some level of social justice, represented by prominent lawyers. For instance, in the case of O. J. Simpson, Simpson was able to circumvent racist strictures in the Nicole Brown Simpson murder trial owing to his ability to secure sophisticated defense lawyers by paying out millions of dollars. Since Black working-class people are victims of a capitalist system, they are unable to achieve social justice and equality because the Constitution of the United States, originally drawn up in 1787, is still geared toward protecting the rights of wealthy white male property owners first and foremost.[104] The historic and globally publicized case of the wrongful imprisonment and sentencing to death of Mumia Abu-Jamal, a journalist and former member of the Black Panther Party in Philadelphia, and his inability to receive a fair retrial despite the flagrantly flawed evidence in the case is yet another instance substantiating the fact that Black people with little financial clout have virtually no chance of receiving justice, even with constitutional provisions.[105]

The guidelines described in the new South African Constitution adopted in May 1996, in which the inalienable rights of food, water, shelter, and health care are assured so long as the state has resources to provide such rights (note that gainful employment is not considered an inalienable right, following typical capitalist societies) are similarly impuissant regarding the country's effort to eradicate the vestiges of economic injustice and social inequality. If the state is bankrupt, for instance, it could deny access to these inalienable rights on the grounds that it does not possess the "available resources." Such denial would be consequently shielded by the Constitution.

Prohibiting racial, gender, and ethnic discrimination in a colonialist-capitalist society does little to transform the socioeconomic conditions in such a society, where capital already wields power and discriminates against workers and the poor. Under a capitalist dispensation, as in South Africa, there is no level playing field between whites and Blacks, between the wealthy and the impoverished, or between the haves and have-nots. Black workers are essentially discriminated against by virtue of their historical role of subjugation by racial capitalism. To argue for nondiscrimination, "color-blindness," and equal treatment of all does not take into consideration that the very structures of South African society are hostile to Black workers and that Black workers are not equal with white industrialists. These respective groups do not possess putatively egalitarian proportions of social and economic power. Black workers are economically and socially at the bottom of society and are disadvantaged in every way. A constitution that outlaws racial and social discrimination in a capitalist society without prohibiting the correlative rapacious practices of capitalist exploitation and curtailing the ownership of resources by a few large conglomerates is an indication of a superficial concern with social and economic justice. Under capitalism, "democratically defined" constitutions that claim class neutrality ineluctably protect the wealthy classes and maintain the subordination of the working class and the underclasses. They legitimate already unjust economic structures and accept their existence as de facto. The United States of America provides the best illustration for the substantiation of this point.[106]

The adoption of the "affirmative action" model in South Africa, as derived from the U.S. political order, has its flaws in many respects, principally because it is seen as a temporary measure as opposed to being a permanent feature of the society that is geared toward redressing the condition of historical racial and economic injustices against Blacks. Moltin Ncholo, a constitutional law scholar, asserts in his critique of the ANC Bill of Rights, which has become a part of South Africa's new Constitution:

> The document [Article 1 on equality] seems to give recognition to one form of equality which is equality in law. It fails to realize that there is both equality in law and in fact. Equality in law deals with matters of substantive and procedural rights while equality in fact deals with real and substantive equality.
>
> The draft further makes reference to positive action in Article 14. In this instance, positive action is equated to the directive principles of State policy. This creates confusion as to the usage of terminology within the field of law. Worst of all, the draft does not state the justiciability of this provision for "positive action." This indicates that insufficient work has been done by both the ANC and the present government in espousing a coherent theory on affirmative action.[107]

The existence of economic and social racism and white supremacy in the United States today, notwithstanding constitutional prohibitions against racist policies and practices, is just one index demonstrating that legal mechanisms are inadequate in redressing fundamental historical injustices.[108] It is also obvious that racism, sexism, and classism continue to persist in the world's most powerful economic nation precisely because of the continuance of the colonialist-capitalist economic system that breeds racial hatred of people of color. This system protects monopoly ownership of wealth by a tiny white capitalist clique, institutionalizes the impoverishment of indigenous, African, and Latino people, and reproduces violence through exaggerated forms of militarism represented by an inflated military, an extensive military-industrial complex, and the proliferation of weapons that enrich the armaments corporations for accumulation of profit. If such practices persist in the United States, where a colonial bourgeois "democracy" prevails, then one is compelled to challenge the new government in South Africa for its emulation of this superpower because the results would be identical: the extraction of wealth for minority whites and Black elites and the continued exploitation of poor Black people, other peoples of color, and all working-class women and men.

Capitalism is an intrinsically racist and anti-Black system principally because it was founded on the backs of Africans who were kidnapped, forcibly shipped, and enslaved in the Americas and on the genocide of other indigenous colored peoples who were conquered by Western Europeans through wars of dispossession and colonization in Africa, Asia, and the Americas. It thrives on cheap and coerced labor. Slavery and colonialism provided the material basis for the emergence of capitalism. During slavery, the value of free and forced African labor was confiscated and expropriated by white plantation owners in the Southern and Eastern United States, South America, and the Caribbean, laying the foundations for the accumulation of wealth by the landed and propertied elites of European descent. The effects of the slave system were multiplier in effect; for instance, fortunes amassed from slavery in England were used to finance banking and insurance enterprises such as Lloyds of London and Barclays Bank and industrial employment and expansion.[109]

During colonialism, indigenous colored peoples, especially African and Native peoples in the Americas, were forced to work as cheap laborers to extract vital natural resources such as cotton, sugar, rice, tea, gold, diamonds, coal, rubber, mahogany, and other materials appropriated by the European colonial powers that dispossessed these peoples. The same system of accumulation followed the abolition of slavery with the spawning of the capitalist system, in which African labor was available cheaply and was once more seized by the burgeoning Northern industrial capitalist class in the United States. Karl Marx confirmed the racist foundations of industrial capitalism when he wrote:

> The discovery of gold and silver in America, the extirpation, enslavement and entombment in mines of the aboriginal population, the beginning of the conquest and looting of the East Indies, the turning of Africa into a warren for the commercial hunting of black-skins, signalised the rosy dawn of the era of capitalist production. These idyllic proceedings are the chief moments of primitive accumulation.[110]

Manning Marable, a leading political scientist, underscores the distinctive confluence of racism and capitalism in the African American experience, verifying our contention of "racialized capitalism" where racism is viewed as intrinsic to the capitalist system:

America's "democratic government" and "free enterprise" system are structured deliberately and specifically to maximize Black oppression. The Capitalist development has occurred not in spite of the exclusion of Blacks, but because of the brutal exploitation of Blacks as workers and consumers. Blacks have never been equal partners in the American Social Contract, because the system exists not to develop, but to *underdevelop* Black people.[111]

Hence, capitalism can never be liberating for *any mass of Black people,* principally because it was built through the bloodshed of African people. No people can expect to realize true independence and dignity by embracing a system that enslaved or exterminated its ancestral forebears. Capitalism is a Eurocentric economic and political system that was originally founded to essentially enrich white ruling classes and is predicated on the predatory principle of raping the resources of colored peoples and forcing such people to work as cheap labor for the industrialization of capitalist empires. It is anti-Black and antiworker at its core because it exists on the foundation of confiscated value of Black people's labor in Africa and the Americas.

Summary and Conclusion

As substantiated in this chapter, the changes that have occurred since 1990, with the beginning of the dismantling of formal apartheid and culminating with the first post-apartheid elections of 1994, were not geared toward empowerment of the Black working-class majority. Rather, they were designed to maintain the hegemony of capital and white ownership of the land and economy of South Africa.

The next chapter analyzes the situation of post-apartheid South Africa, underscoring its neocolonialist typologies in order to underscore what really is at stake in the "new South Africa."

Notes

1. Martin Murray, *Revolution Deferred: The Painful Birth of Post-Apartheid South Africa* (London: Verso, 1993), 8.

2. *Africa Confidential,* Vol. 35, No. 9, 1994.

3. *Southern African Perspectives,* No. 1, 1993 (New York: Africa Fund), 2.

4. *Supplement to the Financial Mail,* June 26, 1998, 272.

5. Mark Swilling and Khehla Shubane, *The Soweto Experience: Urban Popular Movements in "Post Apartheid" South Africa* (Montreal: Centre d'Information et de Documentation sur le Mozambique et l'Afrique Australe [CIDMAA], 1992), 4.

6. Murray, *Revolution Deferred,* 21.

7. Joshua Lazerson's *Against the Tide: Whites in the Struggle Against Apartheid* (Boulder: Westview Press, and Bellville, South Africa: Mayibuye Books, 1994), is an informative text that describes in detail the courageous role that such white individuals played in confronting the apartheid system.

8. Deborah Posel, "Whiteness and Power in the South African Civil Service: Paradoxes of the Apartheid State," *Journal of Southern African Studies,* Vol. 25, No. 1, March 1999, 119.

9. See *Sowetan*, March 20, 1996, and *New Nation*, March 22, 1996, and *Saturday Star*, March 23, 1996, Johannesburg, for the circumstances concerning Malegupuru Makgoba's agreement to step down as deputy vice chancellor and assume the position of research professor at the University of the Witwatersrand. Malegupuru Makgoba's book *Mokoko: The Makgoba Affair: A Reflection on Transformation* (Florida, South Africa: Vivlia Publishers, 1997) documents Makgoba's excruciating ordeal to effect academic transformation at Wits.

10. The works of figures like T.R.H. Davenport (*South Africa: A Modern History*, 4th ed. [London: Macmillan, 1991]), Joseph Lelyveld (*Move Your Shadow* [New York: Random House, 1985]), Johann Van Rooyen *(Hard Right: The New White Power in South Africa* [New York: I. B. Taurus, 1994]), Belinda Bozzoli *(The Political Nature of a Ruling Class: Capital and Ideology in South Africa, 1890–1933* [London and Boston: Routledge and Keegan, 1981]), Leonard Thompson *(The Political Mythology of Apartheid* [New Haven: Yale University Press, 1985]), Colin Bundy *(The Rise and Fall of the South African Peasantry* [London: Heinemann, 1979), Hermann Giliomee and Lawrence Schlemmer *(From Apartheid to Nation Building* [Oxford: Oxford University Press, 1989]), Dan O' Meara *(Volkscapitalisme: Class, Capital, and Ideology in the Development of Afrikaner Nationalism, 1934–1948* [Cambridge: Cambridge University Press, 1983]), Peter Walshe (*The Rise of African Nationalism in South Africa* [Los Angeles and Berkeley: University of California Press, 1971]), and Heribert Adam and Herman Giliomee (*Ethnic Power Mobilized: Can South Africa Change?* [New Haven and London: Yale University Press, 1979]), among other significant authors, are elucidating with regard to issues of South African oppression, yet they overlook the foundational manner in which European colonialism destabilized all of Africa and the world, through slavery and colonialism, a legacy that manifests itself in the pathology of capitalist exploitation and white supremacy and racism today. Although many of these authors rightly stress the primacy of class considerations in the discourse on change in South Africa, European colonialism is treated as a footnote to the principal text on change. A second problem is the peculiarly Eurocentric character of many of these texts, in which Black sources are either seldom mentioned or nonexistent, presuming somehow that the essence of intellectual reflection on South Africa derives from white scholars, a flagrant contradiction considering that Africans cannot be marginalized in any discussion on the future of Africa!

11. Ruth Milkman, "Apartheid Economic Growth and U.S Foreign Policy in South Africa," in Martin Murray, ed., *South African Capitalism and Black Political Opposition* (Cambridge: Schenkman, 1982), 405–406.

12. Robert Fine and Dennis Davis, *Beyond Apartheid: Labor and Liberation in South Africa* (London and Concord, Mass.: Pluto Press, 1990), 290. Although Fine and Davis outline the limitations of both tendencies, they omit the centrality of white supremacy as essential to the development of racial capitalism in South Africa and downplay the formation of the apartheid state as a white colonial state.

13. Anton D. Lowenberg and William H. Kaempfer, *The Origins and Demise of South African Apartheid* (Ann Arbor: University of Michigan Press, 1998), 39.

14. W. H. Hutt, *The Economics of the Color Bar* (London: Andre Deutsch, 1964), Mats Lundahl, *Apartheid in Theory and Practice: An Economic Analysis* (Boulder: Westview Press, 1992), and Walter Williams, *South Africa's War Against Capitalism* (New York: Praeger, 1989).

15. Ihechukwu Chinweizu's classic and trenchant work is unparalleled in its cogency in describing the historical roots of white racism and the legacy of white supremacy, which eventuated in the genocide of indigenous peoples in the Americas and Australia, the extermination of millions of Africans, and the dispossession of people in Asia. See his *The West and the Rest of Us: White Predators, Black Slavers, and the African Elite* (New York: Random House, 1975).

16. This fact is best illustrated with the fact that people of mixed European-African heritage are essentially seen as Black in South Africa in spite of the "colored" designation under apartheid, as well as in the United States, where the "one drop" rule applies, decreeing that any person with the slightest inkling of African ancestry is considered "Black." The 1896 U.S. Supreme Court ruling in *Plessy v. Ferguson* ruled that "Blacks possessed no rights that whites had to respect." These historical facts substantiate the racist grounding of all European behavior in the colonial and industrial eras, regardless of whether these occurred in Africa or the Americas. See, for instance, George Frederickson's work, *White Supremacy: A Comparative Study in American and South African History* (Berkeley and Los Angeles: University of California Press, 1981), and Anthony Marx's *Making Race and Nation: A Comparison of the United States, South Africa, and Brazil* (Cambridge: Cambridge University Press, 1998).

17. See, for instance, William Corbett, Daryl Glaser, Doug Hindson, and Mark Swilling, "A Critical Analysis of the South African State's Reform Strategies in the 1980s," in Philip Frankel, Noam Pines and Mark Swilling, eds., *State, Resistance, and Change in South Africa* (New York and London: Croom Helm, 1988), which explains the multifaceted character of the crisis of the apartheid state as its persistent "reform" initiatives saw unremitting disaster through the 1970s and 1980s.

18. Robert Davies, Dan O'Meara, and Sipho Dlamini, *The Struggle for South Africa: A Reference Guide to Movements, Organizations, and Institutions,* Vol. 1 (London and Atlantic Highlands, N.J.: Zed Books, 1985), 54ff.

19. These financial crises are elucidated in ibid.

20. Murray, *Revolution Deferred,* 29.

21. Quoted from the *Washington Post,* 1990, and cited in a paper by J. Davis, *The Experience of Sanctions in South Africa* (New York: The Africa Fund, 1993).

22. Stephen R. Lewis Jr., *The Economics of Apartheid* (New York: Council on Foreign Relations Press, 1990), 28.

23. Ian Goldin, *External Financial Assistance and Post-Apartheid South Africa,* University of the Western Cape, Economic Policy Research Project, Economic Policy Series No. 9, July 1992, 17.

24. *Post Tribune* (Gary, Indiana), September 10, 1993.

25. Kenneth Grundy, *South Africa: Global Crisis and Domestic Challenge* (Boulder and London: Westview Press, 1991), 71.

26. R. W. Bethlehem, *Economic Restructuring in Post-Apartheid South Africa* (Braamfontein: The South African Institute of International Affairs, 1992), 8.

27. Grundy, *South Africa,* 65.

28. See, for instance, Fatima Meer, ed., *Resistance in the Townships* (Durban: Madiba Publications, Institute for Black Research, 1989), for an illumination of the intensification of Black resistance to the violence of oppression in the urban townships.

29. Tom Lodge et al., *All, Here, and Now: Black Politics in South Africa in the 1980s* (London: Hurst, 1991), 95.

30. Heribert Adam and Kogila Moodley, *The Opening of the Apartheid Mind: Options for a New South Africa* (Berkeley: University of California Press, 1993), 18. One of the weaknesses of this text, as with other works on post-apartheid by white scholars such as Chris Alden (*Apartheid's Last Stand* [New York: Macmillan and St. Martin's Press, 1996]), David Ottaway (*Chained Together* [New York: Times Books, 1993]), Tom Lodge and Bill Nasson (*All Here and Now: Black Politics in South Africa in the 1980s* [London: Hurst, South Africa Update Series, 1992]), Anthony Ginsberg (*South Africa's Future: From Crisis to Prosperity* [London: Macmillan, 1998]), Helen Kitchen and J. Coleman Kitchen (*South Africa: Twelve Perspectives on the*

Transition [Westport, Conn.: Praeger, 1994]), Fine and Davis *(Beyond Apartheid),* Dale McKinley (*The ANC and the Liberation Struggle* [London: Pluto Press, 1997]), J. E. Spence, ed. (*Change in South Africa* [London: Pinter Publishers, 1994]), Guy Arnold (*South Africa: Crossing the Rubicon* [London: Macmillan, 1992]), and Julie Frederickse *(The Unbreakable Thread: Non-Racialism in South Africa* [London: Zed Books, 1990]) is the acute subjectivity and refusal to view issues of race openly and confront South Africa's colonial legacy of racial occupation squarely, a consistent denial that the very notion of race and ethnicity was introduced by European colonial rule. Adam and Moodley, for instance, maintain that "during several centuries of residence the settlers have become indigenous . . . White Africans share the land as citizens legitimately with other groups," which is patently false. Robert Fine and Dennis Davis distort the Africanist tendency in a deeply biased manner on the basis that Africanism does not respect the rights of whites *(Beyond Apartheid,* 149), presuming speciously that "Africanism" does not possess the ingredients for inclusion of diverse tendencies and ethnic backgrounds and that principally Western European liberal democracy does! This unfortunate racial bias has obscured the fundamental socioeconomic contradictions between Blacks and whites historically and has perpetuated a mythology of transition that is at best, superficial and at worst, Eurocentric wishful thinking. The notion of a "nonracial" or "color-blind" post-apartheid society is a case of obfuscation and a mechanism to preserve historical white privilege. See Peggy Macintosh's classic work on white privilege in the United States, which is relevant for South Africa: "White Privilege: Unpacking the Invisible Knapsack," *Peace and Freedom Magazine,* July–August 1989.

31. *Post Tribune* (Gary, Indiana), September 10, 1993.

32. Alistair Sparks, *Tomorrow Is Another Country: The Inside Story of South Africa's Negotiated Revolution* (Wynberg, South Africa: Struik Publishers, 1994), 98.

33. Johannes Rantete, *The African National Congress and the Negotiated Settlement in South Africa* (Pretoria: J. L. van Schaik Academic, 1994), 132–135.

34. Gilbert M. Khadiagala makes this point in "The Frontline States in Southern Africa," in Harvey Glickman, ed., *Toward Peace and Security in Southern Africa* (New York: Gordon and Breach Science Publishers, 1990), 148–149.

35. Archie Mafeje, *In Search of an Alternative: A Collection of Essays on Revolutionary Theory and Politics* (Harare, Zimbabwe: SAPES Books, 1992), 76.

36. Robert Schire, *Adapt or Die: The End of White Politics in South Africa* (London: Hurst, 1992), 125.

37. Kader Asmal, Louise Asmal, and Ronald Suresh Roberts, *Reconciliation Through Truth: A Reckoning of Apartheid's Criminal Governance* (Cape Town: David Philip, Oxford: James Currey, New York: St. Martin's Press, 1996), x. The Truth and Reconciliation Commission was formed by the post-apartheid government as a forum to provide a public account of perpetrators of crimes during apartheid rule, as part of a process of supposed reconciliation between the protagonists and victims of apartheid.

38. Willem De Klerk has stated that in September 1990, following the all-white referendum, his brother F. W. De Klerk had dispatched a National Intelligence Service official to him to convince him that he ought not to talk with the ANC. The former white president had attacked his brother Willem for holding talks with ANC leaders in London in 1989, a year before the historic parliamentary speech unbanning previously banned organizations and freeing political prisoners. See Willem De Klerk's book, *F. W. De Klerk: The Man and His Time* (Johannesburg: Jonathan Ball, 1991), 54–55.

39. T.R.H. Davenport, *The Birth of a New South Africa* (Toronto: University of Toronto Press, 1998), 8, 9. Davenport wrongly assumes the primacy of "individual" initiatives in the

"historic" 1990 announcements of De Klerk. It needs to be borne in mind that oppression is a systemic phenomenon, not the result of individual aberrations or distortions, as the bourgeois capitalist establishment often leads us to believe.

40. Cited in Rupert Taylor and Mark Shaw, "The Dying Days of Apartheid," in David Howarth and Aletta Norval, eds., *South Africa in Transition: New Theoretical Perspectives* (London: Macmillan, and New York: St. Martin's Press, 1998), 17.

41. Mafeje, *In Search of an Alternative*, 104.

42. *New York Times,* September 8, 1993.

43. Mafeje, *In Search of an Alternative*, 65.

44. Ibid.

45. Karim Essack, *Co-option or Transfer of Power in South Africa* (Dar es Salaam: Thakers Limited, 1991), 35.

46. Strini Moodley, "Avoiding the Reform Trap: Black Consciousness in the 1990s," *Indicator South Africa,* Vol. 7, No. 3, Autumn 1990, 18.

47. Mafeje, *In Search of an Alternative*, 77.

48. Ibid.

49. This point was conveyed to the author in a conversation with Canaan Banana, the former president of Zimbabwe, in March 1995 in Harare, Zimbabwe.

50. Vincent Maphai in his article "The Politics of Transition" points out that suspicion about De Klerk's motives were justified considering the regime's repeated failures to keep to its word: the arrest of Philip Kgosana in 1960 after promising to address Black worker grievances; the 1985 bombing of Botswana, Zambia, and Zimbabwe while the Commonwealth Eminent Persons Group was visiting South Africa; and the 1984 Nkomati accord signed with President Samora Machel of Mozambique when Pretoria continued to support the terrorist movement Renamo, despite the peace accord. Found in Vincent Maphai, ed., *South Africa: The Challenge of Change* (Harare, Zimbabwe: SAPES, 1992), 50.

51. Richard Levin and Daniel Weiner, "The Politics of Land Reform in South Africa After Apartheid: Perspectives, Problems, Prospects," in Henry Bernstein, ed., *The Agrarian Question in South Africa* (London and Portland, Ore.: Frank Cass, 1996), 107.

52. The issue of land reform in Zimbabwe will be discussed extensively in Chapter 4.

53. Themba Sono, *African Perspectives, Selected Works No. 4, The Land Question: Healing the Dispossessed and the Surplus People of Apartheid* (Mmbatho, South Africa: Institute for African Studies, and Pretoria: Centre for Development Analysis, 1993), 8.

54. Essy Letsoalo, "Restoration of Land: Problems and Prospects," in Maphai, *South Africa,* 194, 203.

55. Ibid.

56. *New York Times*, November 18, 1993.

57. Letsoalo, "Restoration of Land," 194, 203.

58. Ibid., 216.

59. Ibid.

60. Adebayo Adedeji, ed., *South Africa and Africa: Within or Apart?* (London: Zed Books, Cape Town: SADRI Books, Ijebu-Ode: ACDESS, 1996), 223.

61. *Business Day* (Johannesburg), November 20, 1995.

62. B. Huntley, R. Siegfried, and C. Hunter, *South African Environments into the 21st Century* (Cape Town: Human and Rousseau, 1990), 55.

63. Vinodh Jaichand, *The Restitution of Land Rights: A Workbook* (Johannesburg: Lex Patria, 1997), 32.

64. Khehla Shubane, "Civil Society: Apartheid and Beyond," *Theoria: A Journal of Studies in the Arts, Humanities, and Social Sciences,* No. 79, May 1992 (Pietermaritzburg: University of Natal Press, 1992), 37.

65. David Pallister, Sarah Stewart, and Ian Lepper, *South Africa Inc.: The Oppenheimer Empire* (New Haven: Yale University Press, 1988), 294.

66. See the article "Corporate Giant Charts Tough Future in South Africa," *San Francisco Chronicle,* July 18, 1990, for an illumination of Anglo American's cynical profit motives in supporting "anti-apartheid" activity in South Africa.

67. Pallister, Stewart, and Lepper, *South Africa Inc.,* 38.

68. "Corporate Giant Charts Tough Future in South Africa" substantiates this point of Anglo American's subtle tactics in promoting "change" in South Africa.

69. *Sunday Times* (Johannesburg), January 29, 1995.

70. Davies, O'Meara, and Dlamini, *The Struggle for South Africa,* 65.

71. *New York Times,* June 17, 1993.

72. *Africa Today,* Vol. 1, No. 1, September–October 1995, 33.

73. *Southern Africa Perspectives,* No. 1 (New York: The Africa Fund, 1993).

74. Andrew Donaldson, "Basic Needs and Social Policy," in M. Lipton and C. Sunkus, eds., *State and Market in Post-Apartheid South Africa* (Johannesburg: Witwatersrand University Press, and Boulder: Westview Press, 1993), 273.

75. P. Le Roux, "The Social Democratic Imperative" in L. Howe and P. Le Roux, eds., *Transforming the Economy: Policy Options for South Africa* (Durban: Indicator S.A., University of Natal, and Cape Town: Issue Focus, Institute for Social Development, University of the Western Cape, 1992), 16.

76. Human Sciences Research Council, *A Profile of Poverty, Inequality and Human Development in South Africa, 1995,* cited in a statement by COSATU, NACTU, and FEDSAL (Federation of South African Labor), April 1, 1996.

77. The World Bank, *Reducing Poverty in South Africa* (Washington, D.C.: World Bank, June 1994).

78. Ben Turok, "Development Context for South Africa," in Ben Turok et al., eds., *South Africa: Perspectives on Development* (Pretoria: Institute for African Alternatives, 1994), 97–98.

79. B. Huntley, R. Siegfried, and C. Hunter, *South African Environments into the 21st Century* (Cape Town: Human and Rousseau, 1990), 55.

80. *Business Day,* March 30, 1993.

81. Mafeje, *In Search of an Alternative,* 74.

82. Pierre Beaudet, "The New Terrain," in *Urban Challenges in "Post-Apartheid" South Africa* (Montreal: CIDMAA, 1993), 4.

83. On April 15–16, 1990, leaders of the ANC and the PAC met to discuss the formation of a united front in dealing with the apartheid regime, in which all organizations representing the oppressed but which had not been promoted within the white bourgeois media, such as the Workers Organization of South Africa, the New Unity Movement, and the Azanian People's Organization, would unite to negotiate directly with De Klerk and his representatives. Tragically, the Patriotic Front did not materialize due to philosophical and tactical differences between radicals and moderates from the various organizations. The ANC remained central to the negotiation process and expressed little profound concern over the need for unity of the oppressed prior to discussions with the oppressor. See the section "Trend Towards Unity of the Oppressed" in Essack's *Co-option or Transfer of Power in South Africa.*

84. Essack, *Co-option or Transfer of Power in South Africa,* 60.

85. Shubane, "Civil Society," 37.

86. *Three Years of Destabilization: A Record of Political Violence in South Africa from July 1990 to June 1993,* Special Report 13, Human Rights Commission, Braamfontein, 1993, 3.

87. Ibid., 16.

88. Annette Seegers, *The Security Forces and the Transition in South Africa: 1986–1994,* Africa Seminar, Centre for African Studies, University of Cape Town, March 8, 1995, 39–41.

89. *Urban Challenges in "Post-Apartheid" South Africa.*

90. *Southern Africa Perspectives,* No. 1, 1993.

91. *Focus,* University of Natal, Autumn 1991.

92. *Weekly Mail* (Johannesburg), July 27, 1991.

93. South African Broadcasting Association television news report, August, 17, 1994.

94. Over 21 senior police and military officers from the previous apartheid regime are being currently charged in South African courts for their roles in assassinating Black people in townships around the country during the late 1980s.

95. Shubane's article "Civil Society" raises precisely this question about the assumption that the liberation movement (essentially the ANC) made in arrogating to itself the right to represent the diversity of tendencies in South Africa, especially among the oppressed.

96. This was reported in the *New York Times,* November 18, 1993, and in the *Organizer* (San Francisco), September 1993.

97. The tragedy of the "Colored" vote heavily swaying in favor of the National Party was the result of the tactics employed by the apartheid party, which disseminated racist propaganda depicting an "African" swamping of "Colored" rights, leading to wide-scale tension between these two colonized Black groups. This situation has clearly changed since the June 2, 1999, elections, with the "Colored" vote for the ANC increasing from 33 percent since the 1994 elections to 42 percent, reducing their support for the New National Party, formerly the National Party, from 53 percent of the vote to 39 percent. See the *New York Times,* June 7, 1999.

98. See Lani Guinier's excellent work *Tyranny of the Majority: Fundamental Fairness in Representative Democracy* (New York: Free Press, 1994) for an illumination of the racial redistricting policies in the United States that keep people of color disparate and divided so that collectively they will not garner a majority in areas where these people of color predominate.

99. *New York Times,* September 8, 1996.

100. *Chicago Tribune,* May 9, 1996.

101. It needs to be mentioned that a classic case of genocide continues presently, as over 12,000 Dineh people are being forced to move from their traditional lands, as a result of U.S. colonial policy that pitted indigenous Navajo people against their fellow Hopi, in northern Arizona. The beneficiaries have been Peabody Coal, a huge mining company, which exploited the indigenous people for their labor without taking precautions to protect the Navajo and Hopi against hazardous conditions in the coal and uranium mines. The net effect is that 25 percent of those who worked for Peabody Coal are now dying from radiation cancer. The U.S. government is fully responsible for this outrageous act of genocide yet refuses to recognize the rightful claims of the Dineh people on their own land. Little of this situation has been publicized around the United States. The documentary *Vanishing Prayer: Genocide of the Dineh,* produced by the Big Mesa Support Group (Sol Communications, Malibu, Calif., 1999) describes this horrific tragedy of the late twentieth century.

102. *Metro Times Detroit,* June 23, 1999.

103. *Emerge,* October 1997.

104. See Manning Marable's work *How Capitalism Underdeveloped Black America* (Boston: South End Press, 1983) for a detailed substantiation of the point of Black powerlessness in the face of monopoly capitalism.

105. The recent rejection by the Supreme Court of the United States to grant Mumia Abu-Jamal a retrial verifies the contention that justice and law enforcement is race- and class-biased, and hardly color-blind. The incredible element about Abu-Jamal's incarceration under the threat of a death sentence is that he is being persecuted solely because he was a former member of the Black Panther Party, is a member of the antiestablishment MOVE organization, and is an outspoken critic of injustice in the United States; and the criminal justice system is prejudiced against any militant Black organization and Black voice of liberation. See Mumia Abu-Jamal's classic works *Live from Death Row* (1995) and *Death Blossoms: Reflections from a Prisoner of Conscience* (Farmington, Pa.: Plough Publishing House, 1997) for an illumination of the violent nature of U.S. capitalism and racism.

106. Marable's work, *How Capitalism Underdeveloped Black America*, is one important text that verifies this contention.

107. M. P. Ncholo, *Equality and Affirmative Action: The Ideas of Equality and Affirmative Action in the Context of Bills of Rights with Special Reference to Post-Apartheid South Africa* (Cape Town: Sig Publications for the Community Peace Foundation, 1994), 180.

108. With regard to Native people in the United States, M. A. Jaimes, ed., *The State of Native America: Genocide, Colonization, and Resistance* (Boston: South End Press, 1992), is a comprehensive text that establishes the foundational racist and unjust legal foundations of the United States. Guinier's *The Tyranny of the Majority* confirms the precedence of racial considerations in political inclinations, to the extent that "empirical evidence suggests that race, not class, more often defines political preference" (p. 68).

109. See Eric Williams, *Capitalism and Slavery* (New York: Capricorn, 1966), 98–105, 126–134, for a detailed delineation of the argument that the industrialization of England was financed by the profits accruing from the slave trade.

110. Karl Marx, *Capital: A Critique of Political Economy*, Vol. 1 (New York: International Publishers, 1967), 751.

111. Marable, *How Capitalism Underdeveloped Black America*, 2.

3

Neocolonial Political Economy in South Africa

The new South African government claims that it has earnestly embarked on a path of economic and social transformation through emphasis on growth of the country's capitalist economy, specifically through the GEAR (Growth, Employment, and Redistribution) program, which was preceded by the Reconstruction and Development Program. Both programs were formulated as a result of the ANC's "Discussion Document on Economic Policy," compiled in 1990. The 1990 document stressed that economic growth would be spurred by redistribution of resources, with the product of growth being redistributed and derivatively functioning to meet fundamental needs. Coupled with this distributive plan was emphasis on boosting government spending in the areas of education, job creation, skills building, health care, and housing.

However, through 1991 and 1992, as global market forces ascended to towering heights with the collapse of socialist governments in Eastern-bloc countries, a paradigmatic shift occurred, with the ANC gravitating toward calls for intensified foreign investment and tightened fiscal disciplinary policies. At the 1992 National Conference of the ANC, the slogan "Growth Through Redistribution" evaporated and the rhetoric of "economic growth" took precedence.[1] During the ANC's Policy Conference in May 1992, the organization pledged to challenge monopoly capital in its economic policy document, which read:

> The concentration of economic power in the hands of a few conglomerates has been detrimental to balanced economic growth in South Africa. The ANC is not opposed to large firms as such. However, the ANC will introduce anti-monopoly, anti-trust and mergers policies, in accordance with international norms and practices, to curb monopolies and the continued domination of the economy by a minority within the white minority and to promote greater efficiency in the private sector.[2]

The current situation in the South African economy certainly seems a far cry from these statements on monopoly capital in 1992. The ANC has yet to engage even in the introduction of such promises.

The ANC-led government today asserts that it is necessary to water down criticism of the excesses of the white minority in terms of ownership of land, industries, and wealth, so that foreign investors will not be dissuaded from investing in the country, a policy that thereby generates jobs for the Black unemployed and serves as a catalyst to the beginning of Black economic prosperity. In line with this objective, the government has embarked on a program of massive privatization of the state-controlled sector of the economy, with capital inflows moving in concert with privatization measures, urged by the U.S. government.[3] It will be cogently demonstrated in this chapter that the posture of accommodation by the government of white ownership of the country's resources is essentially taken to appease capital and concomitantly permit the enrichment of the country's growing Black ruling elite and the emergence of the Black bourgeoisie.[4] It is now palpable that a neocolonial situation has come into existence in post-apartheid South Africa/Azania, a situation in which an indigenous petit-bourgeois has assumed the reins of political governance while essentially servicing the needs of Western capital and imperialism, making it more difficult to isolate the lines of liberation and those of exploitation, as is the case with much of "post-independent" Africa. As Amilcar Cabral, the late revolutionary leader of the Partido Africana do Independencia da Guine e Cabo Verde who led the people of Guinea-Bissau to freedom from Portuguese colonial imperialism, put it:

> In the neo-colonial situation, the more or less accentuated structuring of the native society as a vertical one and the existence of a political power composed of native elements—national state—aggravate the contradictions within that society and make difficult, if not impossible, [for] the creation of as broad a united front as in the colonial case. On the one hand, the material effects (mainly the nationalization of cadres and the rise in native economic initiative, particularly at the commercial level) and the psychological effects (pride in believing oneself ruled by one's fellow-countrymen, exploitation of religious or tribal solidarity between some leaders and a fraction of the mass of the people) serve to demobilize a considerable part of the national forces.[5]

Although South Africa/Azania is significantly dissimilar to Guinea-Bissau of the 1970s in the context of history and political economy, Cabral's point is persuasive, reinforcing the view of the current South African political economy as neocolonial as opposed to neoliberal, as recent critics have described it.[6] South Africa/Azania's colonial history cannot afford to be obscured in any theoretical critique of its recently instituted post-apartheid dispensation, owing to colonialism's historical rupturing of continental Africa, including South Africa/Azania, and that legacy reaching into present-day Africa.

Neocolonialist Capitalism and the Black Elite Class

The new government in South Africa is fully conscious that the country is being used as a springboard for maintaining the hegemony of Western capital and penetration into the rest of Africa in areas of mining, tourism, finance, banking, energy resource development, and agriculture, given its knowledgeable economic and political advisers.[7] It is aware of the extractive role of South African corporations in depleting Africa's mineral wealth and exploiting and impoverishing its people in the process.

South African Breweries now wields substantial ownership of brewing industries in Botswana, Lesotho, Swaziland, Tanzania, and Zambia. Shopping chains like Shoprite-Checkers and Pick 'n Pay have expanded into Zimbabwe, Kenya, Mozambique, Namibia, Botswana, and Zambia. South Africa's Standard Bank has existing operations in thirteen African countries and its mining corporations have penetrated Angola, Ghana, Burkina Faso, Mali, Senegal, Zambia, and the Democratic Republic of the Congo.[8] These conglomerates have poured substantial investments into non-African countries, too, such as Indonesia, Australia, Chile, Venezuela, and Ecuador, to reap mammoth profits.[9] Old Mutual, South Africa's largest financial institution, acquired Albert E. Sharp Holdings in Birmingham in the United Kingdom in September 1998, escalating the value of its British assets to over $20 billion (R100 billion).[10] South Africa and the European Union have recently concluded a sweeping trade agreement that "will open up South Africa's market to 86 percent of EU goods over a 12-year period, while opening the 15 EU economies to 95 percent of South African goods over a 10-year period," a move that will significantly undercut neighboring South African economies, as they will lose up to 15 percent of fiscal revenues as tariffs are abolished under the new agreement.[11] SASOL, the domestic South African oil production company, has cut lucrative deals to "convert unmarketable sources of natural gas to environment-friendly diesel with Qatar in the Gulf, Norway's giant Statoil and now with U.S. oil company Chevron in Nigeria."[12]

The post-apartheid South African regime has even expanded weapons production and export through the parastatal corporation Armscor, continually seeking licenses from foreign military industrial corporations to manufacture military equipment in South Africa. Israel, for instance, provided arms supplies to South Africa worth $300 million in 1987 and $100 million in 1988, and South Africa supplied R1.8 billion ($600 million) worth of weapons to other countries in 1989. South Africa shipped artillery shells to Iran in exchange for oil and spare parts for Phantom F-4 fighter aircraft.[13] Armscor exports military equipment and weapons to over fifty governments around the world. The South African deputy minister of defense, Ronnie Kasrils, a former freedom fighter against the apartheid machine, made the following statement regarding the armaments industry in post-apartheid South Africa:

> South Africa's defense industry annually earns substantial amounts of foreign exchange, since it is one of the country's largest exporters of manufactured goods. Worldwide, the armaments trade is a big and lucrative business. . . . South African exports of defence

equipment amounted to R886 million in the 1993–1994 financial year and are expected to reach the one billion rand mark this year. South Africa exports defence equipment to 50 countries . . . With the lifting of sanctions, South Africa expects to increase its sales substantially. Its preference is for niche markets in areas such as artillery and armoured vehicles. The country cannot compete with the quantities of weaponry or levels of sophistication of arms proffered by the major industrialized nations.[14]

Ironically, much-needed local investment is diverted abroad, and instead of requiring such companies to invest heavily domestically, the new government looks desperately to foreign capital for investment in the South African economy. South Africa is now the latest addition to the states of Africa that appears to be headed in the direction of neocolonial political economies.

However, on the domestic front, the Government of National Unity in South Africa cannot hide behind the façade of elevating the living standards of the domestic Black population while depending on "foreign investment" and loans from the International Monetary Fund and the World Bank (borrowing $850 million a few years ago).[15] In 1995, the South African government expected to borrow approximately $284 million in 1996, $663 million in 1997, $142 million in 1998, and about $855 million in 1999.[16] The former finance minister of South Africa, Chris Liebenberg, announced in late 1995 that South Africa would appoint two permanent advisers, one at the IMF and the other at the World Bank to represent "English-speaking" African countries, typical of the European settler-colonial mode of dividing Africans on the basis of colonization by the British and the French.[17]

South Africa does not need to depend on foreign investment as the principal avenue of generating a viable economy and look to the IMF and the World Bank for its economic salvation when it has enormous material resources and its most precious resource, millions of Black people, including hundreds of thousands of skilled men and women. Such dependence on foreign investment will permit unbridled extraction and acquisition of the country's natural resources, as has occurred under apartheid. Although authors like Anthony Ginsberg impishly espouse foreign investment as the panacea for South Africa's economic hardships, they are compelled to acknowledge that there is no history of foreign investment in South Africa providing a boost to domestic manufacturing ability and creating jobs for the unemployed, since the bulk of the billions of dollars that have poured into South Africa following the 1994 elections has flowed into stock and bond market portfolios.[18] Nonresidents invested R74 billion ($14 billion) in equities and bonds between January 1996 and the end of March 1998.[19] Capital investment has been in middle-class business, hotels and tourism, construction, electronics, food, and mining, since capital on principle invests only in areas where it can rake in large profits.[20] Billions of profits were made at the cost of the blood and sweat of Black workers, encouraging a wholesale looting and transnational capitalist ownership of the country at the price of impoverishing the Black working-class majority.

In 1995, the Johannesburg Stock Exchange had a capitalization worth of R1 trillion ($200 billion) and represented the tenth largest in the world.[21] Astoundingly, it exceeds the value of the stock exchanges in the rest of Africa combined by over two and one-half times. Yet the inordinate wealth that accrues from this cartel of commerce and trade quintessentially benefits the wealthy European, North American, and Japanese capitalist establishments with the minuscule residues trickling down to South Africa's vast Black working-class majority. Further, it is the Black working class in South Africa, like the working-class masses of the rest of Africa and the colonized world, that continue to remain long-term economic slaves to the designs of these global capitalist agencies by living on borrowed money and "aid," reducing Africa's ability to become economically independent and financially free from the shackles of Western colonialism and capitalism. The burden of paying back the exorbitant rates of interest on these international loans falls on the backs of the Black working class, resulting in decreased incomes and an escalating impoverishment of Black workers, as we have seen in Ghana, Zambia, Kenya, Zimbabwe, Chad, Egypt, Malawi, Senegal, and so many other parts of Africa and the "Two-Thirds World," where the structural adjustment policies (SAPs) of the IMF and World Bank loans have caused the deaths of millions of Africans.[22] In 1996, African nations paid out more than $24 billion in interest on debt payments, draining the various governmental coffers of much-needed revenue for developing infrastructure and human services.[23] Inadequate health care, increased disparities between the poor and the rich, devalued currencies, dismantling and curtailing of educational institutions, undermined indigenous production, induced declines in wages of urban workers, frozen subsidies on basic food products, exacerbated poverty conditions, and rising infant mortality rates because of the malnutrition of children are all provoked by the policies of the IMF and the World Bank.[24]

Former South African state corporations such as Safcol, the forestry company; Alexkor, the diamond company; Telkom, the telecommunications company; Abaokor, the abattoir company; and two divisions of Denel, the arms company, have been privatized. The Airports Company of South Africa, a former state corporation, has been heavily privatized, with substantial holdings by Italy's Aeroporti di Roma, which has already acquired the rights to build Kilimanjaro Airport in East Africa.[25] Homenet, the country's largest real estate group, has developed partnerships with the Portuguese rental estate company, Saviotto and Esaguy, to market South African properties in Europe, so that today some 20 percent of housing sales in some suburbs of Cape Town are to European buyers.[26] The push toward massive privatization is coming from the capital sector, predicated on the pretext that only capitalist-run organizations are efficient and cost-effective. The beneficiaries in almost all of these instances of privatization have been domestic white companies and their European partners or a paltry number of rich Black business tycoons.

For South Africa, Africa's wealthiest country, to travel down this path of economic dependence through mortgaging the nation's resources and privatizing state assets to supposedly "reduce the national debt" is to perpetuate neocolonialism by offering handsome profits to foreign capitalists and undermine the interests of the working-class masses.[27] It is geared toward protecting the white minority population's position of economic privilege and prosperity by providing a cushion against some of the reforms that may entail minor economic adjustments for whites. Concurrently, it *permits* a limited number of Black people to reap a larger, albeit meager, portion of the colonially-sown, capitalist-baked pie, so as to accord the system moral legitimacy.

As part of its strategy of co-opting Black people into the bourgeoisie, the capitalist class is actively involved in disseminating its message of "foreign investment" dependence and "economic growth" throughout the media establishment. The March 16, 1995, issue of the *Star* newspaper in Johannesburg (Egoli), for instance, had a special supplement entitled "Pioneers of Black Business," in which one can find captions such as "Ready to claim a real stake in the economy," "Here's living proof that assistance is given to those who try on their own," "Moving to Africa," "Go-ahead woman takes tough mining world by storm," "Headhunter with high standards," "Small company aiming high," and the like. On the back page of the same newspaper, a picture of Jay Naidoo, the then minister without portfolio and formerly the general secretary of COSATU, is shown, with the caption "Drawing in the foreign investors" in bold print. These depictions are part and parcel of the capitalist system's strategy of promoting individual middle- and upper-middle-class Black success so that the world will be deceived into thinking that the "new South Africa" is a Black economic success story. Policies of promoting organizations like the Small Business Development Corporation in South Africa, like its counterpart in the United States, the Small Business Administration, is part of the racist ideology and practice that is endemic to capitalism: institutionalizing the ownership of *large business* by wealthy whites and arrogating the ownership of *small business* to the province of Black people.[28]

The move to include more Black people within the capitalist economic system in South Africa is designed to convey the specious impression that Black people are becoming significant beneficiaries of the new post-apartheid system.[29] It is directed toward creating a tiny and affluent Black middle class that will eventually come to legitimate the system of capitalism. During the days of late apartheid, the white regime employed a similar strategy, what came to be referred to as the Total Strategy:

> The Total Strategy aimed to win allies for the regime by encouraging the emergence in urban areas of a stratum of Africans with considerably higher incomes and standards of living than the majority of African people. As a result of various "reforms," the regime argues, this group of people would be given a stake in the system and drawn away from revolutionary activity. The regime hoped that it would win support for the capitalist system from this group, and that its members would have a moderating influence on the liberation movement.
>
> Big business, which had long urged the state to adopt such a policy, now gave it active support.[30]

This Black elite class in turn is now being used by the post-apartheid regime to legitimate its moral credibility and economic policies, since the latter is increasingly facing a crisis of moral bankruptcy in the eyes of the Black working-class majority.

The subject of class is decisive and instructive for any critique of the evolution of the political economy of South Africa. The Black working class was the most victimized and exploited by the apartheid system; it thus will be the most critical element in the radical transformation of post-apartheid capitalism. Under the existing post-apartheid dispensation, the principles of a broad interclass movement are espoused by the new government.[31] This notion of a broad "rainbow nation" may sound attractive, yet it overlooks both the reactionary role that the liberal white establishment plays and the class configurations and contradictions that have begun to emerge prominently within the Black community in recent years.[32]

The minuscule Black middle class, particularly the class of Black commercial entrepreneurs represented by a tiny group of merchants, traders, and financial executives, conjoined with the burgeoning Black professional elite (doctors, lawyers, local political appointees, public administrators, and managers), have contrived to lend credence to the reforms instituted by the preceding apartheid oligarchy. This class has functioned to mitigate the severity and grave effects of apartheid while couched in the comfort of their plush homes furnished with many modern gadgets available from Western technology and transported in their sleek European automobiles that cost tens of thousands of dollars. This class has regurgitated an ambiguity about apartheid, expressing the view that apartheid was discriminatory on racial grounds and was *principally* problematic because it excluded Black people from its benefits, not for its fundamentally exploitative nature. The Black elite generally desired being co-beneficiary with whites from apartheid capitalism and were envious of whites because the latter were successful in their economic exploits and functioning within the apartheid system.

The colonial apparatus that governed South Africa during the apartheid era needed the countenance and support of this tiny privileged elite to brutally suppress the vast working class and the underclasses. Colonial legislation such as the Local Black Authorities Act, which sanctioned the existence of community councils and administrative bodies in Black urban areas, facilitated the development and protection of this paltry Black elite. John Saul and Samuel Gelb describe this phenomenon of Black elitism, effectuated and cultivated by the racist regime to stratify and polarize the large Black majority and consequently heighten existing class disparities, all measures that are essential for the perpetuation of minority colonial rule:

> A first aspect here is the encouragement of a black trader class. This has meant the lifting of a number of regulations limiting their ability to accumulate (for example, being allowed to own only one business). In addition Riekert [a commission formed to look at Black worker grievances] recommended that African employees be given a claim equal to that of other employees in relation to the allocation of labor supplies. Other legislative amendments have opened new market opportunities for this group; thus the authority granted to the community councils will have an economic payoff over and above its political implications, some of the township powers—allocation of trading sites, provision of some township services—promising a possible platform (albeit a small and flimsy one) for private accumulation by the small group of African traders around them.[33]

This elitist class creation is part of the system's effort to maintain its capitalist superstructure with the goal of obtaining a level of support from Black people for laissez faire economics and the rhetoric of the market.[34] The function of the apartheid state's Riekert Commission, established in 1979, was ostensibly to study the issue of Black workers' grievances, part of the ruling class's strategy in fragmenting the oppressed Black majority working class.

> In outlining the Riekert proposals for the permanent urban Blacks we must keep in mind that this amounts to a process of *trying to create* a Black middle class, a petty Black bourgeoisie (that in the eyes of the regime) will hopefully come to identify with the capitalist system. Class formation is normally a slow and uneven process in capitalist society, and there is no guarantee that this last-moment effort by the white ruling class will succeed.[35]

The Riekert measures thus constituted a deceptive and desperate attempt on the part of the white minority regime to polarize rural and urban Black people by granting limited "rights" to the latter while denying the former even a semblance of human rights. Consequently, the promulgation of the Riekert plan represented an insidious effort to perpetuate white rule through a two-pronged strategy of pronouncing limited "race reforms" while essentially undermining Black workers' struggles for justice and freedom. Sheena Duncan, an activist with the Black Sash, a liberal white women's support organization in South Africa, captured the essence of the Riekert legislation when she stated:

> The Riekert report is a very clever and highly sophisticated piece of work which will probably result in a longer period in which the status quo will be maintained through the creation of a relatively small African privileged group which may serve as a buffer against unrest. In the interim, dreadful human suffering in the homelands and problems experienced by migrant and non-permanent workers will increase.[36]

As is prototypical of capitalism in the United States, for instance, where the system *breeds* and *tolerates* an attenuated Black elite class to obtain legitimacy,[37] so too in South Africa this minuscule Black elitist class is *allowed* to exist, being subordinate to and functioning in the interests of white capital in South Africa. The African American social critic and scholar bell hooks explains that these class contours have affected Blacks so that different Blacks experience the racism of the same oppressive system in various ways:

> Employing a critique of essentialism allows African-Americans to acknowledge the way in which class mobility has altered collective black experience so that racism does not necessarily have the same impact on our lives. Such a critique allows us to affirm multiple black identities, varied black experience. It also challenges imperialist colonial paradigms of black identity which represent blackness one-dimensionally in ways that reinforce and sustain white supremacy.[38]

The Black petit bourgeoisie has no independent existential social and economic basis; it behaves according to the whims and fancies of the capitalist establishment.[39] Groups such as the Urban Foundation, initiated by white financier Anton Rupert on the supposed philanthropic grounds of providing assistance for the procurement of houses for a tiny Black middle class through the Small Business Development Corporation, are part of the economic strategy of appeasing Black antagonism toward racial capitalism.[40] Such institutions are very similar in character to the Black Enterprise establishment in the United States, which represents the core of Black commercial activity but is in fact an ineffectual economic cog within America's mammoth capitalist financial empire.[41]

The rhetoric and implementation of "race reform" by the white authorities in South Africa functions in concert with the ploy of generating an economically advantaged Black elite. As Saul and Gelb point out, this pattern of class fabrication further exacerbates within the Black community the existing class tensions between the attenuated middle class and the vast working class and the underclasses:

> A more developed initiative in this respect is the ending of the "petty apartheid" of racial segregation in (some) social areas (restaurants, trains and buses, park benches and the like). Though nominally targeting all blacks (and always packaged internationally in this way) this actually reinforces class polarization since only the middle class can afford to enjoy the newly opened hotels, restaurants, first class trains and the like in white areas. Such a growing gap between the cultures and lifestyles of the different black classes will no doubt be further widened as such facilities directed towards class-defined markets become more common in the townships themselves as part of their increasing commercialization . . .
>
> . . . A second important front here is the cultivation of appropriate "middle class values" among the urban black population, attempting to shape in this way their self-definition as a group.
>
> Significantly, this is likely to be relevant even as regards working-class blacks, who can be expected after all to have already imbibed a strong dose of capitalist culture by virtue of their eating into the "modern" economy (at least to the extent that this has not been countered by alternative forms of ideological work).[42]

The purpose of the white ruling class in nourishing this Black elitist class is to "encourage political support for capitalism while demobilizing Black militancy and to facilitate the expansion of the Black consumer market"[43] while serving the designs of the Western industrialized capitalist countries.[44] It was with this plan in mind that Anglo American, South Africa's largest conglomerate, sold its subsidiary, African Life Insurance, to a Black consortium headed by Anglo director, Don Ncube, and for the same reason, Sanlam, a huge insurance conglomerate, sold Metropolitane Life to a group headed by Nthatho Motlana, a Black medical doctor and former anti-apartheid leader, following the elections in 1994.[45]

Interestingly, there are indications that some members of this Black elitist class are becoming somewhat uncomfortable with the façade of Black mobility under such fanciful slogans as "the African renaissance and Black economic empowerment." Nkosi Phathekile Holomisa, an ANC member of Parliament and president of the Congress of Traditional Leaders of South Africa, describes the elitist character of these concepts in reflecting on the future prospects for his sister, Babalwa Holomisa, a recent graduate in social work from the University of Transkei:

> In the cities she is likely to be invited to forums where these concepts—the African renaissance and black economic empowerment—are discussed by an exclusive, privileged, upwardly mobile black elite. She will find out about the advantages of joining one or another empowerment consortium whose aim it is to gain access to the public enterprises that government is selling in terms of its macro-economic policy . . .
> . . . She will soon realize, however, that this particular campaign for black empowerment has nothing to do with uplifting the condition of the entire black SA family. It has more to do with enriching a few individuals who will be joining the ranks of the white middle class. The vast majority of blacks will continue to provide labor as before without experiencing any material changes to their lives.[46]

Although Holomisa is an ANC parliamentary representative, he laments the fact that a paltry number of Blacks are becoming economically privileged, including such notables as former ANC general secretary Cyril Ramaphosa, who is now deputy chairman of New African Investments Limited (NAIL) and director of Jonnic, a private corporation; and PAC leader Dikgang Moseneke who has left parliament to co-head NAIL. Black Consciousness leader Peter Jones has now embarked on a lucrative venture with businessman Sam Dube, promoting tourism in the Western Cape, and former head of the South African Broadcasting Corporation Zwelakhe Sisulu moved on to join the Denel Corporation and is now an executive with NAIL.[47] Recently, Nthatho Motlana and Jonty Sandler, executive directors of NAIL, indicated that they would be leaving the company. Both received payments of R40 million ($6.7 million) as a farewell payoff.[48]

Union leaders who previously denounced capitalism as exploitative and unjust, such as Jay Naidoo, former leader of the Congress of South African Trade Unions, and Mbhasima Shilowa, outgoing general secretary of COSATU and the premier of Gauteng Province, have now been lured and co-opted by capital's designs in South Africa.[49] These leaders, together with some other historic figures in the anti-apartheid struggle, have become quite wealthy, in some instances multimillionaires, and have embraced the new "businessocracy" emerging with lightning speed in the "new" South Africa.[50]

Today, Black-owned businesses on the Johannesburg Stock Exchange are worth close to $8.6 billion.[51] In almost of all these ventures, white corporate monopolies have sold shares to Black stockholding individuals, the majority of whom do not possess the capital wherewithal to compete in a system where financial capital reigns supreme. The result has been the dismal performance and failures of numerous Black commercial ventures such as New Age Breweries. Khehla Mthembu, the chairman of New Age Breweries, conceded that Blacks were significantly disadvantaged in entering the stock market because they were moving into "a monopolistic market with limited resources."[52] Salukazi Hlongwane, the chairperson of another Black corporation, Nozala Investments, notes the sheer hegemony of white financial ownership and the fact that the bulk of the returns from investments in Black companies accrues to white entrepreneurs, where "the economic wealth generated by the company—through ownership of 95 percent of the shares of the company, is enjoyed largely by white financial institutions."[53] Similarly, JCI, a mining house heralded as the beginning of Black ownership of South African mining interests a few years ago, collapsed. Wiseman Nkuhlu, the former head of JCI, explained that the company was unable to compete with the large white mining monopolies on the mining market.[54] Black companies such as New African Investments Limited; Merchant Bank, which has a capitalization worth of R5 billion ($830 million); Real Africa Holdings; and P. Q. Africa have experienced immense difficulties in staying afloat within the circles of financial and commercial conglomerates, principally because of crushing debt burdens.

> Many BEE [Black Economic Empowerment] transactions have required innovative control and funding structures to make them possible. Many of the transactions have necessitated the participants taking on high levels of debt. Should the underlying investments not perform sufficiently to allow the debt to be serviced, some of the control structures could collapse.[55]

In the aftermath of the stock market crash of July 1998, the companies that were hardest hit were the new Black corporations that had arisen as a result of the Black Economic Empowerment program. The market capitalization worth of twenty-eight Black-controlled firms plunged one-third from R68.7 billion to R48.3.[56] The steepest declines were recorded by Maxtelm Infiniti and AM Moola. Metlife, the insurance giant, lost 58 percent of its value, losing R8.5 billion in the process. New African Investments Limited similarly dropped 58 percent of its capitalization value. Even though many of these fledgling companies appeared to have gotten off to a good financial start, the aura of prosperity was short-lived. The turbulence in the stock market forced many of these new Black companies into further indebtedness, underscoring their acute vulnerability to market forces, unlike the historical white capital conglomerates that are well cushioned during such financial downturns.

The picture of Black corporate business is certainly not as sanguine as many have been led to believe in the global media. The condition of Black corporations is acutely tenuous and the face of prosperity does not reveal the actual predicament of indebtedness and virtual financial insolvency. This situation is akin to that of an individual possessing a luxurious mansion and an expensive automobile without disclosing that the house and car were purchased through credit cards and, in fact, belong to the financing bank. This façade of financial prosperity obtains for most members of the newly created Black elite in post-apartheid South Africa, many of whom are extremely vulnerable to banking and financial pressures from the white corporate establishment. Of the 25 billion South African rands that pass through the country's banking system each day, almost all of the financial capital belongs to the white banking institutions.[57] The central problem, however, is that like their white counterparts and investment partners, the fundamental principle motivating these Black companies—such as the Black Management Forum Investments, for instance—is profit and seeking multimillion-dollar deals that enrich a few individuals, while deluding the Black working class into believing that capitalist investment is the key to the extrication of the Black poor from poverty.[58] Nothing could be further from the truth.

If anything, the disparities between the tiny Black elite and the vast Black working class and the underclasses have grown phenomenally, a by-product of which is the booming wave of street crime. The gap between wealthy and poor Black people is now equivalent to that between whites and Blacks, and while one-fourth of the wealthiest 20 percent of households are Black, the incomes of the poorest 40 percent of Blacks have steadily decreased over the past two decades.[59] This trend represents a continuation of the effects of the "race reforms" of the 1970s and 1980s; although the income of Africans increased from 29.0 percent to 36.3 percent during the 1980–1990 period, it was essentially more skilled Blacks who benefited and the less skilled who suffered.[60] Black empowerment is a façade, a ruse by white capital to legitimate its existence by enriching a few Black elites. Ashwin Desai, a South African sociologist, reflecting upon the financial acquisitions of the Black nouveau riche as represented by Cyril Ramaphosa, Mzi Khumalo, and Nthatho Motlana, correctly recognizes this political sham when he observes:

> What is ignored by the media and themselves is that they conduct their business in the same way as the white man does and often at his behest. What actually happens is that for every Motlana that is empowered, thousands of black people become disempowered. Motlana might take over a mine or two, but thousands of black miners are retrenched by him to make the operation a financial success. And because it suits the black elite to label their ascension into monopoly capitalism as racial empowerment, it covers up these aspects.[61]

The paltry group of elitist Black individuals are empowered at the cost of impoverishing the vast majority of Black workers, and corporate greed is made to look respectable and signify an index of "success" in post-apartheid South Africa.

The swelling ranks of these new financially wealthy Black elites come with tremendous social costs, particularly as the media explodes with the hype about crime being the most formidable social problem in the "new South Africa." The reason for rampant crime rising to unprecedented levels, including violent crime (South Africa now has the dubious distinction of having the third-highest murder rate after the United States and Swaziland), is the expanding gap between the haves and the have-nots.[62] After the author and businessman Phinda Madi was held up by a car hijacker, he recalled the words of the hijacker, "I want your car because nidla nodwa, which loosely translated means 'because you eat alone.'"[63] What this statement reveals is the consciousness by the Black underclass of the Black elite's benefiting lucratively from South Africa's "free market" yet selfishly allocating all benefits for themselves. The Black underclass views the behavior of the Black elite as predatory and shameless because members of the elite live in opulence without a care about the poor. The "criminal" element among Blacks in urban townships is the inevitable result of a capitalist economy where the elites "lawfully" make millions from the system, and the generally unemployed members of the underclass view it as their right to "unlawfully" engage in crime to make whatever they can in a rip-off system where individual greed is the norm.

Black Working-Class Responses to the Post-Apartheid Economy

The trend of unions investing in the stock market and other sectors of the commercial economy has produced wealth, but only for a tiny sector within the union organizations, particularly union leaders like Marcel Golding, the former head of the National Union of Miners (NUM); former COSATU general secretary Sam Shilowa, who is married to millionaire businesswoman Wendy Luhabe; and Johnny Copelyn.[64] Shilowa has indicated that "former trade unionists do not need to be poor" and that "socialism is not about poverty."[65] It is evident that many of these former trade union leaders have become phenomenally wealthy through cutting financial deals with the large conglomerates and through the engineering of personal investments in the capitalist economy, while the rank and file of Black workers, particularly rural workers, struggles to eke out a decent existence with meager poverty earnings.

The continuance of a capitalist system in post-apartheid South Africa, minus its overt racist ideological trappings, is the Western powers' fervent desire. This solution is material anathema to the largest sector of the disadvantaged population in South Africa, Black workers.[66] The organized Black working class, represented by militant trade unions such as the National Union of Miners; the National Union of Metalworkers of South Africa (NUMSA); the Amalgamated Clothing, Textile, and Allied Workers Union of South Africa; the Engineering and Allied Workers Union; the Glass and Allied Workers Union; the Paper, Pulp, Wood and Allied Workers Union; the Commercial, Catering, and Allied Workers Union of South Africa; and many others have expressed their rejection of privatization and labor-hostile legislation through industrial strikes and other mass actions, such as the November 1991 anti-VAT strike and the August 1992 strike that involved over 4 million workers following the Boipatong massacre.[67] The ANC-COSATU-SACP tripartite alliance may not be as solid as external impressions convey.[68]

COSATU has rejected the government's GEAR program following the state's failure to implement a substantive Reconstruction and Development Program in the immediate aftermath of the 1994 elections. COSATU snubbed the "obsession around deficit reduction" and stated that the government "should avoid binding itself to specified deficit targets without first assessing their impact on service delivery, ability of the State to extend services to poor communities and developing the institutional capacity of the state."[69] The SACP also warned that the government's GEAR program would concentrate power and wealth in a nonracial elite composing 30 percent of the population, while the remaining 70 percent of the population would remain in poverty. The SACP's Economic Transformation document declared that it rejected "all approaches suggesting that essential developmental transformations should be postponed until growth has been achieved."[70]

At a gathering of the Congress of South African Trade Unions, a coalition organization representing some 1.5 million workers and about 600 trade unions, held in 1989, the question of a workers' charter and a workers' political program was raised.[71] This issue was elevated in the context of the current wide-scale discussion around the Freedom Charter, the document representing the African National Congress's blueprint for a post-apartheid South Africa. Moses Mayekiso, formerly a dynamic trade unionist leader of the National Union of Metalworkers of South Africa, who had spent almost three years in political detention for fighting apartheid oppression and who recently resigned from an appointment in the new government, echoed the need for a working-class political program at that meeting beyond the workers' charter that the Freedom Charter proposes. He insisted that Black workers play a decisive and central role in the reformulation of a new constitution for the liberated South Africa. He articulated the sentiments of South Africa's grassroots Black workers vividly when he underlined the differences between the idea of a workers' charter and that of a workers' political program:

> The workers themselves want to build political programs that would be workerist. If you say workers' charter, that means that workers themselves should have their own political program. We believe that this is not true. We believe that the *working class* must have a political program, not the workers. The workers' charter can never answer things that trouble the workers, like the rights to strike, negotiate and a living wage. Then the working class political program must answer the question, what is the future society? What is the political setup of the future society? So the charter and our program are two different things. (emphasis mine)[72]

Although Mayekiso may have compromised his erstwhile radical position today, his vying for a working-class *political* program then, and the envisaged axiological role that the Black working class in South Africa ought to play in recasting the politico-economic destiny of South Africa, is illuminative for all those struggling against post-apartheid capitalism. The Black working-class community is the sector that creates South Africa's wealth, finances its industrialization, and maintains the nation's existence and well-being.[73] It is thus logical to have this class assume the position of major player in molding a nonracial and participatory, democratic and socialist South Africa/Azania.[74] Solomon Mlambo, a COSATU activist, articulated the role of the Black working class as the vanguard capable of ushering in genuine democratic socialism and an independent South Africa, since it is this class that faces the brunt of apartheid capitalism and stands to benefit most from a radical restructuring of the colonial-capitalist system:

Because the working class is the most forceful and consistent fighter for even limited democracy, it is the only class capable of guaranteeing concrete "bourgeois democratic rights" for itself and the oppressed petty bourgeois. However, the working class is interested not only in bourgeois democracy, but far more importantly in socialism (i.e., the planned economy, state ownership and control of the means of production, distribution and exchange, and a state monopoly over foreign trade).[75]

Mlambo implies a radical democratic socialism that has yet to be concretely determined by the Black working class. The point that he makes regarding the working class being in the forefront of South Africa's political struggle is well taken, however. It is this class, reinforced by the praxis of Black women and youth, that propelled South Africa into the phase of post-apartheid extant today. As a consequence of this analysis, Mlambo proposes a very significant movement that he describes as the United Front. He argues that a united front led by the *Black working class* is the path to the realization of veritable people's economic and political democracy, as opposed to a populist, interclass front with the objective of bourgeois democracy, which is in vogue in contemporary South Africa. He states:

> As opposed to the united front, the populist conception of class alliances (i.e., the popular front) is *not* based on agreement around a practical program of action between the working class and petty bourgeoisie with the working class maintaining its independence and freedom to criticize the vacillations of the middle class.
>
> Instead the popular front robs the working class of its political independence and organization, and replaces its historical mission (striving for socialism), with the struggle for us to "win the moral high ground" (JODAC: Workers in Progress 53, p. 38). The petty bourgeois ostensibly joins hands with the working class within the popular front, but in reality it keeps its own hands free, to manipulate "the question of post-apartheid economic structures and precise shape of democratic institutions" (SACP: WIP 50/51, p. 55). Popular front or "class alliance politics" calls for class partnership, a loose arrangement in which it is unclear which class is to be the senior and which the junior partner. In fact, popular frontism bends over backwards to satisfy the petty bourgeois (and bourgeois) and placates them by not offending their liberal sensibilities with socialist demands. . . . That is why popular frontism and class alliance politics is reformist and anti-socialist. It fails in practice to bind the petty bourgeois to a practical agreement on a specific issue to defend or advance the working class struggle.[76]

Mlambo's insights here, though somewhat disconcerting to accept for persons from more bourgeois sectors of the resistance, are nevertheless perceptive and elucidating as regards the contradictory nature of class interests in South Africa. It underscores the reality that is obvious in the view of the oppressed Black working class and underclass communities: that there are class antagonisms extant within the Black community and even more accentuated chasms between the liberal white middle-class sector and the militant sections of the Black working class.

The post-apartheid ANC-led government claims that the path of transformation it is pursuing in the post-elections era is that of rapprochement among various sectors of South African society to achieve change with a minimum of conflict among the various classes, organizations, and tendencies so as to prevent a chaotic and destructive transition to "freedom." In June 1990, after being released from prison earlier in February, ANC leader Nelson Mandela made the umbrella character of the ANC patently clear when he asserted:

> The ANC has never been a political party. It was formed as a parliament of the African people. Right from the start the ANC has been a coalition, if you will, of people of various political affiliations. Some support free enterprise, others socialism; some are conservative, others are liberals. We are united solely by our determination to oppose racial oppression. That is the sole thing that unites us. There is no question of ideology as far as the odyssey of the ANC is concerned, because any question approaching ideology would split the organization from top to bottom.[77]

This broad interclassist and open-ended posture on the part of the ANC has demonstrated the political naïveté of the organization: anticipating an amorphous unity of the national population without understanding the riveting divide between workers and the bourgeoisie, between capitalists and socialists. Although the ANC's populist-based support was heavily generated within working-class circles, represented by organizations like COSATU and the SACP prior to 1990, it opted to break this alliance by aligning itself with capital after 1990, even at the cost of denying the legitimate material aspirations of the Black poor. At the present juncture, it is willing to threaten the pro-working-class elements, with independent legislative action, regardless of unionist sentiments, demonstrated, for example, in the instance of Mandela's rebuke of working-class interests at a convention of COSATU in 1998. The interclassist inclination of the ANC and its portrayal of South Africa as a diverse "rainbow nation" that needs to be united in the new post-apartheid era is politically weak, precisely because of the bourgeois character of the "rainbow nation" and the failure to be accountable to the vast Black working-class majority. In a related though significantly different context, the Rainbow Coalition formed by Jesse Jackson in the mid-1980s in the United States suffered a similar shortcoming, owing to its dependence on populist rhetoric and little working-class political mobilization. Manning Marable contends in this regard:

> The Rainbow movement was not a socialist movement, but it was a social justice movement in which many revolutionary nationalists, communists, Marxists, and democratic socialists freely participated. It had tremendous potential.
>
> The major weakness of the Rainbow Coalition was its failure to consolidate a democratic, membership-based organization with elected leadership accountable to all members. Jackson favored a charismatic, populist-style of Black leadership, which inevitably destroyed his own organization after 1989.[78]

The ANC-led government in South Africa thrives on a charisma akin to this, hoping to sway the population through popular charisma and diction, the bulk of it devoid of substantive means for ushering in economic justice for Black workers. In actuality, there can be little common ground between the-custodians of monopoly capital who are bent on accumulating profit and the interests of Black workers. This class of workers consists of women and men who are struggling to earn a decent living wage with basic rights to a proper and complete education, adequate health care, humane housing conditions, and the pursuit of opportunities for the development of their human potential in all respects, accorded in a climate of social and economic justice.

One important factor that needs to be noted is that the rural areas of South Africa, such as the Eastern Cape, Northern Transvaal, Northwest Province, and Kwazulu-Natal, which are the poorest regions of the country, have been direly neglected in the new government's plan of restructuring the political economy of the country. The focus of the bulk of investment initiatives and incentives has been in historically more prosperous regions of the country: the Gauteng area, where the mining industry is concentrated, and the Western Cape.

The rural areas, where up to 60 percent of South Africans/Azanians live, are largely untouched by the wave of post-apartheid "reforms," particularly in areas of land redistribution, health care, and indigenous agricultural support. Black rural workers have not been drawn into the circle of politicians and bureaucrats responsible for shaping the future Azania. In essence, they have been viewed as irrelevant to the discourse on transformation, simply because of their indigenous rural and peasant experiences. This is precisely the area in which the new government has failed to understand the process of fundamental change as rooted in the lives and cultures of the indigenous rural community. It has devoted its energies to urban "economic development" and has viewed the metropolitan cities as the central loci of transformation, just as it views the indigenous rural population as needing to "graduate" to urban levels of development, productivity, and modernization. At a meeting of the national Rural Convention in Bloemfontein in late April 1999, a discussion document noted that

> half of South Africa's population lives in rural areas, with about 70 percent living in dire poverty. Rural development continues to receive lower priority than urban development. The potential of rural development to generate wealth and livelihoods is underestimated. The government's spatial development initiatives are being planned and implemented with little regard for the effects on individual rural communities. Many of the causes of poverty are the result of the private sector's interests in short-term profits at the expense of longer-term social goals.[79]

The Rural Convention document criticized the government's failure to construct and implement rural development initiatives that are geared toward women's development, farmworkers, land and tenure reform, agricultural support, water and the environment, education and training, health care and social security, and infrastructure supports. This critique becomes trenchant when one considers that at the ANC Policy Conference of 1992, where the ANC's economic policy was officially adopted, it was declared that "[t]he ANC will redirect government expenditure on housing, infrastructure, education, health, and social welfare to ensure equality for all South Africans, *especially rural people*" (italics mine).[80]

With accentuation on models of urban industrial development, this has implied an orientation of modernization as defined by Westernization that ineluctably produces and caters to bourgeois urban elites. It is for this reason that there has been no fundamental change in the arena of governmental initiatives on socioeconomic transformation, since the essence of development has been urban, European-oriented, capitalist-based, and elitist.

Michael Lipton's analysis in the 1970s of the skewed nature of development paradigms in most postindependence African countries that favored urban metropolises to the detriment of investment in rural production applies to the contemporary post-apartheid context in South Africa.[81] Although over three-fourths of Africa's working population was engaged in some form of agricultural production during the 1960s and 1970s, only 10 percent of capital investments were earmarked for rural agricultural development.[82]

The same trend obtains in South Africa and in most other countries throughout Africa today. Implicit within this disposition of the elevation of urban-based industrialization over against a developmental focus on rural areas, where the bulk of African populations reside, is the assumption that rural indigenous forms of production are subsistence economies and would be inevitably and gradually overtaken by the commodity-exchange system as introduced by Western capitalism.[83] This policy signifies a fundamental flaw in the posture of the post-apartheid regime, in that it overlooks and neglects one of the most important segments of the colonized and dispossessed population, the rural peasantry. According to Cabral, it is only through the dynamic cultural interaction between the urban intelligentsia and the rural peasantry that a revolutionary synthesis can be born that is able to include the materialization of radical transformation, a critical observation that warrants extensive citing:

> The leaders of the liberation movement, drawn from the "petty bourgeoisie" (intellectuals, employees) or from the background of workers in the towns (laborers, drivers, salaried workers in general), having to live day by day with the various peasant strata, among the rural populations, come to know the people better. They discover at its source the wealth of their cultural values, (philosophical, political, artistic, social and moral) acquire a clearer awareness of the economic realities of the country; of the difficulties, sufferings, and aspirations of the mass of the people. The leaders realize, not without a certain astonishment, the wealth of spirit, the capacity for reasoning and clear statement of ideas, the facility for comprehension and assimilation of concepts on the part of populations who only yesterday were forgotten if not despised and regarded by the colonizer, and even by some nationals, as incompetent beings. The leaders thus enrich their culture—they cultivate the mind and free themselves from complexes, strengthening their capacity, to serve the movement in the service of the people.

On their side, the mass of workers and in particular, the peasants, who are generally illiterate and have never worked beyond the confines of the village or region, in the contact with other categories shed the complexes which constrained them in their relations with other ethnic and social groups. They understand their situation as determining elements of the struggle, they break the fetters of the village universe to integrate gradually into the country and the world; they acquire an infinite amount of new knowledge, of use to their immediate and future activity within the framework of the struggle; and they strengthen their political awareness, by absorbing the principles of national and social revolution postulated by the struggle. They thus become fitter to play the decisive role as the principal force of the liberation movement.[84]

In the South African case, the indigenous rural peasantry has been disregarded by the ruling establishment and viewed as irrelevant to the shaping of political and economic change because they are African, often illiterate, and provincial in their outlook. The normatization of the future South Africa is seen as most responsibly defined by Western-educated Black and white technocrats, bureaucrats, and financial advisers, people who are viewed as the sole custodians of utilizable and relevant knowledge for socioeconomic transformation, with the indigenous rural population expected to conform to the didactic designs formulated by this class of the urban intelligentsia. It is a one-sided movement, with the urban petit-bourgeoisie "experts" doing all the talking and the rural peasant folk doing all the listening. The new post-apartheid reality has little tangible grounding in the social and cultural milieu of the indigenous rural community, particularly agrarian workers. It is precisely for this reason, as Cabral so cogently illustrates in his analysis of the liberation struggle in Guinea Bissau, that material changes proposed by the new ruling class in South Africa are basically cosmetic in character and ultimately reproduce a Eurocentric, urban-based, and elitist-served status quo. Ironically, this arrogance and chauvinism generally demonstrated by South Africa's indigenous ruling elite replicates the same religio-cultural haughtiness toward Africans shown by the European missionaries who infiltrated the country in the 1800s and 1900s.

The anti-working-class nature of the new South African government is also evident in its xenophobic attitude and policy toward African workers from the neighboring countries, most of whom bore the brunt of the brutality of the South African military when it engaged in attacks on ANC forces in Botswana, Mozambique, Lesotho, Swaziland, Zimbabwe, Angola, Zambia, and Tanzania. The clampdown on "illegal migration" by South African immigration authorities, akin to the swoops made by the Immigration and Naturalization Service on Latin American people in the United States, has resulted in the arrests and "deportation" of thousands from Zimbabwe, Mozambique, Zambia, and Angola. The victims of these mass deportations are almost entirely Black. The same xenophobia does not apply to bourgeois European migrants, who are seen as skilled and useful technocrats in the expansion of the post-apartheid economy.[85]

The South African regime has demonstrated its neocolonial character in this regard, as it totally ignores the fact that the apartheid economy was founded on the backs of Black workers, particularly miners, many of whom originated from neighboring African states. The description of these workers as "illegal aliens" coming in to "take South African jobs" is intrinsic to the posture of national chauvinism of the South African regime, identical to that practiced against Mexican agricultural workers in the United States, who are repeatedly harassed as "illegal aliens" in border states such as California, Texas, Arizona, and New Mexico, even though the Mexicans are descendants of the original Native peoples of the Americas.

Xenophobic attitudes have precipitated attacks by South Africans on people from the Frontline States living in places like Alexandra Township outside Johannesburg, yet another tragic result of the suppression of the working classes and the product of intra-working-class divisions issuing from the aura of the "booming capitalist economy."

The "Free Market" Economy: South African Style

The new post-apartheid government claims that it cannot introduce radical change measures because such action would destabilize the local economy and cause untold hardship. It thus touts the program of "economic pragmatism," which is essentially neoliberal Keynesian economics. One could argue that South Africa has not had an *authentic* economy since the first European colonialists came to Africa, in the sense that the economy of the colonial era through the present has quintessentially been a capitalist-owned and run one, predicated on the cheap labor of the conquered indigenous population. The majority of the Black population has never truly benefited from the colonially originated and subsequently capitalist-driven economy.

The government has also defended its transformation program by indicating that it made some significant "gains" in 1997, such as increasing the supply of clean and accessible water from 700,000 to 1.3 million people; built and upgraded more than 500 clinics; made 421,000 telephone connections and 400,000 electricity connections; is feeding 4.9 million primary school children a year; constructed 400,000 houses and allocated 700,000 subsidies for housing; ensured over 60,000 loans at "the upper end of the subsidy market"; and passed the Skills Development Bill and the Employment Equity Bill.[86]

Although one may concede that these are important developments and that the envisaged transformative changes of the ANC need more time and many more resources before they can take effect, concomitantly, Black working-class critics perceive these measures as perfunctory because they are designed to appease the Black working-class population as opposed to *empowering* the working class and the underclasses. The perfunctory character of the "reforms" is noted on the grounds that they do not address the fundamental needs of the Black poor, since they are concerned with a small fraction of the population. The majority of the poor are left to eke out an existence in whatever way possible. In fact, they bespeak cosmetic and incremental reforms because they are conducted within the same old capitalist economic system, where profit accumulation is still the motive of all private companies involved in housing, telecommunications, construction, and so on and presuppose the principles of the same market system.

Further, as previously noted, many of the houses built by the new government are defective in construction and structure (90 percent of them), and many are simply too small. They are defective because construction contracts are given to companies that are ill qualified and use the cheapest route and products possible to construct these houses. Many families are still overcrowded in these tiny two-room matchboxes.[87] Low-income residents still need to take out loans to supplement the subsidies for the houses or for extensions to the buildings. Others sell their homes and use the monies for other needs such as paying for children's books and other family expenses. Transformation is not about appeasing the poor or adding defective houses to impoverished, polluted, and resource-strapped ghettos like Alexandra and Soweto, Umlazi and Khayelitsha; it ought to be about constructing new neighborhoods in qualitatively different areas where Black people can live in dignity and free from the filth and squalor of apartheid slums.

Similarly, the entire school system, involving curricula, teacher training, school construction, and book availability, needs to be revamped. Giving school lunches to hungry children is important for satisfying hunger, but one must ask right at the outset why it is that children are hungry when they come to school and what needs to be done to transform that condition of structural poverty.

The government's rhetoric on job creation and job training is empty and devoid of substance; it has no radical program to force private enterprise to create such jobs or train skilled workers because it does not have the power to radically restructure and reorient the capitalist-controlled economy. If the government did have the *power*, it would engage in revolutionary transformation as opposed to the gradual, incremental, and piecemeal reforms that it has produced. In short, the post-apartheid government does not possess power to effect genuine and radical transformation; the ANC disempowered itself by depending on the negotiated settlement that resulted in foundational compromises and powerlessness on the part of the new post-apartheid government.

It is clear that short of any revolutionary transformation in South Africa, there will be no fundamental change, because the assumptions of the old capitalist apartheid system would still be tacitly operational, as is the case in the current post-apartheid order. As Karim Essack points out, "Socialists believe that it is not possible to use structures of capitalist societies to serve the workers and peasants. These have to be destroyed and new structures should arise to serve the new structures in power."[88]

The ANC-led government continues to work within capitalist structures because of economic and material benefits that accrue to a tiny segment of its leadership clique, while portraying itself as a benevolent institution that will eventually deliver on its plethora of promises to the Black poor. As in the mold of other neocolonial regimes in Africa, politics has been made cheap by the unrestrained capitalist ethos in post-apartheid South Africa.

In the current post-apartheid economy, the single–sex hostel system of apartheid still remains, in which over 300,000 Black male miners are forced to live away from their families for forty-eight weeks each year. The new government claims that it cannot immediately dismantle the subhuman immigrant labor hostel system because it would cause economic havoc and wreck the Southern African regional economy. Black mine workers still live in humiliating conditions with filthy kitchens and fetid putrescent toilets with overflowing sewage and endure the denial of individual privacy, justified under the pretext of "economic necessity and stability."[89]

The new government also contends that it does not have the economic resources and lacks the capital to engage in massive investment in areas of rural development and education, health care for the poor, job training, and housing, hence its dependence on private enterprise. Yet, as has been demonstrated in the preceding section, the problem is not that South Africa lacks financial wherewithal but rather that the vast portions of wealth are controlled by a white capitalist clique and locked into the hands of a few. This same group dictates economic policy to the government, which has capitulated to the designs of capital, since its ruling members benefit lucratively from this unjust arrangement. South Africa is not a poor country by virtue of the fact that white capitalists make billions of dollars each year, all deriving from the impoverishment of the Black working class and rural peasant majority.

The South African post-apartheid economy, as with all other capitalist economies, operates on the principle that the large conglomerates determine the prices of goods and services and consequently decide the standard of the cost of living in South Africa. The prices of goods are inflated way beyond the *cost of producing* such goods, so the industrial conglomerates can make gargantuan profits and offer lucrative salaries to commercial entrepreneurs and business executives, the large majority of whom are white, with a sprinkling of Black faces. Prices of locally manufactured goods are generally priced way beyond the reach of the average Black working-class consumer.

Companies are driven first and foremost by profit motives as opposed to satisfying material needs for workers. The raw materials needed for making cement, for instance, are found locally, and items such as cement (mortar) are manufactured locally, yet the prices for these goods are exorbitant. South Africa is distinctive within Africa and the underdeveloped world in that it has independent manufacturing capacity. It does not need to depend on imports for the fundamental running of its economy because most of the needed raw materials are found in the country. The government keeps prices of goods inflated on the grounds that labor costs have risen as a result of protracted union activity over the 1980s and 1990s, such as in the mining sector. However, most Black workers, even those unionized workers who were successful in increasing wage scales significantly through industrial strikes and other resistance actions, still earn incomes that barely keep up with the escalating cost of basic food, housing, clothing, and other essential consumer goods.

Black workers generally possess a variety of industrial skills and are able to build homes, manufacture and assemble automobiles, manufacture agricultural and food products, erect high-rise buildings, pave roads and construct bridges, electrify residential areas, and print books, yet all these skills are performed in the service of the wealthy elite. The bulk of the technical work in South Africa, from mining to industry to agriculture, is done by Black workers. Black miners in South Africa earn far less and work much harder under less safe conditions than their counterparts in Australia, the United States, Germany, and Britain. Yet the cost of purchasing a decent home in South Africa runs into several thousand dollars, making it cost-prohibitive for the average Black worker who is essentially responsible for producing the nation's vast wealth. The distribution of wealth is so obscenely skewed that most Black workers who work on massive construction sites, building thousands of plush homes and offices for middle-class and upper-middle-class people, will never be able to afford the very buildings they were responsible for erecting, a painful irony of capitalist socioeconomic strictures.

In the South African context, because of a large skilled workforce, the costs of vital goods can be dramatically reduced so that they become accessible to Black workers. The country is capable of manufacturing vital utilitarian items domestically, as opposed to purchasing them cheaply from places as far away as South Korea and Indonesia, where workers are exploited for their cheap labor by transnational conglomerates. South Africa engages in wasteful industrial production as part of its intrinsic membership within the global capitalist economy, as one sees in the United States, for example, where several brands of automobiles, electrical equipment and appliances, pharmaceuticals, clothes, and so forth are manufactured at artificially expensive prices, making those goods only accessible to the wealthy and upper middle classes. The plethora of brand products that one sees in Western industrial societies exists for the sole purpose of whetting the appetite of the populace so that primary consumers are created as opposed to conservationist producers. The end result of this "right of choice" defined by market principles is billions of dollars' worth of profits for parasitic transnational corporations, while most in the working classes are deprived of access to these high-priced goods because of low earnings that are defined by the owners of the market—global capital.

South Africa continues to be a sycophantic nation that apes the perversity of Western European industrial capitalist civilization, now promoted by the ruling Black and white elite. With this orientation comes an implicit anti-indigenous working-class culture, absurdly reflected, for instance, in the recent move by a Pretoria automobile company to sell 500 sports cars at the cost of $25,000 each to the United States. To add insult to African injury, the cars are to be dubbed "the Shaka."[90] These startling facts convince us that the capitalist basis of the South African economy, in addition to being exploitative of Black workers, is also *Eurocentric* and *racist*. Capital's obsession with profit compels it to commodify and prostitute indigenous African history and culture.

Land, Housing, and Economic Dependency

On the land question, we see that the contradictions inherent in so-called independent Africa continue to manifest themselves in South Africa. In Kenya, the bulk of the most productive land that is used by commercial farmers is owned by the European minority, and commercial institutions are preponderantly owned by wealthy Asian businessmen. In Zimbabwe, where similar apartheid-type conditions prevail, 800,000 peasant families lived on 16 million hectares of land at the time of independence in 1980, and 6,000 rich white farmers lived on 14 million hectares of land. Today, 4,000 white commercial farmers still control almost half of the productive arable land (called large-scale commercial, or LSC) and account for two-thirds of farm production.[91]

In the South African context, too, it is clear that the question of land reclamation for the millions of Black people dispossessed after three and a half centuries of European colonialism will not be addressed under the current dispensation, since the colonization of South Africa by Europeans never surfaced at any stage during the negotiations. The ANC has been very ambiguous on the subject, with former ANC secretary general Cyril Ramaphosa declaring during the heydays of CODESA, when a negotiated settlement was being reached between the ANC and the white National Party, "that restitution would be fundamental to ANC land policy, but warning of the danger of wrecking the country if the issue was not handled *sensitively*" (italics mine).[92]

"Sensitive" in this context refers to the anxieties by the white population that the lands that they had acquired through theft of Black land would be returned and redistributed. However, given the accommodationist posture of the new South African government, such fears on the part of the settler population appear to be groundless. Consider the ambiguity in the following statement on land restitution during the CODESA discussions: "Nothing in this section shall preclude measures aimed at restoring rights in land or compensating persons who have been dispossessed of rights in land as a consequence of any racially discriminatory policy (where such restoration or compensation is *feasible*)" (italics mine).[93]

The glaring question that many commentators on the South African situation omit to openly recognize is that *all* lands acquired by Europeans in Africa occurred through illegal and discriminatory means. The statement on possible compensation for land dispossession, contingent upon feasibility, is profoundly superficial. It is revealing to record the new government's putative continuation of the policy of forcing "squatters," or "informal settlers"—people who are determined to reclaim their ancestral lands—to move. In June 1994, several hundred homeless Black people who attempted to occupy their lands, stolen from them by white "developers," were forced off by police authorities.[94]

Although there is now a Land Claims Commission that the government formed to receive land claims for compensation and grant restitution to people who can furnish evidence of dispossession (which is often impossible in written form), there are several clauses that protect rich white farmers. The government has declared no intention of radically eliminating these laws, even though it claims to have abolished the Native Land Act of 1913, which allocated the most unproductive 13 percent of the total land area of South Africa for Black occupancy. It still has the Subdivision of Agricultural Land Act No. 70 of 1970, which forbids "the fragmentation of lands into units, which in the opinion of the Department of Agriculture, are 'uneconomic.'"[95]

The government has initiated cosmetic measures to convey the false impression that it is committed to land redistribution. The Commission on the Restitution of Land Rights was established under the interim constitution, which operated until the final constitution was approved in 1996. In November 1994, the South African Parliament passed the Restitution of Land Rights Act, which provided for the "restitution in rights of land in respect of which persons or communities were dispossessed under or for the purpose of furthering the objects of any racially based discriminatory law."[96] Claims had to be lodged between May 1, 1995, and April 30, 1998, a deadline that was then extended to December 31, 1998. Lodging a claim entails "a difficult and complex process" of about fourteen successive measures, broken down into the registration, notification, research, mediation, referral, and monitoring stages.[97] Over 22,000 people have filed claims to the Restitution of Land Claims Commission.[98] Given the complicated process, only five claims have been finalized in a space of three years.[99] At the rate that the Land Claims Commission is conducting its work, "it will take many centuries to deal with the backlog."[100] Obviously, the Black people filing the claims will not be alive by the time their claims are settled. Noting that 3.5 million Black people have been dispossessed since 1960 and the cumbersome character of the claims process, it is extremely unlikely that most Black people, particularly members of the rural population, will come forward because most will feel that it is a futile exercise, which the neocolonial system anticipates and accepts.

The government is mouthing the empty rhetoric of "land reform" echoed by repressive landlord-supporting regimes in Latin America and Asia. It is not committed to restoring the land of dispossessed African people as a minimum precondition for establishing independent Black agricultural producers. A Ministry of Land Affairs spokesperson, Helmut Schlenter, declared that 30 percent of private and state land will be redistributed over the next five years.[101] The import of this declaration, however, must be derived from what was not stated: The remaining 70 percent of private and state land will continue to remain in the hands of wealthy white farmers or industrialists. South Africa appears to be headed on the neocolonial path of the Congo and Kenya, particularly in the area of land restoration to the vast African peasantry. Land reform, let alone redistribution, is a façade for the following reasons:

A. The ANC-led regime will not expropriate the bulk of the best arable and most productive lands belonging to white commercial farmers and industry because it does not have the power to do so and such action contravenes its ostensible "reconciliation" policy. The Ministry of Land Affairs has indicated that 26 percent of the surface area is state land, about 32 million hectares, of which 1 million hectares of rural land is available for redistribution.[102] It needs to be noted that the amount of private (white commercial agricultural and industrial land) is not mentioned, nor is its size area explained or discussed in the Green Paper on South African Land Policy. The inference is that the remaining 74 percent of South Africa's land, which constitutes the best and most productive lands, is not included within the discussion on land reform, essentially retaining and protecting white-minority monopoly of South African land.

B. The new government will need to pay compensation for whatever land is taken from whites as part of the constitutional provisions, and it does not have the financial resources to execute such reform, prompting possibilities of dependence on Western financial agencies for the very limited appropriations of land that it will pursue. This will significantly retard and restrict the process, as was the case in Zimbabwe. The Department of Land Affairs has established a maximum of R15,000 ($2,500) "per beneficiary household, to be used for land acquisition, enhancement of tenure rights, investments in internal infrastructure, and home improvements according to beneficiary lands."[103] This limitation indicates that the value of lands acquired will be of the lowest premium, considering this absurd financial amount.

C. The land-reform program is conditioned by market considerations, as part of the overall market philosophy of capitalism, such as the GEAR program in progress, which implies that the substance of the land-reform program will be in the urban areas with the objective of creating a bourgeois class of landowners, with little or no earnest attention to the need for viable land redistribution among the vast rural peasantry. Most rural residents will continue in essence to remain impoverished and be prevented from developing sustainable agrarian economies.[104] The "willing buyer and willing seller" policy adopted by the state in this regard gives volitional character to land reform, and when the sellers are wealthy and the buyers poor, this will place the buyer in an invidious market position.[105] Following capitalist market principles and employing them in the land-reform program replicates economic inequalities and contradicts a program designed to assist those who are poor and victims of the market system.

D. There can be no land-reform program that is substantive, equitable, and just where there is "reconciliation" between wealthy white land-owning farmers and the landless Black peasantry, as the Land Affairs Green Paper by Derek Hanekom, former Minister of Land Affairs suggests. Neither is it a "finite resource which binds all together in a common destiny."[106] Such views are chimerical and do not enunciate the truth that land redistribution is a painful process for those who acquired the land through illegitimate means (colonialism) and land possession/dispossession is and will continue to remain at the heart of the South African racial, political, and economic conflict. Neither can there be a "non-racial approach to urban and rural development," as the Urban Foundation absurdly suggests.[107]

E. Half of South Africa's population still lives in rural areas, with over 70 percent living in acute poverty.[108] In some instances, such as at the Cape Flats, twelve families that were dispossessed received R953,000 as a form of monetary compensation.[109] Generally, Black people living in rural areas are totally untouched by the government's land-reform program, which appears to be orchestrated for political consumption rather than tangible and sincere land return. There has been no earnest attempt to retrieve rural land owned by white farmers that was taken from Blacks a century ago or to redistribute land to the desperately poor and landless rural folk. The government has made grants of R16,000 ($2,667) to certain small individual Black farmers, which is a woefully inadequate amount for any form of viable rural production. The former minister of agriculture and land affairs, Derek Hanekom, has conceded that the government has failed to create rural jobs with the land-reform program and that it has invested very little in land reform.[110] Clearly, land return and redistribution to the Black poor majority is not a serious priority for the ANC-led government. Land redistribution is essentially mythic rather than real, similar to the scams of "land reform" that one observes in places like Brazil, El Salvador, and Nicaragua in Latin America, where the rich still own the best lands and the majority of the impoverished peasantry ekes out an existence on the most unproductive and overcultivated of lands.

In the area of housing, similar promises to the 7 million Black homeless remain just that—promises devoid of practice. The government's former Reconstruction and Development Program responsible for housing construction appears to generate more form than substance and is more cosmetic than transformative. In Gauteng Province alone, there is a backlog of 1.5 million houses, and to address this backlog, the government would have to build 150,000 houses annually for ten years. An additional problem is the fact that 45 percent of those in need of housing earned below R800 ($160) per month in 1995, well below the poverty line.[111] In Gauteng, thousands of homeless people formed the National Homeless Organization of South Africa in the mid-1990s and engaged in mass mobilization against draconian measures of evictions of so-called squatters, an ugly racist term that needs to be exorcised from South Africa's political lexicon ("squatters" are essentially African people who are now reclaiming their ancestral lands pillaged by white colonial settlers).[112]

In September 1995, the Gauteng provincial government issued a Supreme Court order that denied rights to Black people who were homeless from occupying vacant land in Alexandra East and threatened some 280 families with eviction.[113] It is further ironic to learn that land sites have been identified in Gauteng for the proposed construction of a new prison, while the government claims that it is excruciatingly difficult to make more land available to the Black landless in the region.[114] One is compelled to ask the recurring question, in light of poor and homeless Black people being subject to such evictions and censure, among other oppressive violations: *Is apartheid really dead?*

A crucial area of housing that warrants further criticism is the government's promise to upgrade conditions in the overcrowded ghettos and townships and provide electricity, sewage, and water to make life more comfortable in these unhealthy environments. The South African regime has no program or plan of action to liberate people from the unsanitary and humiliating conditions of the townships, in other words: to begin dismantling townships and single-sex hostels.[115] The government is unwilling to initiate the radical process of massive reconstruction of homes for Black people in decent and economically productive areas (where most whites live), so that the Black working class can live in dignity and with adequate physical space and land. At this stage, the government plans on housing are obviously to maintain the status quo and to tinker with peripheral elements of these abominable living conditions for Black people and luxurious lifestyles for whites by encouraging a token number of Black elites to live in exclusive white suburbs. In the sphere of mortgage lending, for example, yet another goal (such as the ANC's initial promise to build 1 million homes in five years, when the state built only 878 houses from March 1994 to March 1996)[116] has been scuttled: Fewer than 1,000 low-income earners had received loans from the country's four big mortgage lenders, totally off the anticipated goal of financing 50,000 homes by 1996 that the lenders agreed upon with the government.[117]

Even more troubling is the government plan to settle twenty white families from South Africa in the lush Congo Niare Valley as part of an agreement reached between the South African and Congolese governments.[118] Another group of farmers is scheduled to settle in Zaire and Mozambique, for similar colonial-agricultural purposes. The amazing fact is that the South African government, which claims that it is a post-apartheid organization, is protecting the interests of wealthy white farmers who would not only displace other African farmers and people on the continent but also ship African fruit produce for export to Europe when such produce needs to feed indigenous people on the continent. Such measures, promoted by the South African government "to bring prosperity and jobs to Central Africa" and supported by the European Union and the World Bank, point up the government's conviction that the needs of the Black working classes in South Africa are subordinate to protecting the economic desires of the white-farmer population.[119]

When the government is faced with many of the questions raised here, it typically responds by stating that it is aware of the gross injustices extant in South Africa today as a result of historical oppression and that the antidote to the country's socioeconomic ills of dispossession, homelessness, impoverishment, and disparities between the wealthy and the poor is the Reconstruction and Development Program, recently renamed the Growth, Employment, and Redistribution Program. This program is supposed to reverse the effects of injustice as created by the apartheid-capitalist system and is equipped with a budget of a few billion rands. Since the elections of April 1994, this program at best addresses the needs of a small segment of the Black community in a symbolic way and at worst forms a smokescreen for the fundamental contradictions between wealthy white capitalists and the economically depressed vast Black working class.

The Reconstruction and Development Program, interestingly, was also formulated in Zimbabwe under the auspices of the Zimbabwe Conference on Reconstruction and Development, shortly after the realization of that country's independence in March 1981, with the explicit objective of mobilizing "international financial resources through grants and soft loans" to help finance the Transitional National Development Plan, in which a major objective was to redistribute wealth.[120] The "reconstruction" results in Zimbabwe, after sixteen years of independence, are negligible: The economy and most productive lands are still held in the hands of white settlers and a few Black elites. In examining the details of the South African Reconstruction and Development Program, one finds that the document is astute in analyzing elements of the South African political economy and society. For instance, it accurately states in the introduction, that in South Africa

> The economy was built on systematically enforced racial division in every sphere of our society. Rural areas have been divided into underdeveloped Bantustans and well-developed white-owned commercial farming areas. Towns and cities have been divided into townships without basic infrastructure for Blacks and well-resourced suburbs for whites
> . . .
>   Segregation in education, health, welfare, transport and employment left deep scars of inequality and economic inefficiency. In commerce and industry, very large conglomerates dominated by whites control large parts of the economy.[121]

Yet no mention is made of the *capitalist* system that is responsible for the economic disparities and impoverishment of Black people in South Africa. The entire document describes six rudimentary principles: an integrated and sustainable program; a people-driven process; peace and security for all; nation building; linking reconstruction and development; and the democratization of South Africa.[122] It describes in detail the range of basic needs that the majority of Black South Africans lack, such as land reform, housing services, water and sanitation, energy and electrification, telecommunications, transport, a clean environment, nutrition, heath care, social security and social welfare, education and training, arts and culture, sports and recreation, and youth development. The Reconstruction and Development Program set its goals for the five years following the elections of 1994, promising to "redistribute a substantial amount of land to landless people, build over 1 million houses, provide clean water and sanitation to all, electrify 2.5 million new homes and provide access for all to affordable health care and telecommunications"[123] and proposing "to distribute 30 percent of agricultural land within the first five years of the program."[124]

According to the new government, all of the objectives outlined above would be financed through a restructuring of the national budget, introducing a fair tax structure based on a system of progressive taxation on income and raising funds through the private sector. The foundations of the capitalist system that protect large private enterprise and the wealthy elites will remain intact according to this schema of reconstruction. It takes simple arithmetic and basic understanding of economic theory to point out that one does not generate substantial income in any moderate-size economy through taxation. Income in South Africa can be harnessed through expanded agricultural production, intensive manufacturing production based on natural resource utilization, and export of vital materials and products output that refurbishes the domestic-based economy, which then can be used to finance infrastructure development such as roads, hospitals, schools, and factories in urban *and* rural areas. Workers' incomes need to be at a level sufficient to support families' essential needs: adequate food, decent shelter, health care, social security, and recreational facilities. In any *just* economy, the right to employment, health care, and education must be inalienably assured and protected by the state. Progressive taxation may be a liberal approach toward appeasing the lower income segment of society by conveying the impression that national tax policies are against the rich, but in essence it protects the right of the wealthy corporations to earn billions. Even if multimillionaires are taxed heavily, their income base is still several hundred times that of the working class. The disparities between the poor and the wealthy may see minor reductions under liberal capitalist policy that institutionalizes progressive taxation, but such policy by no means undermines the fundamental position of the ownership of wealth in capitalist society by a tiny minority elite, which in the South African situation is essentially white capitalists and their managerial associates. Taxation is fundamentally ineffective in undoing the effects of economic disparities in a capitalist economy precisely because it presupposes the legitimacy of an unjust economic system that pays lucrative salaries to the upper and middle classes and poverty-level wages to the remaining working-class population.

Tinkering with the budget, as the government did in 1995 when it declared that more money would be invested in education and health care, while earmarking R10 billion ($2 billion) for defense and R28 billion ($5.7 billion) for servicing the country's national debt, is no fundamental departure from apartheid. The new Black government ought not to pay for debts accruing from the era of apartheid anyway, especially if it is owed to international banks and financial institutions that benefited lucratively from the slave system of apartheid during its heyday. When the annual national South African budget was drawn up in March 1995, the total amount allocated was some R135 billion ($27 billion). The R20 billion earmarked for reconstruction and development did not provide the 1 million homes that the ANC promised in five years. Since the apartheid-capitalist budgetary structure remains intact under a post-apartheid government, with peripheral modifications and adjustments, it logically follows that no radical change will be produced. Actually, under the capitalist dispensation, where living costs are determined by the market and international capitalist values, it is estimated that about R90 billion ($18 billion) will be needed to provide significant rehabilitation provisions to even begin to address Black community impoverishment, funds that the new government does not have. Prior to 1994, white bureaucrats and other profiteering renegades stripped the South African economy of R70 billion ($14 billion in 1994) in anticipation of the coming of Black rule, leaving the government's coffers bankrupt.

In the final analysis, the Growth, Employment, and Redistribution Program functions as a cover and is a ploy to conceal the innate contradictions between capitalist economic power and the vast Black working class. It is a ruse encouraged by global capital because it protects the interests of the conglomerates and wealthy whites, while conveying the specious impression to the world that the new South African government is committed to the grassroots Black working class.

No fundamental economic transformation and justice can occur under a capitalist dispensation, and it is criminal for the new government of South Africa to deceive the Black working class into believing that houses for the homeless will be built, education will be free and available to all, including the Black poor, that land will be given to the landless and dispossessed, and the economically distressed will be able to enjoy security through expanded economic growth and foreign investment. The GEAR program will not be able to realize its goal of authentic "reconstruction" or "development" because it is enshrined within a capitalist framework that is foundationally unjust and because it depends on private enterprise, the very sector that benefited from the exploitation of Black workers.

The IMF has offered to provide loans of $1 billion to the country's Reconstruction and Development Program. Given the blood-stained hands of the IMF as it forces women and children in the rest of Africa into the ranks of poverty though its diabolical Structural Adjustment Policy, the South African government demonstrates its neocolonial collaborationist designs by viewing such "assistance" as benign. One cannot expect radically transformative and positive results if one is beholden to the very hands that engineered the chains that enslaved Black people. Chango Machya w'Obanda makes this point abundantly clear as he excoriates neocolonial regimes in Africa for catering to the needs of capitalism's imperialist interests and mortgaging Africa's people and resources to the very powers that subjugated Africa, a critique that applies to the "new South Africa":

> The national bourgeois governing class, having turned its back on the interior, on the people, and on the real facts of its underdeveloped country, now has to look to the former colonial power, foreign capitalists and donor agencies as the source of the survival of the regime. This implies that the governing class has abandoned its responsibility to the people. Experience shows that the real masters of the country become the World Bank and the IMF. As we have already noted, independence was mere sham. Because as Onimode points out, the old masters, the direct colonial officials, who left in the 1960s, merely handed over to the rule of the multinational corporations. These have now handed over to IMF and the World Bank which "represent increasing multilateralism, decreasing direct visibility of imperialism and the current multilateral recolonization of Africa" (Onimode, 1988, p. 280).[125]

In the same vein, Samir Amin reminds us of the blatant hypocritical crocodile tears shed by the IMF when it comes to poverty and underdevelopment in the Two-Thirds World, while concealing its very generation of such evils:

> Adjustment policies unilaterally imposed on the weakest partners [of the global economy] (the Third World and the Eastern Bloc) fulfill this requirement for management of the crisis. They are not mistakes or aberrations produced by following an absurd ideology. The IMF did nothing to prevent the excessive borrowing of the 1970s because the rising debt was very useful as a means of managing the crisis and *the overabundance of idle capital* which it produced. The logic of adjustment now being carried out requires, therefore, that the free mobility of capital prevail, even if this should cause demand to contract because of reductions in wages and social spending, the liberalization of prices

and elimination of subsidies, devaluation, etc., and thus bring about a regression in the possibilities of development. The ritual statements made by these institutions, which in practice, place management of the crisis over every other consideration, the tears that they shed over the plight of the "poor," their incantations in favour of "stimulating supply," are nothing but rhetoric, and there is no reason to believe them sincere or find them credible. (italics mine)[126]

Even the IMF and the World Bank acknowledged in a recent report that the current debt-relief program entitled the "Heavily Indebted Poor Countries Initiative" will not assist poor countries in extricating themselves from indebtedness and becoming economically viable.[127] In fact, this initiative will further impoverish the poor peasantry of the underdeveloped Two-Thirds World. The demand that formerly colonized countries continue to transfer billions of dollars to Western banks and governments in payment for past debts is shameful, immoral, and inhumane; it does not consider the fact that the Western world essentially reaped its riches from the backs of enslaved Africans and other indigenous people of color in Asia and the Americas, whose confiscated labor and natural resources over the past five centuries have never been compensated. What revolutionary Africans and other oppressed people need to demand from the amnesia-ridden West are *reparations* for 500 years of genocide, slavery, and land and resources theft.

For Black and other colonized people to place their economic trust in Bretton Woods institutions[128] like the IMF and the World Bank is akin to accepting an internal investigative report recently released by the CIA disputing that it was involved in supporting drug dealers associated with the Contras in Nicaragua during the 1980s, which eventually brought drugs by the ton into the United States.[129] This bizarre rationality is tantamount to asking a fox to guard the chickens, as Malcolm X would say. Mohamed Babu, the exiled Tanzanian political economist, describes this strange phenomenon as "seeking a solution to the status quo from within the status quo!"[130]

The critique of the dependence on foreign investment and "liberalized" capitalism in the policy and thinking of the economic planners in the new government is not entirely isolated. Organized labor has also sharply challenged the false notion of the "engine" of economic growth that the government hails as the swiftest way of creating jobs for the millions of Black unemployed and extricating the Black underclass from the confines of poverty, even though it is not as intensely radical concerning capitalism as the Black Consciousness formations are. In an April 1, 1997, statement, COSATU, NACTU, and the Federation of South African Labor (FEDSAL) declared:

Market deregulation, or the "rule of the free market" is often put forward as the main engine to develop the society. In theory, the immediate freeing of markets would allow capital, labor, and other resources to flow to the areas where they can most productively be used. In practice, it is not so clear, nor so simple. Societies which have deregulated their markets have not necessarily grown fast.

Many market failures—particularly the failure to allocate resources to education and training, or to investment in infrastructure and big capital projects—have led to poorly performing economies. The premature and uncoordinated removal of tariffs, without attempts to address structural weaknesses of local industry, has wiped out employment. The failure to have financial market regulation has led in some countries to the collapse of major financial institutions, and of public confidence in the financial system. The system itself has often been inadequate in allocating capital to new economic activity. Markets have been inadequate in responding to the social needs of human beings—in setting decent wages and fair standards, in protecting the poor and marginalized, in correcting imbalances of wealth and inequality, and in addressing the problems of exploitation.[131]

Although this segment of organized labor does not appear to be explicitly calling for a socialist alternative because of its general political alignment with the ANC, it is clear that the ideology and idolatry of the market in a capitalist society like South Africa is under attack by these organizations. The new government would do well to heed their prophetic and veritable words.

Finally, as we reflect on the aftermath of the Pan-Africanist working-class movement in the South African liberation struggle, we would be remiss in not alluding to the experiences of many women and men from the African Diaspora in the United States who played a leading role in the anti-apartheid–Free South Africa Movement during the late 1970s and throughout the 1980s. This heroic and courageous community must be saluted for its tenacious efforts in advancing the tenets of the South African struggle, particularly with regard to the call for economic sanctions against the apartheid regime. One can say without a shadow of doubt that barring the involvement of these sisters and brothers, things might have been quite different on the current South African political landscape, and perhaps the release of members of banned organizations like the ANC, PAC, and the Communist Party could have been delayed.

Many Black people from the United States flew to South Africa for the country's first nonracial elections in April 1994. Today, many are returning to South Africa to become involved in a variety of initiatives that are concerned with the economic rebuilding and social reconstruction of South Africa. These are important steps within the Pan-African movement, particularly in bridging the experiences of two communities of Africans on either side of the Atlantic who have strikingly common histories of enslavement, colonialism, and racism. However, a caveat needs to be sounded so that the euphoria that many Africans experienced in the United States (or Turtle Island, as the Native people call it) during the South African elections is sobered by the realization that Africans in South Africa and the Americas are engaged in an anticapitalist and anti-imperialist struggle. These Africans thus need to see the elections as part of the tactical designs of the imperialist nations to subvert the Black movement of radical transformation of South Africa's system of colonial-capitalism.

The naïveté that many Black people from the United States reflect needs to be replaced with a new consciousness of the fact that the real enemies of Black people—

capitalism and imperialism—have not yet been defeated in the South African liberation struggle and, in fact, have been couched in new forms, this time under the guise
of a "post-apartheid" South Africa and "the new South Africa." The struggle for
racial and economic justice for the people of African descent in the United States
and the indigenous Indians in that land can only be realized through the elimination
of American capitalism, colonialism, and imperialism. The United States functions
as an imperialist monster that thrives on the blood of the labor of the colored workers and slaves of the world, consistent with its genocidal history of the chattel enslavement of Africans and the decimation of the Native peoples of North America.
Just as the indigenous Indians need to have their stolen lands returned to them for
authentic peace and harmony in the United States and the rest of the Western hemisphere, so too the stolen lands of South Africa/Azania need to be returned to the
African people, specifically to the working classes.

"Black working-class cultures of resistance" as implied in this context of struggle
refers to the all-encompassing character of Black resistance that militates against the
tripartite evils of racism, classism, and sexism, all intrinsic to the cog of colonial oppression and capitalist exploitation imposed on African people. The operating assumption here is that oppression is a composite phenomenon that is organically constituted and maintained by the interwoven nature of the threads of oppression by
race, capital, and gender. In essence, capitalist exploitation is integral to racial oppression, and all capitalist systems are both racist and sexist in character.

Black working-class cultures of resistance are revolutionary in formation when
they attack structures of racism, classism, and sexism in a multipronged yet simulta-
neous movement of liberation. However, they reflect reactionary trappings when
they assume regressive cultural tendencies such as the proscription of women's mobility and the denial of access to critical and decisive positions of leadership of
women by Black working-class males. A classic example of this tragedy is observed in
the makeup of the Zimbabwean government, following the protracted guerrilla war
waged by ZANU, in which many female cadres sacrificed their lives and families to
advance the liberation struggle *(chimurenga)* only to be later disappointed to discover
that they were viewed as incapable of shouldering "male" responsibilities following
independence in 1980.[132] There are few women who have been accorded significant
decisionmaking power in the Zimbabwean government because many male parliamentarians and legislators have reverted to previous male-biased practices in their
overall sociopolitical dispositions. Protests by women erupted in Harare in May
1999, as a result of a court ruling that declared that customary laws will continue to
prevail, under which "only men can inherit and all family members are subordinate
to the male head of the family."[133]

One thus needs to clarify the meaning of Black working-class culture in its radical
proportions, particularly given the rising phenomenon in post-apartheid South
Africa, where leaders previously associated with Black working-class resistance and
"socialism" have now openly and uncritically embraced the capitalist system by taking on lucrative positions in the private sector. Membership in the Black working
class does not necessarily and inevitably translate into revolutionary consciousness.

Summary and Conclusion

The ideals of Pan-Africanist struggle can only be attained and advanced through the praxis of working-class Black people in Africa and in the African Diaspora, not through the empty rhetoric of the Black bourgeoisie who benefit from capitalist economic systems. Black people in South Africa now need to come to grips with a situation that many of us had not predicted a decade ago: the emergence of a Black majority government that is not primarily committed to the liberation of the Black working classes from the yoke of monopoly racial capitalism but rather toward titivating the very system that was responsible for the impoverishment and misery of the Black masses. In a very real manner, the post-apartheid situation in South Africa has several similarities with other "post-independence" situations in Africa, a subject that will be considered in the next chapter, given South Africa's intrinsic connection to the rest of the African continent.

Notes

1. Cited by Nicoli Nattras, "Economic Reconstruction and Development in South Africa," in *Reaction and Renewal in South Africa* (New York: St. Martin's Press, 1996), 133. Although Nattras describes this change in ANC economic policy and direction, he is typical of the neoliberal capitalist tendency and views the initially proposed posture of the ANC of "altering the structure of demand to tilt the balance towards the poor" as negative. This liberal reformist status-quo position advocates that the South African economy ought to remain firmly ensconced in the hands of white capital and their petit-bourgeois Black and white allies (p. 132) or else economic "chaos" would result, "chaos" in this instance implying loss of wealth to the capitalist class.

2. Patrick Cull, compiler, *Economic Growth and Foreign Investment in South Africa* (Cape Town: IDASA, 1992), 3.

3. The U.S. State Department, in its annual survey in early July 1998, rated South Africa as an "attractive" destination for U.S. businesses but cautioned that slow privatization would hamper capital investments in the country; cited in *Business Day,* July 9, 1998.

4. Hein Marais, *South Africa: Limits to Change: The Political Economy of Transition* (London and New York: Zed Press, and Cape Town: University of Cape Town Press, 1998), 147.

5. Amilcar Cabral, *Unity and Struggle: Speeches and Writings* (New York: Monthly Review Press, 1979), 132.

6. See, for instance, Marais, *South Africa: Limits to Change*, 1998.

7. Ibid., 125.

8. Ibid., 137.

9. Ibid., 124.

10. *Star Business Report*, September 14, 1998.

11. *Africa Recovery,* United Nations Department of Public Information, Vol. 13, No. 1, June 1999.

12. *Supplement to the Financial Mail*, June 26, 1998.

13. Cited by Sean Archer, *Defence Expenditure and Arms Procurement in South Africa*, Center for African Studies, University of Cape Town, March 22, 1989. Many of these statistics are taken from the Stockholm International Peace Research Institute Yearbooks of 1983 and 1987.

14. Ronnie Kasrils, "The Future of South Africa's Defence Industry: The Government's Perspective," in William Gutteridge, ed., *South African Defense and Security in the 21st Century* (Brookfield, Vt.: Dartmouth Publishing Company, 1996), 121–122.

15. *Citizen* (Johannesburg), October 5, 1995.

16. *Business Day* (Johannesburg), October 10, 1995.

17. Ibid., October 11, 1995.

18. Anthony Ginsberg, *South Africa's Future: From Crisis to Prosperity* (London: Macmillan, 1998), 184.

19. *Financial Mail,* June 19, 1998, 42.

20. Marais, *South Africa: Limits to Change*, 126.

21. *Citizen* (Johannesburg), December 8, 1995.

22. Julius O. Ihonvbere's article "Political and Economic Restructuring in Africa: Constraints and Possibilities for the 1990s," in Mulugeta Agonafer, ed., *Africa in the Contemporary International Disorder: Crisis and Possibilities* (Lanham, Md.: University Press of America, 1996) is instructive in this regard, particularly pp. 363–364 and 374, where the excruciating effects of World Bank and IMF policies in Africa and the incredible levels of indebtedness crippling African economies are discussed. "Two-Thirds World" replaces the pejorative notion of "Third World" and refers to Africa, Asia, and Latin America, where two-thirds of the world's people live.

23. *Business Day* (Johannesburg), April 18, 1996.

24. *World Press Review,* May 1993.

25. *Star Business Report* (Johannesburg), September 14, 1998.

26. *Citizen* (Johannesburg), September 14, 1998, 24.

27. *Star* (Johannesburg), December 8, 1995.

28. *Star* (Johannesburg), March 16, 1995.

29. See, for instance, John Saul's work in this connection, *Reconstruction and Resistance in Southern Africa in the 1990s* (Trenton, N.J.: Africa World Press, 1993), 118–121.

30. John Pampallis, *Foundations of the New South Africa* (London: Zed Books, 1991), 282.

31. Saul, *Recolonization and Resistance*, 121.

32. See Sam Nolutshungu, *Changing South Africa: Political Considerations* (Manchester, U.K.: Manchester University Press, 1982), 116–145; Baruch Hirson, *Years of Fire, Years of Ash: The Soweto Revolt: Roots of a Revolution* (London: Zed Press, 1979), 5–7; Bernard Magubane, *The Political Economy of Race and Class in South Africa* (New York: Monthly Review Press, 1979), 347–349; Gail Gerhart, *Black Power in South Africa* (Berkeley: University of California Press, 1982), 108–111; "The Transkeian Independence," *South African Students Organization (SASO) Newsletter* 6, March–April 1971; Robert Fatton Jr., "Class, Blackness, and Economics," in *Black Consciousness in South Africa: The Dialectics of Ideological Resistance to White Supremacy* (Albany: State University of New York Press, 1986); Stanley Greenberg and Herman Giliomee, "Managing Class Structures in South Africa: Bantustans and the Underbelly of Privilege," in Irving Leonard Markovitz, ed., *Studies in Power and Class in Africa* (New York: Oxford University Press, 1987); and Martin Murray, *South Africa: Time of Agony, Time of Destiny: The Upsurge of Popular Protest* (London: Verso, 1987), 225–235.

33. John Saul and Stephen Gelb, *The Crisis in South Africa,* rev. ed. (New York: Monthly Review Press, 1986), 132–133.

34. See Chapter 4, "Racial Capitalism," in David Mermelstein, ed., *The Anti-Apartheid Reader* (New York: Grove Press, 1987).

35. Ken Luckhardt and Brenda Wall, *Working for Freedom: Black Trade Union Development in South Africa Throughout the Seventie*s (Geneva: Program to Combat Racism, 1981), 59.

36. Sheena Duncan, "The Effects of the Riekert Report on the African Population," *South African Labor Bulletin*, Vol. 5, No. 4, November 1979, 74.

37. See Manning Marable's excellent delineation of the Black capitalism myth in *How Capitalism Underdeveloped Black America* (Boston: South End Press, 1983), 158.

38. bell hooks, *Yearning: Race, Gender, and Cultural Politics* (Boston: South End Press, 1990), 28.

39. C.L.R. James's article "The West Indian Middle Classes" in Fred Horde and Jonathan Lee, eds., *I Am Because We Are* (Amherst: University of Massachusetts Press, 1995), captures the dependent and sycophantic role of the Black middle classes in the Caribbean, an observation I would argue also applies to South Africa.

40. Saul and Gelb, *Crisis in South Africa*, 133.

41. Manning Marable discusses the fact of the Black corporate core in the United States being so financially small within corporate capitalism that its total worth would be less than the liquid assets of Mobil Oil in the 1980s. See his *How Capitalism Underdeveloped Black America*, 158. Although the Black corporate core may have significantly increased in the 1990s, it is still a drop in the bucket in the world of U.S. transnational capitalism.

42. John Saul and Stephen Gelb, *The Crisis in South Africa*, 135–136.

43. Ibid., 137.

44. See, for instance, the article "The Blacks in the Gray Flannel Suits" regarding the emergence and frustrations of the new Black middle class in *Newsweek*, May 13, 1996.

45. *Africa Confidential*, Vol. 35, No. 9, 1994.

46. *Financial Mail*, June 26, 1998.

47. *Tribute*, July, 1998.

48. *Mail and Guardian* (Johannesburg), April 23–29, 1999.

49. Ibid., May 7–13, 1999.

50. The term "businessocracy" was coined in *Tribute*, September 1996.

51. *Arizona Daily Star*, April 25, 1999.

52. *Tribute*, July 1997.

53. Ibid., January 1997.

54. *Supplement to the Financial Mail*, June 26, 1998.

55. Ibid.

56. *Mail and Guardian*, September 11–17, 1998.

57. *Financial Mail*, March 22, 1996.

58. *Sowetan* (Johannesburg), September 14, 1998.

59. *Africa Confidential*, Vol. 35, No. 9, 1994.

60. I. Abedian and B. Standish, "The South African Economy: An Historical Overview," in Iraj Abedian and Barry Standish, eds., *Economic Growth in South Africa: Selected Policy Issues* (Oxford and Cape Town: Oxford University Press, 1992), 15.

61. Ashwin Desai, "All the Baas' Wear the Same Clothes," *Towards Democracy*, Journal of the Institute for Multi-Party Democracy, Vol. 7, No. 1, 27.

62. *Arizona Daily Star*, April 25, 1999.

63. *City Press* (Johannesburg), September 13, 1998.

64. *Mail and Guardian*, September 11–18, 1998.

65. Ibid.

66. The promotion of racial capitalism, free from overt racist legislation in South Africa, was mediated through the U.S. policy of "constructive engagement" under Ronald Reagan, articulated in Chester Crocker's (U.S. assistant secretary of state for African affairs during the Reagan term in office) article "In Defense of American Policy," in Mermelstein, *The Anti-*

Apartheid Reader, 346–350. For a detailed role of U.S. policy in Southern Africa, see William J. Pomeroy, *Apartheid Imperialism and African Freedom* (New York: International Publishers, 1986); James North, *Freedom Rising* (New York: New American Library, 1985); and Phyllis Johnson and David Martin, eds., *Frontline South Africa: Destructive Engagement* (New York: Four Walls Eight Windows, 1988). Cornel West also echoes this view in his reflections on witnessing apartheid in South Africa in "On Visiting South Africa," in *Prophetic Fragments* (Grand Rapids, Mich.: William Eerdmans, 1988), 111.

67. Martin Murray, *Revolution Deferred: The Painful Birth of Post-Apartheid South Africa* (New York: Verso, 1994), 159.

68. Alan Ward, "Changes in the Political Economy," in F. H. Toase and E. J. Yorke, eds., *The New South Africa: Prospects for Domestic and International Security* (London: Macmillan, and New York: St. Martin's Press, 1998), 49. Although the objectives of this book in terms of the interests of the Black working class are questionable, owing to the militaristic origins of the writers and editors, it is nevertheless clear that the political unity among the three leading players on the South African political stage, namely the ANC, COSATU, and the SACP, is indeed fragmentary.

69. *Financial Mail,* June 19, 1998.

70. Ibid.

71. In resolving to formulate a workers charter, the Amalgamated Clothing and Textile Workers Union of South Africa noted that "the Freedom Charter, while being a historic document which raises many issues fundamental to all oppressed in South Africa, is no substitute for clearly spelt out protections of minimum worker rights which ought to be included in any genuine democratic constitution for South Africa if it is to enjoy the respect of organized workers"; cited in *South African Labor Bulletin*, Vol. 14, No. 2, June 1989, 54.

72. Moses Mayekiso, interview with *South African Labor Bulletin*, cited in the *New African* (Durban, South Africa), July 7, 1989.

73. For a delineation of the historical role of Black workers in political and economic struggle in South Africa, see Ken Luckhardt and Brenda Wall, *Organize or Starve: A History of the South African Congress of Trade Unions* (New York: International Publishers, 1980), and "The ANC and Workers Organizations, 1919–1928," in *A History of the ANC: South Africa Belongs to Us* (Harare: Zimbabwe Publishing House, and Bloomington: Indiana University Press, 1988). See also Stephen Dain, "Democratic Trade Unions," chap. 7 in *Apartheid's Rebels* (New Haven: Yale University Press, 1987), which elucidates the participatory character of democracy within the Black trade unions in South Africa.

74. At the historic workers' summit of August 1988, where workers' organizations from both trade union federations, COSATU and NACTU, were represented, the issue of a workers' labor relations amendment was raised once again. The workers, cognizant of the fact that even the African National Congress was under pressure abroad to portray a mediocre image on the question of workers' demands and a socialistic economy with workers making decisions, were determined to articulate their workerist views unequivocally and unabashedly. See Karl von Holdt, "The Summit: A Step Towards Unity," *South African Labor Bulletin*, Vol. 14, No. 1, April 1989, 27.

75. Solomon Mlambo, "Popular Front or United Front?" *South African Labor Bulletin*, Vol. 13, No. 8, 1989. See an article discussing the trade union coalition around the tensions of the "united front approach" and the "Africanist–working class" tendency, represented by COSATU AND NACTU, respectively, in "Conferees Endorse Unity Call," *Africa News*, December 25, 1989.

76. Mlambo, "Popular Front or United Front?"

77. Cited in Rich Mkhondo, *Reporting South Africa* (London: James Currey, and Portsmouth, N.H.: Heinemann, 1993), 127.

78. "Being Left: A Humane Society Is Possible Through Struggle," interview with Manning Marable, South End Press, June 1998.

79. *Mail and Guardian*, April 23–29, 1999.

80. Cull, *Economic Growth and Foreign Investment*, 3.

81. This posture is described, for instance, in Charles Becker and Andrew Morrison's article, "Public Policy and Rural-Urban Migration," in Josef Gugler, ed., *Cities in the Developing World* (Oxford: Oxford University Press, 1997), 92.

82. Wayne Nafziger explains this point of urban bias in post-independence governments of Africa and the resultant chasms between urban and rural development and the accompanying disparities between urban elites and rural peasants in *Inequality in Africa: Political Elites, Proletariat, Peasants, and the Poor* (Cambridge: Cambridge University Press, 1988), especially in chap. 11.

83. Irving Markovitz's work *Power and Class in Africa: An Introduction to Change and Conflict in African Politics* (Englewood Cliffs, N.J.: Prentice Hall, 1977), for instance, is one elementary yet informative analytical text on this subject, particularly p. 76.

84. Cabral, *Unity and Struggle*.

85. The article "Meet SA's Strange New 'Racists'" (*Sunday Times Insight*, September 13, 1998) underscores the gravity of this anti-Black xenophobia, reflected, for instance, at an automobile manufacturer where unionized workers opposed the hiring of a Zambian but had no qualms about employing an Australian and a Scotsman at the same time.

86. Thabang Makwetla, "Challenges Facing the ANC in Parliament in 1998," *Mayibuye: Journal of the African National Congress,* March 1998.

87. The author interviewed people living in such new houses in Joza Township, outside Grahamstown, and in Alexandra, outside Johannesburg, in July 1997 and July 1998, respectively, and the inhabitants expressed sentiments of disgust and dejection at the minuscule size of the houses, the outdoor toilet, the poor ventilation, and the low quality of materials used to construct the houses.

88. Karim Essack, *Co-option or Transfer of Power in South Africa* (Dar es Salaam: Thakers Limited, 1991), 78.

89. For a description of the sordid and inhumane conditions under which migrant miners are forcibly subject, see Shula Marks and Neil Andersson's article, "The Epidemiology and Culture of Violence," in N. Chanbani Manganyi and Andre du Toit, eds., *Political Violence and the Struggle in South Africa* (London: Macmillan, 1990). The author himself visited worker hostels in Thlabane, in Rustenburg, and at Umlazi, in 1989 and 1991 and saw the horror of the hostel system firsthand.

90. *Business Day* (Johannesburg), March 26, 1996.

91. Tor Skalnes, *The Politics of Economic Reform in Zimbabwe: Continuity and Change in Development* (London: Macmillan, and New York: St. Martin's Press, 1995), 150.

92. Doreen Atkinson, "Insuring the Future: The Bill of Rights," in S. Freedman and D. Atkinson, eds., *The Small Miracle: South Africa's Negotiated Settlement* (Johannesburg: Ravan Press, 1994), 139.

93. Ibid.

94. News broadcast, South African Broadcasting Corporation, June 15, 1994.

95. Peter Moll, *The Great Economic Debate* (Johannesburg: Skotaville, 1991), 142.

96. *The Commission on the Restitution of Land Rights* (CRLR) (Braamfontein: Human Rights Committee, 1996), 3.

97. Ibid., 4.

98. Marc Wegerif, *Poverty, Land, and Rural Development in South Africa: Poverty Hearings Background Paper* (Johannesburg: SANGOCO Publications, 1998), 9.

99. *Commission on the Restitution of Land Rights,* 7.

100. Wegerif, *Poverty, Land, and Rural Development,* 9.

101. *Business Day* (Johannesburg), February 28, 1995.

102. *Our Land, Izwe Lethu, Ilizwe Lethu,, Shango Lashu, Tiko Ra Hina, Naga Ya Rona, Live Letfu, Naha Ya Rona, Naga Ya Rona, Inarha Yethu, Ons Grond, Green Paper on South African Land Policy,* Department of Land Affairs, February 1, 1996, 65.

103. Ibid., v.

104. Ibid., 25.

105. Ibid., 25.

106. Ibid., 1.

107. *A Land Claims Court for South Africa? Exploring the Issues: An Executive Summary* (Johannesburg: Urban Foundation Research, September 1993), 1.

108. *Mail and Guardian* (Johannesburg), April 23–19, 1999.

109. Ibid., May 21–27, 1999.

110. Ibid., April 30–May 6, 1999.

111. *Business Day* (Johannesburg), November 29, 1995.

112. Ibid., January 5, 1995.

113. Ibid., September 22, 1996.

114. *Star* (Johannesburg), December 5, 1995.

115. Ibid., October 11, 1995; see also *Star*, October 30, 1995, which describes this level of upgrading in Alexandra Township outside Johannesburg.

116. *Business Day* (Johannesburg), February 24, 1995.

117. Ibid., October 6, 1995.

118. *Star* (Johannesburg), August 4, 1995.

119. *Business Day* (Johannesburg), November 23, 1995.

120. *Socio-Economic Review, 1980–1985* (Harare, Zimbabwe: Ministry of Finance, Economic Planning and Development, 1985), 8. The relationship of the Zimbabwe Conference on Reconstruction and Development to the Transitional National Development Plan is discussed in Lawrence Shuma's *A Matter of (In)Justice: Law, State, and the Agrarian Question in Zimbabwe* (Harare, Zimbabwe: SAPES Books, 1997), 53–54.

121. *The Reconstruction and Development Programme* (Johannesburg: Umanyano Publications for the African National Congress, 1994), 2.

122. Ibid., 4–6.

123. Ibid., 8.

124. Ibid., 22.

125. Chango Machya w'Obanda, "Conditions of Africans at Home," in Tajudeen Abdul-Raheem, ed., *Pan Africanism: Politics, Economy, and Social Change in the Twenty-First Century* (New York: New York University Press, 1996), 50.

126. Samir Amin, *Capitalism in the Age of Globalization* (London: Zed Books, 1997), 21.

127. Robert Naiman, "G7: Drop the Debt or Stand Aside," in *Sunday Journal* (Washington, D.C.: Preamble Center), June 13, 1999.

128. "Bretton Woods" institutions are named after the location, Bretton Woods, New Hampshire, where the question of postwar reconstruction of Europe was discussed by the Western imperialist powers in 1944.

129. *Arizona Daily Star* (Tucson), January 30, 1998.

130. A. M. Babu, "The New World Disorder—Which Way Africa?" in Abdul-Raheem, *Pan Africanism*, 91.

131. Statement on *Social Equity and Job Creation, the Key to a Stable Future*, released by COSATU, NACTU, and FEDSAL, April 1, 1996.

132. The subject of women's oppression and revolutionary struggle will be taken up in Chapter 6.

133. *Mail and Guardian* (Johannesburg), May 21–27, 1999.

4

A Pan-Africanist/Black Working-Class Critical Perspective on "Independent" African Political Economies

It is important to illuminate the politics of contemporary independent Africa, particularly in order to understand South Africa's political economy, since South Africa is an intrinsic part of Africa, even though many have been led to believe that South Africa is part of the West. Ihechukwu Chinweizu reminds those of us from Africa, including academics, that we are not a part of the West, even though we often attempt to believe so.[1] From the vantage point of conscientized African people, South Africa, or Azania, is an integral nation of the African continent, with the history of the country being indissolubly linked with historical developments in the rest of the continent. Ossified and atomized thinking and conceptualization of South Africa as a disparate and distinctive political and economic entity apart from the rest of Africa is the legacy of colonialist and imperialist divisions of Africa.

South Africa and "Independent Africa"

Black South Africans have been constantly indoctrinated and continue to be vitiated by the perversity of a colonialist analysis of the South African situation that holds that Black people there are better off than in other African nations in the northern, western, and eastern parts of the continent, primarily because we live in a land that was built and "developed" by the prolonged presence of Europeans and thus have had some of the wonders of European civilization rub off on our experiences. Then, too, we are taught that we are different because we have a peculiarly different history from the rest of the continent—that of apartheid. True, the level of humiliation that we suffered under apartheid and the effects that we continue to suffer from today are unprecedented when one examines the history of colonial forces in other parts of Africa. Yet we cannot claim to be living in a post-apartheid era or under radically different socioeconomic conditions from the rest of the continent when the foundational structures of white supremacy, colonial domination, and capitalist hegemony persist, similar to the Western colonial control of the rest of Africa through monetary, commercial, material, military, and cultural means.

Two important points need to be articulated here. First, our historical and contemporary experiences in South Africa are akin to those of our sisters and brothers in Ghana, Nigeria, Kenya, Tanzania, Angola, Senegal, and other parts of Africa, at least in substance. Kwame Nkrumah underscored this point as he laid out the groundwork for African unity in *Africa Must Unite:*

> But I have also, as an African and a political being drawn into the vortex of African affairs out of my dedication to the cause of Africa's freedom and unity, sustained an indelible impression from the experience of my continental brothers [and sisters] under other colonial rulers.
>
> Their history of colonialist subjection differs from ours only in detail and degree, not in kind. There are some who make fine distinctions between one brand of colonialism and another, who declare that the British are "better" masters than the French, or the French "better" than the Belgian, or the Portuguese or the white settlers of South Africa, as though there is virtue in the degree to which slavery is enforced. Such specious differentiations come from those who have never experienced the miseries and degradation of colonialist suppression and exploitation. More frequently they are apologists for the colonialism of their own country, anxious out of jingoistic patriotism to make a case for it.[2]

The vast majority of the African people who are urban workers and rural peasants continue to experience oppression and exploitation—by militarism, by the dictates of transnational corporations, and by greedy and wealthy landowners. These classes are economically exploited and commercial and wage slaves who are dispossessed from their lands and forced to suffer illiteracy, homelessness, joblessness, and malnutrition, from Ethiopia to Senegal, from Egypt to Zimbabwe. These neocolonial conditions of subjugation, impoverishment, class divisions, and economic injustice signify the ugly legacy of colonialism, under which a tiny segment of African society, the bourgeoisie, benefited from a system of servitude and material extraction that bled Africa's wealth and resources. As distinguished political economist Samir Amin notes, these preconditions dictate continuation of underclass subjugation,

> since the entire under-developed world and large regions of Africa suffer already from a shortage of land where it is possible for farmers to grow crops. This land shortage has favored and reinforced the monopoly of landlords, creating a class of peasants without land or obliged to work land which does not belong to them. . . . Colonisation created a class of minor functionaries confined for a long time to inferior positions, postmen,

teachers, clerks, etc. Considering the state of backwardness of the peasant masses and the almost total absence of a proletariat, this group of minor functionaries was for a long time almost the only one to have an anticolonialist conscience. The rural traditional chiefs and the national bourgeoisie of businessmen and plantation owners only thought of using colonisation to enrich themselves at the people's expense.[3]

Even though Amin was reflecting upon the colonial period and newly indepen-dent movements in Africa during the 1960s, his analysis applies to the neocolonial plight that plagues most of Africa today: The wealthy benefit and grow richer, at the cost of impoverishing the vast peasant and proletarian majorities. The widening gap between the haves and the have-nots is a product of the historical colonial economies transplanted and eventually entrenched in Africa. Wayne Nafziger writes:

Colonial policy also contributed to today's agricultural underdevelopment: (1) Africans were systematically excluded from participation in colonial development schemes and in producing export crops and improved cattle. British agricultural policy in Eastern Africa benefited European settlers, and ignored and discriminated against African farmers, pro-hibiting Kenyans from growing coffee; (2) Colonial governments compelled farmers to grow selected crops and work to maintain roads; (3) Colonialism often changed tradi-tional land-tenure systems from communal or clan to individual control. This created greater inequalities from new classes of affluent farmers and ranchers, and less secure tenants, sharecroppers, and landless workers.[4]

Today, the transnational capitalist conglomerates have replaced the colonial gov-ernmental powers of fifty years ago. They have functioned to contribute to

a displacement of indigenous production or a general "denationalization" of domestic industries, and as a corollary, they displace indigenous entrepreneurial groups from ac-tive participation in the productive activities of their own economies. They [critics] ar-gue that multinationals displace indigenous production because they either buy off ex-isting indigenous import-substitution industries or simply outcompete them, forcing them to close down.[5]

The "new world order" propagated by Western political leaders is precisely about the expansion of the gap between the rich and poor and intensifying the accumulation of capital by the conglomerates, making this obsessive extractive behavior appear more respectable under the guise of "economic growth."

In 1995, Black people in South Africa on average earned less than 13 percent of what whites earn, and 25 percent of African households earned less than R300 ($75) per month.[6] When 40 percent of Black people are currently unemployed, when 1 in 3 Black children suffers from stunted growth because of malnourishment, when only one doctor is available to every 10,000 rural men and women, when 2–3 million people suffer from malnutrition, when only 20 percent of African households have a faucet inside their homes, when 60 percent of Black people have no electricity, and when 20 percent of Black people live in shantytowns without basic human facilities such as toilets and bathrooms, then these conditions are indices of underdevelopment and are not fundamentally different from the underdevelopment found in Burundi or Mali or Zaire.[7]

I concede that there may be degrees of exploitation, suffering, and oppression, with people perhaps worse off in one part of Africa than another because of historical conditions and economic circumstances; however, the common denominator for us on the continent is that we are all essentially economic slaves of the system of Western European colonialism, neocolonialism, and capitalism. Kwame Nkrumah's well-known asseveration that Ghana's independence was indivisibly interwoven with the total independence and unity of Africa certainly has relevance in viewing and analyzing the current unfolding of events in South Africa.[8]

The experiences of Black people in South Africa must thus be viewed within the continuum of Western European colonial aggression and domination of the entire African continent. It is absolutely true that those boundary disputes in Africa that have become more acute since the political independence of the early 1960s were the aftermath effect of "artificial boundaries" carved out and imposed upon the new leaders by the departing colonial powers.[9] The history of the artificial boundary dates back to the nineteenth century, when the European scramble for Africa became so intensified that the Berlin Conference was called in 1884, during which the African continent was partitioned without due regard to national and socioeconomic groupings. Each European power controlled and administered its own share of the slice of the African "prize."[10] It is only distorted thinking and ideology that locates South Africa either "above" or outside Africa and the same that construes a territory called "sub-Saharan Africa."[11] European colonialism needs to separate North Africa from the rest of Africa so that imperialist control and division of Africa on ethnic, geographical, economic, and regional lines can be clearly established. These imperialistic demarcations by colonial ideologues and academics alike must be abolished and vehemently opposed. Africa is one entity—from Cape to Cairo, from Mauritania to Madagascar.

Second, it is important to overcome being consumed with the unabashed, hideous, and naked violence of the externalities of apartheid that were manifest in racial categorization, discrimination, and deprivation and to examine what lurked below the surface of apartheid. The apartheid colonial ideology was masterfully artful in that it managed to inject repugnance for apartheid in our veins, to the extent that we viewed the principal obstacle to Black humanity as being apartheid, or apartness between Black and white. Apartheid is now abolished, racial classification has been dismantled, and other racially discriminatory statutes have been struck off the law books. Yet oppression and exploitation have neither disappeared nor even been mitigated.

The problem resides in the false analytical perceptions of oppression that have been imposed on Black people in South Africa. We were made to view our problem in South Africa as racial apartness, and the deprivation of a democratic vote. However, as Black people who have been historically victims of colonialist, neocolonialist, and capitalist oppression, we need to see the apartheid beast for what it really was and is. Apartheid was the ideological veneer for the concealment of the racially enslaving and exploitative character of capitalism, as the logical sequence to the establishment of settler-colonialism.[12] The denial of the fundamental human right to participate in a democratic dispensation under apartheid was just one element within a totalitarian political and economic machinery of repression. We Black South Africans have become so mesmerized by the bourgeois capitalist myth that the vote is the avenue of attaining Black political and economic power that we have become blinded and mystified by the casting of a ballot. In the true spirit of Malcolm X, we should understand the superficiality of the ballot and the compelling inevitability of the bullet, when it comes to confronting and overcoming colonial oppression and the tenacity of the tentacles of Western European and now East Asian capitalist hegemony in South Africa in particular and in Africa in general.[13]

To mention a parallel context similar to that of South Africa, in the United States, Black people are still not economically and politically free even after the passage of the Voting Rights Act of 1964 and the Civil Rights Act of 1965. The "State of Black America" issued by the Urban League in early 1996 evinces a profoundly disturbing picture for Africa's descendants in the United States. It reports that a Black baby is thrice as likely as a white baby to be born to a mother with no prenatal care; a Black infant is twice as likely to die in the first year of life as a white baby; a Black child's father is twice as likely to be unemployed as the father of a white child and if two parents earn an income, the Black parents' income will be 84 percent of what the white parents earn; a Black child is 40 percent more likely to lag behind in school progress than a white child; a Black youth is twice as likely as a white youth to be unemployed; a Black college graduate possesses the equivalent opportunities for employment as a white high-school graduate. In 1994, only one-fourth of all Black men between the ages of twenty and twenty-four earned enough to support a spouse and one child, and more than 50 percent of all Black children lived under conditions of poverty.[14]

Desmond Tutu, the archbishop of Cape Town and the 1984 Nobel Peace Prize Laureate, went overboard when he described the act of voting in South Africa for the first time on April 27, 1994, and declared, "Man, it was like falling in love!" Since falling in love requires another person, it remains unclear who Tutu was in love with at the time of his exclamation. The retired cleric later criticized the ANC-led government for engaging in unnecessary and extravagant excesses and "getting on the gravy train." If this love relationship was based on the singular act of voting for the first time in colonially occupied South Africa, then the relationship has not lasted as long as many had anticipated. The euphoria has been short-lived and behooves us to more accurately describe the act of casting one's ballot for the first time as infatuation.

The number of Black people interested in voting in the municipal elections on November 1, 1995, rapidly dwindled following the 1994 national election.[15] After voting for Nelson Mandela in April 1994, many felt that little had changed regarding the conditions of deprivation, poverty, and squalor for the Black masses of women, men, youth, and children in the urban ghettos and rural villages. Four years after the country's first democratic election, many Black people in South Africa had clearly become disillusioned with the false and inverted concepts of a "new South Africa" and a "post-apartheid" South Africa. In an interview conducted in the township of Joza outside Grahamstown in the Eastern Cape in July 1997, when asked about whether Black people were free from the scourge of oppression, an elderly woman replied: "Yes, we are free . . . to build more shacks."[16]

Even though a number of poor Black people voted for the ANC in large numbers in 1999, it is clear that the sentiment of support for the ANC is equivocal and is in fact a desperate measure, as many Black people believe that the ANC-led government would make things better if it was only granted more time. Although skeptical, the Black masses do not firmly believe that there is an alternative; though they may subscribe to more radical political inclinations like the Pan-Africanism represented by the Pan-Africanist Congress of Azania and Black Consciousness represented by the Azanian People's Organization and the Socialist Party of Azania, they hold little hope for these parties because they do not believe that these radical formations have even the slimmest chance of winning any current elections.

Some Neocolonial Political Economies
in "Independent Africa"

As victims of Western European fraud, deception, colonialism, genocide, land dispossession, material theft, slavery, and repression, we as South Africans need to be more astute in our analysis of the ballot. There are practically no situations on the African continent, or on any continent for that matter, where the using of one's hand to cast a ballot has won economic liberation of the working class and underclass masses. Western colonial powers have indoctrinated us with the specious belief that if we were to have multiple party elections in the various African countries that European colonialism created, we would then live in peace and prosperity and benefit from enjoying the fruits of Western European capitalist civilization. Adebayo Adedeji, the renowned Nigerian economist, confirms the paradox of elections and participatory democracy:

> It is needless to add that good governance is not necessarily guaranteed by multi-party elections, although elected governments should be better motivated than authoritarian ones to practise it. In a country where low-intensity democracy prevails, however, the quality of governance may not be much different from that prevailing under a military dictatorship. In fact, it can be worse because such regimes are not only illegitimate and coercive, but also divisive and polarized, pursuing a zero sum political game where winners take all and losers are not only excluded from power but run the risk of being exterminated.[17]

"Democracy" is the buzzword that is used to stupefy us into accepting any form of inhumane society so long as elections are held. There is not one country on the African continent where elections have been held and the masses of the people have been simultaneously set free on the path of deliverance from the manacles of illiteracy, poverty, economic exploitation, military terror, land dispossession, barely existent health care, and personal insecurity.

Angola and Mozambique held elections in 1992 and 1994, respectively, pressured to do so by the imperialist powers following the dismantling of the Soviet Union as a gesture demonstrating that the two African nations were committed to Western-style democracy. Yet nothing substantive has emerged to indicate that these countries have changed dramatically for the better in terms of socioeconomic conditions. True, there are more goods available in stores, more luxury cars, more expensive houses built, new supermarkets called Supermarcado LM (Maputo used to be called Lourenco Marques, or LM, under Portuguese rule) in Mozambique's capital, Maputo—quite a facelift compared to the situation a decade ago.[18] The only problem is that these fancy items are generally out of reach for the average person because of exorbitant prices that only upper-middle-class people can afford. Mozambique, like many other African countries, has been forced to open its economy to foreign capitalist ownership, so that "Portuguese, South African, and British companies are taking over farms, factories, shops and tourist facilities that their compatriots abandoned in panic twenty years ago, just after independence."[19] Peasants, even children, in the North, are forced to pick cotton to make a living for their families and sell it to monopoly buyers. The result is an enrichment of foreign capitalists and the local commercial and business elite and the impoverishment of the vast urban and rural working classes, all a result of free-market economics and World Bank and IMF loan dependence. Incredibly, with the return of Portuguese merchants and financiers, Mozambique appears to have been thrown back into the era of colonialism. The Catholic Church of Mozambique has satirically described the IMF and the World Bank as "bosses of Mozambique," since so much of policy, including that of education, agriculture, and health care, is dictated by these two Bretton Woods institutions.[20]

Bowing to pressure from the Western capitalist powers on the part of many African countries has borne no political and economic fruit for the masses because Western imperialism's intentions in Africa are essentially exploitative: keeping African societies destabilized through military coups and dictatorships or benevolent capitalist-leaning oligarchies that ensure the continued siphoning of Africa's minerals, industrial raw materials, energy and agricultural resources into Western European and American national coffers. Nigeria, Sierra Leone, and Liberia, where the most brutal of military regimes have reigned and reign, are just three classic cases in point.

In Nigeria, the military has ruled the country consistently (with the exception of minor intervals) since independence, violating every human right and subjecting the masses to consistent terror and death. The Nigerian military is the historical product of the Niger Constabulary, established by the British colonists in 1872 to protect those Africans who collaborated with the British against the wrath of the colonized African masses. The regional instability of Nigeria was assured with the powerful northern part of the country accorded economic precedence over the underdeveloped South by British colonial structures and policies, provoking an ongoing regional and ethnic conflict. The Nigerian lawyer B. O. Nwabueze, once critical of the government but who later became minister of education in the former Babangida regime, confirms this assertion: "There could therefore be no other explanation for the refusal of the British to alter the boundaries of the North than that it was a deliberate design to foist Northern domination upon the country. It has quite rightly been described as 'one of the greatest acts of gerrymandering in history.'"[21]

It is for this reason that Nwabueze contends that federalism in Nigeria has been ineffective in fomenting unification of Nigeria's diverse population and proposes a single unified state in which diversity would be configured. Diversity, according to him, cannot be accorded primacy at the cost of sacrificing unity following the levels of political polarization in Nigeria's short-spanned postindependence history.[22] Additionally, the Nigerian Constitution protects the state and ensures that state machinery will perpetually maintain responsibility for determining allocation of the fruits of economic production. Since Nigeria is a neocolonial state, this disbursement is earmarked for the tiny wealthy elite. Ironically, the constitution declares, under Section 16 (1) that the state shall "[c]ontrol the national economy in such a manner so as to secure the maximum welfare, freedom and happiness of every citizen on the basis of social justice and equality of status and opportunity."[23] It is not too difficult to ascertain why the state is constantly bankrupt: because it functions to provide security to its ruling members and the rich classes of Nigeria, who are constitutionally assured of their continuity in power as a carryover from the colonial period.

Civilian rule following independence was short-lived in Nigeria, with the Azikiwe government toppled by the first military coup in 1966, led by General Ironsi, six years after Nigeria declared "independence" from the British. Ironsi was subsequently overthrown later that year by Yakubu Gowon, another military leader, who ruled till he was ousted in a bloodless coup in 1975. General Olusegun Obasanjo assumed the position of commander in chief from 1976 to 1979, when Shehu Shagari was elected president.[24] In 1983, the elected civilian regime of Shehu Shagari announced that new austerity measures would have to be introduced as a result of the falling price of oil, which peaked in 1980. The consequence was a peaceful overthrow of the Shagari regime by Mohammed Buhari, a U.S.-trained military general who had also headed the Nigerian National Petroleum Corporation at one time. Buhari promised to clean the country of economic corruption, as have so many military tyrants who have ruled Nigeria, with the only result being the wealthy continuing to prosper and the military gaining access to excessive power that licensed violent abuse of civilians. On August 27, 1985, Buhari was overthrown in a relatively bloodless coup by another military figure, Ibrahim Babangida. Buhari deferentially began to pay off the country's "debt" to the IMF, a debt that accrued from the money borrowed to finance "development" projects at exorbitant rates of interest that essentially ended up in the pockets of the greedy economic elite and the erratic military leadership. Buhari is being sued by Gani Fawehinmi, Nigeria's most famous civil rights lawyer, for squandering money in the process of moving the government from military to civilian rule, but Buhari has refused to accept court papers on the grounds that a head of state is immune from being prosecuted as a civilian.[25]

The IMF was fully aware of this state of graft; it permitted such corruption because it thrives on the economic fraudulence of the neocolonial regimes of Africa, such as Nigeria, making billions of dollars from the interest on "bank lending." The African working-class masses are forced to pay for this greed of the wealthy, further spiraling into the ranks of poverty as incomes drop because of IMF policies. Privatization and deregulation of the economy in Nigeria as required by the IMF and World Bank, on the grounds that state enterprises were inefficient and the closed economy was incompatible with economic growth, was specious on both accounts. Not all state enterprises were inefficient, and the irony of privatization was that "it was the same public officers who recklessly managed these enterprises that were asked to buy them for the purposes of 'efficient management and profit'" at absurdly low cost.[26] Whereas Buhari favored paying the IMF on Nigeria's indebtedness, Babangida severed all borrowing ties with the IMF and offered to pay a maximum of 30 percent on foreign debt service. He announced new austerity measures, devalued the naira so that it is at its lowest international value today (86 naira are now equal to $1.00; in 1979, 1 naira was worth $0.80), and dismissed thousands of civil servants as part of the ostensible campaign to clean up government. Fuel prices and prices of other commodities skyrocketed as incomes for the working-class majority dropped.[27] Babangida encouraged the development of a political class that would support him in power, while he permitted its members to "pilfer and accumulate wealth."[28]

In 1992, elections were held under Babangida's rule and a wealthy businessman, Mashood Abiola, won the majority of the vote. Abiola himself was a dubious figure, having enriched himself significantly through lucrative oil and commercial government contracts and having been a close personal friend of Babangida[29] (Abiola was the candidate favored by the United States during the elections). Babangida nullified the elections results, detained Abiola for a period, repressed growing worker and student strikes, clamped down on the press, and declared that he was installing a new leader in the country, General Sani Abacha.

Abacha ruled with an iron fist, forcing millions of Nigerians to live under one of the most brutal military systems in recent history, with their basic human freedoms curbed, most civil liberties suspended, and the voices of protest of the working classes suppressed. Police and armed terror became the norm of the day. The wealthy elite, many of whom are either middlemen in the oligopolies of imperialist oil-expropriation networks or part of the favored status of senior military bureaucrats, benefited lucratively from this tragic situation of dire oppression and dehumanizing impoverishment and continue to deplete the resources of the country for their own material and financial gain.

Nigeria's catastrophic predicament today is principally attributable to the legacy of European colonial greed in Africa, known as neocolonialism, and is manifest in the unbridled extraction of oil by mammoth Western oil companies like Shell and British Petroleum that are bent on draining the precious energy resource at low cost to drive the engines of Western industrial expansion. These oil companies are so obsessed with accumulating wealth through profit maximization that they are willing to work with any governmental authority, even ones as rabidly violent and inhumane as the Nigerian junta. The blind eye that Shell Oil was willing to turn during the imprisonment and subsequent execution of well-known writer and activist Ken Saro-Wiwa and several others in late 1995 is one dramatic illustration of this fact. The death of these environmental leaders is a continuation of the unholy alliance between the greedy oil corporations and the unconscionable military, evidenced in the devastation of 1981. Gesiye Angaye states:

> Oil spillages, land, water, and air pollution have begun to constitute a very serious damage to the security of life and property of the inhabitants of petroleum-producing areas. There has also been serious disruption of economic life in erstwhile fishing and agricultural areas because of massive pollution and destruction of fishing grounds and farmlands. The oil spillage in early 1981 which caused extensive damage to human and wildlife in Sangana, Koluamama, Akassa and Brass in Rivers State is a typical case—neglect of agriculture, and a consequent rise in food prices ... the oil industry has enriched a few at the expense of the masses. What exists is a "dubious relationship between the foreign businesses, the local middlemen and the comprador state officials" as the commercial triangle. If a contract is awarded to the foreign businessman, the state official is usually rewarded with a payment arranged by the local middleman or go-between. There is no doubt that a few private and public functionaries have made some easy fortunes through the petroleum industry.[30]

Regardless of which regime has ruled Nigeria since independence in 1966, the tiny African elite has become progressively more wealthy and the peasant masses have grown increasingly poverty-stricken. Under military regimes, oppression and terror is compounded: Impoverishment is consistent, compounded by wide-scale military terror against the civilian working classes and a sense of enhanced fear and insecurity among the masses, uncertain about whether tomorrow will be the day of death at the hands of trigger-happy soldiers.

The recent situation appears to be somewhat hopeful, in that following the sudden and mysterious death of the dictator Sani Abacha in June 1998, the military leader who succeeded him, General Abdulsalami Abubakar, ruled for a year, held civilian elections, and Olusegun Obasanjo was elected to the presidency. Abubakar declared new austerity measures in the budget compilation for 1999, announcing that over $3 billion was unaccounted for and the minimum wage would be reduced from 5,200 naira to 3,000 naira.[31] Oil prices were raised by 127 percent, prompting national protests, all this in a country that ranks among the top ten oil producers in the world. It has since been learned that the late General Sani Abacha had quietly accumulated a fortune of some $3 billion by pilfering state funds, of which $750 million has now been returned to the government by his family, leaving $2.3 billion missing.[32]

Although Obasanjo has functioned as a leading international statesperson as a member of the Eminent Persons Group that visited South Africa urging the repeal of apartheid in the late 1980s, it needs to be borne in mind that he is a former military general. Since his ascendancy to the presidency, he has purged the military of scores of former senior military officers. However, the latest news from Nigeria indicates that no fundamental transformation of the skewed socioeconomic foundations of Nigerian society, which favored the military and commercial elites, will occur under Obasanjo's tenure as president.[33] The ongoing protests by oil workers, teachers, students, and other sectors of the working class continues, in the hope that the new regime will take heed. The fundamental disparities between the tiny wealthy Nigerian elite and the vast impoverished African working class and underclasses will persist under the new dispensation, akin to that of post-apartheid South Africa. The probability of any working class–biased transformative measures by the new regime are acutely remote.

The masses seek justice and equitable access to vital resources such as education, skills training, economic empowerment, decisionmaking in public policy, and the right to indigenous religiocultural practices, as they are the legitimate producers of wealth in the country. Western European–style elections in Nigeria, in the rest of Africa, or elsewhere in the Two-Thirds World, for that matter, have not produced justice for Africa or for any working-class masses around the underdeveloped segment of the globe. If anything, owing to their quintessentially corrupt character, these elections have been part of the opiate of imperialist ideology that compels repressive African regimes to emulate the West. Ebenezer Babatope (once radically critical, but who later ironically became the minister of transport under Abacha's regime) laments this fact about Nigeria's imitation of the United States and its economically exploitative ramifications: "But our sovereignty as a nation is being insulted. Nigerians are being humiliated by all the characters of our political life who have been dancing to the music of the Americans . . . Nigeria is now a clear extension of America and Nigeria is playing the ball well to the acclamation of all Americans."[34]

It is no coincidence for instance, that U.S. president Jimmy Carter visited Nigeria in 1978 as a follow-up to the visit of Nigeria's head of state General Obasanjo to the United States in 1977. These visits represented a prelude to the events that led to the 1979 "fall" of the shah of Iran, one of the staunchest allies and oil suppliers for the United States. The United States needed another major oil-producing nation to supply it with millions of barrels of oil for its fastidious oil-consumption needs. Nigeria, as one of Africa's major producers, was viewed as the replacement for the subsequent "fall" of the shah. (About 50 percent of Nigerian oil is exported to the United States today.) Several members of the Nigerian House of Representatives visited the United States in 1980, as part of a "democracy-learning" experience to pave the way for the installation of a U.S. inspired "democracy" that would be presented to the world as the success of the American "democratic" system. The nefarious character of oil-extracting capitalism that has destroyed Nigeria's land fertility and precious forests and, most important, has impoverished the vast working-class majority of its people, was concealed during this "international democratic exchange."

In Sierra Leone, the tragedy of thousands of Africans having been killed in the civil war between the former regime of Valentine Strasser and the Revolutionary United Front (RUF) headed by Foday Sankoh is yet another instance of the legacy of neocolonialism, where African figures—armed, equipped, countenanced, and propped by the imperialist powers—assumed the role of inflicting terror on the vulnerable African masses. The mysterious character of the RUF, which clearly did not have any constructive program for the economic transformation of Sierra Leone under which the working class would be material beneficiaries but merely aspired toward possessing political power and economic domination, underscores the unconscionable character of neocolonialism. Note that while the civil war raged in Sierra Leone throughout the 1990s, a war in which thousands of Africans were forced to flee their homeland to neighboring countries, diamonds were still being extracted by the large transnational corporations, including the South African global conglomerate De Beers, alongside the mining of platinum and gold. From the early 1930s, diamond mining was the exclusive prerogative of the Sierra Leone Trust Company, formed by the Consolidated African Selection Trust from the Gold Coast, which was actually a subsidiary of Selection Trust, a British company. Similarly, gold mining was operated by Maroc Limited, the Balia Gold Mining Company, and the Sierra Leone Development Company, all British companies.[35] This legacy of foreign capitalist control, mediated through international imperialist agencies such as the World Bank and the IMF, continues to ravage Sierra Leone's political economy today.[36]

Paradoxically, the South African government assigned troops to Sierra Leone to fight alongside those of Strasser and to protect the lucrative diamond-mining interests of De Beers. Elections occurred in Sierra Leone in 1995, and Ahmed Tejan Kabbah assumed the leadership of the country. Many Western political pundits claimed that things would return to normalcy, citing the reinstatement of civilian government as a positive sign. Certainly, one might have taken comfort in the fact that the government had been demilitarized. Yet what did this mean for the illiterate and impoverished masses of Sierra Leone who were earning starvation wages working as factory workers in mostly Lebanese-owned factories or European-owned mining industries? Admittedly, "normalcy" implied demilitarized capitalism, with the tiny clique of African commercial elites growing wealthier and the African working class become poorer.

The civilian government of Kabbah was short-lived, however. Sierra Leone was once more thrust into political terror in April 1997, when Kabbah was overthrown by a military leader, Johnny Koroma, and was forced to flee the country. Since that time, forces of the Economic and Monitoring Group of West African States (ECO-MOG), led by Nigeria, attacked Koroma's forces, causing devastating damage and extensive loss of life in the country. Koroma was finally defeated in March 1998, and Kabbah returned to rule Sierra Leone. The incredible irony was that the tyrannical military regime of Nigeria militarily intervened in a neighboring African country,

Sierra Leone, to establish democracy! Abacha obviously did not view his actions of "liberating Sierra Leone" while inflicting terror in his native Nigeria as contradictory, a logical posture of neocolonialism in Africa. In July 1999, an agreement was signed between Kabbah and Foday Sankoh, the leader of the RUF who had been imprisoned in Nigeria for two years, granting the RUF four government cabinet positions and according the equivalent of a "vice presidency for mines" to Sankoh.[37] All of this occurred in the aftermath of thousands of innocent civilians having been killed by the thuggish bandits of the RUF, with scores of children having their hands hacked off. Now a new government unity prevails, with almost half of the ruling regime consisting of a group of rogues and murderers bent on securing power at the cost of destroying Sierra Leone's civilian population.

In Liberia, as the ongoing conflict between the forces of Charles Taylor and those of Roosevelt Johnson, who was secretly whisked by the United States out of the country on May 2, 1996, raged throughout the 1990s, it was evident that little stability would be achieved through the "leadership" of these power-hungry figures. This country, which was founded by the American Colonization Society in 1787 to repatriate freed Africans who had been enslaved in the United States, will continue to face its crisis of political uncertainty and socioeconomic disarray so long as capitalist multinational corporations need and continue to extract the natural wealth of the country, specifically rubber.

Firestone, the U.S. rubber company, is one of the largest commercial enterprises operating in Liberia. Firestone has played a pivotal role in Liberia's economy since the company was permitted by the Liberian government to lease over 1 million acres of land over a ninety-nine-year period for the extraction of rubber, utilizing several thousand Africans as cheap labor for the extractive process. Firestone supplied over 65 percent of the U.S. rubber needs at one time and represented one of the bastions of the superpower's commercial interest.[38] Liberia was selected by the United States for principal rubber production so that the United States could reduce its dependence on Britain for rubber production as well as avoid paying the British cost of production.

The leaders of the heavily armed political factions in Liberia have often been trained either in European, U.S., or Israeli military and naval academies and persist in power with the full sanction of many of the Western capitalist powers. Liberia is also the center for a huge U.S. satellite communications network, serving to provide the basis for continental surveillance by U.S. intelligence operations, and it provides the outlet for a major channel for the Voice of America radio station.

Elections in Liberia in 1997 resulted in the seating of Charles Taylor as president. Taylor essentially represents a continuation of neocolonial domination in the country and is supported by U.S. corporate interests. There will be no radical shift in Liberia's economic predicament. The country will continue to exist as a neocolonial dependency, with Firestone and the local bourgeoisie becoming richer and the masses suffering from poverty, economic exploitation, illiteracy, and social insecurity. The future portends a hodgepodge of farcical elections, in which a range of individuals, all beholden to Western imperialism and capitalism, will vie for political leadership of the authoritarian and despotic type of government that the country has had since its founding. Stephen Hlophe asserts in this regard:

> The struggles for politico-economic hegemony among the West Atlantic chiefdoms led not only to periodic oligarchic exploitation of less fortunate groups, but greatly complicated the class relationships which subsequently emerged in this region, namely: those between the Africans and the settlers in the nineteenth century; the Americo-Liberian power structure and the Indigenous-Liberian technocratic stratum since 1944. A careful observation of the present Americo-Liberian power "struggles" against the indigenous Liberian technocrats shows an adoption of the same type of strategies which existed in the nineteenth century, when Masonic Craft and family affiliation were heavily used for ascendancy and clique-boundary maintenance. The Indigenous-Liberians, on the other hand, have tended to slip into their pre-colonial (pre-1820) styles of lineage and chiefdom organization, whenever an opportunity of intra-class advantage was availed to them by the Americo-Liberian "power structure."[39]

Military regimes signify the crystallization of autocratic and neocolonial dictatorial rule and derive their legitimacy and sustenance from the powerful Western governments that proffer military training and aid, ruling as "police" states that terrorize local populations into forcibly succumbing to being dominated by the powerful. They breed graft and exploitation and violate all human rights, particularly those of the poor. Adebayo Adedeji verifies the nefarious character of military regimes in Africa:

> Little wonder that countries that have long been under this type of government (Nigeria and Zaire are now the classic examples) are groaning under gross mismanagement, massive capital flight, corruption, unproductive uses of resources, poor resource mobilization, distorted priorities and lack of basic human rights. The authoritarian imposition of unpopular and irrelevant structural adjustment programmes in Nigeria and elsewhere in the 1980s was possible only because of lack of democracy and popular participation.[40]

Elections have been held in Côte d'Ivoire, Malawi, Zambia, Algeria, Benin, Senegal, and Namibia over the past few years. In Côte d'Ivoire, the decrepit regime of Konan Bedie, which practiced the same oppressive and exploitative capitalist system that his predecessor, Felix Houphouet Boigny, did, was overthrown in a bloodless coup in December 1999 by Robert Guei, a military leader, who has promised to hold elections in July. Bedie is now being challenged by opposition party members to account for an oil deal with the former apartheid regime, in which it was believed he received between 28 and 35 cents per barrel of oil in a contract worth $43 million.[41] Bedie has pegged the local currency to the French franc, as did other former regional French colonies such as Senegal, Benin, Niger, Mali, the Central African Republic, and Gabon. The dominance of the French is apparent in the fact that French military advisers continue to have a direct influence on many of these countries of "Francophone Africa."

The Côte d'Ivoire government negotiated an agreement with the IMF in 1989 and again in 1991. It received six World Bank loans between 1989 and 1993 and in 1994 agreed to an Enhanced Structural Adjustment Facility (ESAF) with the IMF. The preconditions for the IMF program were: devaluation of the local currency, extended privatization, retrenchment of the civil service, price decontrol, labor-market deregulation, and open trade policy. These measures have been foisted on all countries of the underdeveloped world that have requested IMF and World Bank funding. The net result of the Côte d'Ivoire–IMF contract and the ESAF was economic devastation in 1989–1993. Between 1988 and 1995, the percentage of people earning less than $1 per day increased from 17.8 percent of the population to 36.8 percent.[42] Abidjan, the flamboyant capital of Côte d'Ivoire, saw its urban poverty escalate from 5 percent to 20 percent. Per capita spending on education declined by 35 percent, and all people were required to pay health care fees, crippling the poor. The incidence of stunted growth as a result of malnutrition among children rose from 20 percent in 1988 to 35 percent in 1995.

Clearly, the government of Côte d'Ivoire cannot be held accountable for the legacy of neocolonialism that permits the West to continue to strangle Côte d'Ivoire and other African nations that have entered into agreements with the IMF and the World Bank on an ESAF package, with the anticipation of financial relief. Yet the Côte d'Ivoire government, like the Ugandan government, which continues to work with the IMF and the World Bank, is fully aware that the conditionalities imposed by these imperialist agencies have caused untold suffering on Africa's poor and further impoverished them and in most instances have increased Africa's indebtedness rather than relieving it.

In the case of Uganda, which is being heralded as the "success story of the 1990s" by IMF and World Bank policymakers, what goes unmentioned is the tremendous human cost involved in fulfilling IMF and World Bank conditionalities and obligations. Liberalization of trade, specifically cash crops, as demanded by the IMF for extending loans to Uganda, resulted in a mere 4 percent per capita private-income increase between 1988 and 1995.[43] The downside of the privatization effort in Uganda is widely suppressed in the bourgeois media: Over 350,000 workers were laid off in the public sector, most ending up in dire economic straits since they were ill equipped to enter the private sector. Health care allocations in real terms decreased between 1989 and 1994.[44] Cost sharing in health care, introduced by the Ugandan government and insisted upon by the IMF, resulted in low-income people being unable to pay for health services and exacerbating illness in most instances, resulting in local hospitals closing, since these institutions depended on people making their payments to keep these medical services open. After less than one year of receiving IMF relief, Uganda is once again in an economic quagmire because of the significantly reduced coffee prices on global markets, which represent the mainstay of government revenue. Uganda's embracing of the global market system and its acceptance of privatization of vital state industries has "benefitted the government and corporate interests," a group of Uganda nongovernmental organizations reported in 1997.[45] Ugandan military commanders and businessmen have even extended their interests to the lucrative business of diamond smuggling in the Congo, with their sights on Mbuji Mayi, considered the diamond-mining capital of the Congo.[46]

In Gabon, an oil-rich country that has one of the highest per-capita incomes in Africa yet where workers still earn wages below the cost of living, elections were held in December 1998. Omar Bongo and his party, the Parti Democratique Gabonais, were once more returned to office as president and ruling party, respectively. Gabon signifies the façade of electioneering, following the introduction of multipartyism in 1990, since the ruling party consistently wins, and the opposition parties, this time the Rassemblement National des Bucherons, are too weak to present any formidable challenge. Typical of these elections, as in most others in Africa, sexist practices persist, and women were used only to promote the candidates of the ruling party, campaigning on behalf of the repressive president, Bongo.[47]

In Benin, Mathieu Kerekou, a dictatorial so-called Marxist leader, won the elections of early 1996, ousting Toglo, with little hope for any radical transformation of the country's commitment to working-class Black people and their uplifting. In Niger, President Ibrahim Bare Mainassara, who had come to power through overthrowing a democratically elected government in January 1996, was assassinated in a hail of gunfire by his presidential guard. Major Mallam Daouda Wanke assumed the role of head of state and chairman of the National Reconciliation Council. Mainassara was not very popular in Niger, considering that he had conducted fraudulent elections in 1996, winning 52.2 percent of the vote under very suspicious conditions.[48] His administration was flawed from the start, with detention of opposition leaders, protracted demonstrations by striking workers, and an economy in collapse. In February 1998, the military revolted, and in April, over 40,000 government workers went on strike, insisting that they receive eleven months' back pay.[49] Wanke has promised that the tenure of military rule will be short and that he intends to return the country to civilian rule. Only time will tell whether there is any truth to his statement.

In Malawi, Bakili Muluzi, yet another oppressive dictator like the preceding Kamuzu Banda, has now assumed leadership of the country, maintaining the same neocolonial and capitalist policies as Banda. In Zambia, Frederick Chiluba, the current president, came to power after Kenneth Kaunda was defeated in the last elections almost nine years ago; workers have repeatedly gone on strike to protest the government's escalating anti-working-class policies. Many expected a more prolabor stance from the government, given that Chiluba is a former trade union leader, yet his service to international capital via intimate ties to the IMF and the World Bank reveal his corrupted profile.

Meanwhile, in Algeria, a bloody civil war continues, with the French supporting the bankrupt Algerian regime under Abdelaziz Bouteflika, a successor of the military ruler, Zeroual, in defiance of the elections of 1992 in which the Islamic Salvation Front was on the verge of declaring a victory but was prevented from assuming power by the ruling regime. More than 100,000 people have been killed by government and rightist violence over the past five years. Bouteflika has decreed that the Islamic Salvation Front will be banned from participating in Algerian politics and has clamped down on opposition groups, exacerbating an already polarized and bloody situation.[50] The Organization of African Unity (OAU) has claimed that it is helpless to intervene to bring a halt to the carnage. Neocolonial genocide continues in Algeria with no solution in immediate sight, even as oil is still pumped for mass export to France.

In the Horn of Africa, a tiny new state, Eritrea, emerged as an independent nation in April 1993, after the Eritrean People's Liberation Front declared nationhood. Initially, the Islamic and non-Tigrayan Eritrean Liberation Front began to resist the regional domination of the Ethiopian state.[51] The achievement of Eritrean independence signaled a victory following a thirty-year war against successive Ethiopian regimes. In the aftermath of the overthrow of the military leader Mengistu Haile Miriam in 1991, the Eritrean liberation organizations forged a united front to establish a new country and, derivatively, a new government in Ethiopia. Although there was optimism expressed about the fraternal ties that both Ethiopia and Eritrea could develop so that both these fractured nations could jointly assist each other in reconstruction, these hopes were short-lived. In early 1998, the military forces of the two impoverished countries exchanged fire, and the ensuing conflict escalated into a fierce war, with aircraft bombing resulting and thousands killed on both sides. Tensions between these fragile countries continue to mount. The tragic eruption of military conflict reinforces the painful condition of nation-states in the Horn of Africa. Mohammed Hassan, a professor of history at Georgia State University, laments the tenacity of military repression in this part of Africa, which he describes in an article entitled "The Curse of Military Dictatorship in the Horn of Africa":

> No words can express the bitterness of the peoples of these countries, who were promised democracy and hoped to shape their societies by democratic experiment but were subjected to live under the worst forms of dictatorship for years. The Sudan started independent existence in 1956 under the Westminister form of parliamentary democracy which lasted only for two years. Somalia started her independent existence in 1960 under a multi-party democracy which lasted up to 1968, with the exception of a few short years. The Sudan has been misruled by the military. From 1969 to January 1991, Somalia suffered under a military dictatorship. From September 1974 to May 1991 Ethiopia too suffered under a military dictatorship which presided over the destruction of the people, their economy and their country. For almost two decades the Horn of Africa has been in the grip of a vast deepening economic, political and social crisis. The curse of military dictatorship, drought, warfare and man-made famine combined with inept economic policies and corrupt leadership, have caused, and still cause, the death of large numbers of people. They will continue to cause the deaths of many millions more in the years to come if warfare is not stopped in Southern Sudan, Southern Ethiopia and Somalia and if criminal neglect of food production and mismanagement of the economy is not reversed in these countries.[52]

It needs to be emphasized that Eritrea was the product of Italian colonialism in 1890, which cut across the homelands of several cultural groups in the region, as is the case with European colonialism throughout Africa. The Afar, Beja, and Tigray communities were separated by colonialism from their indigenous roots in the Sudan and Ethiopia.[53]

Although domination of one cultural group by another is intolerable anywhere in the world, it is most particularly so in Africa. It is imperative that African cultural community groups struggle together to find innovative ways of overcoming their differences and living within the same umbrella nation. Africa's economic and political future depends on the ability of Africans to unite across colonial boundaries, abolishing these in the process and establishing new integrated regional governing entities that are connected to other regional bodies throughout the continent. The condition of neocolonialism and the legacy of colonialism must be extirpated with a new disposition and program that hatches continental unity. Africa, as a continent of historical cultural, religious, and linguistic diversity, must become revolutionized by recovering and reclaiming its indigenous collective traditions that stressed intercommunity unity and coexistence. The politicization of ethnicity by neocolonial regimes throughout Africa for the purpose of fragmenting and polarizing the working classes and the marginalized must be forcefully arrested by new revolutionary identities and movements. Places like the Horn of Africa; the Sudan, which is engaged in a thirty-year civil war between "Christian Southerners" and "Muslim Northerners"; Rwanda and Burundi, where community tensions between the Batutsi and Bahutu still persist following a politicized massacre of both groups by neocolonial military forces aided by Belgium and France; and Kenya, where the poor are pawns in a capitalist game using "ethnicity" as a pretext, all need desperately to undergo revolutionary transformation.

In Rwanda and Burundi, the colonial Germans and Belgians favored the minority Batutsi people against the majority Bahutu people, granting the former social and some educational access and generally denying the latter such opportunities, even though both nations had lived peacefully and shared a common language, religion, and land and had even intermarried for generations.[54] Mohamed Sahnoun, former UN special envoy to the Great Lakes region in Africa, cautioned against analysis that views the problems in Rwanda and Burundi solely in terms of ethnicity, explaining that the rural population density is the highest in the world following Bangladesh and that "the struggle is above all for resources, for land."[55]

The incessant conflicts over "ethnicity" are, in most instances, really the struggle over access to material resources and the deprivation of employment, education, and health care. In fact, "ethnicity" was introduced into Africa and other parts of the underdeveloped world by European colonialists, predicated on personal predilections and with no scientific basis. Antoine Lema points out that the notion of "ethnicity" in Rwanda and Burundi was derived from the English colonial agent J. H. Speke, who upon meeting members of the Batutsi developed a personal theory that they were "all Galla of Ethiopian origin—close to the Middle East" and since that time "the 'Hamitic' 'peoples' in Central Africa were created."[56] Archie Mafeje reminds us, too, that "ethnography has nothing to do with race or ethnic origin but rather with learnt habits" and that contrary to what many anthropologists assume, "people who speak the same language may be divided into a number of independent chiefdoms," substantiating the multiple and fluid character of sociocultural identity.[57] In a rebuttal to the neatly demarcated categories of "ethnic affiliation" constructed by Western anthropologists, Mafeje observes:

> First, it is conceivable that the same people can in different epochs be subject of different ethnographies. In that even they might be referred to by different names. Second, and most important for our purposes, is the possibility that, owing to circular migration within a given area, the same generic stock even within the same epoch might account for ethnographic variation. This might be a result of adaptation to varied ecological conditions or contact with other peoples.[58]

It is essential to note that many of the "ethnic" and fratricidal wars in Africa involve victims who are without exception poor and are often manipulated by military and ruling class elites. Kofi Hadjor, an African political scientist, refutes the ostensible problem of "tribalism" in Africa, citing the revolutionary leader Amilcar Cabral's critique of the distortion of "ethnicity":

> You may be surprised to know that we consider the contradictions between the tribes as secondary ones . . . We consider that there are many more contradictions between what you might call the economic tribes in the capitalist countries than there are between the ethnic tribes in Guinea . . . Structural, organizational and other measures must be taken to ensure that this contradiction does not explode and become a more important contradiction . . .
>
> Our struggle for national liberation and the work done by our own party have shown that this contradiction is really not so important . . .
>
> As soon as we organized the liberation struggle properly the contradiction between the tribes proved to be a feeble secondary contradiction.[59]

At the heart of "ethnic" conflicts in Hadjor's view is the question of the growing disparities between the haves and the have-nots. Hajor contends that Africa has essentially two "tribes":

> the exploiter and the exploited, the rulers and the ruled. African political leaders have succeeded in obscuring the central fact of life. By fomenting discord between different ethnic groups, workers and peasants have set against each other, instead of fighting the exploiter. Tribalism has become a major ideological weapon for perpetrating exploitation.[60]

Hadjor urges the formation of a working class and peasant political party that is drawn from diverse ethnic communities to stave off neocolonialist "divide-and-conquer" strategies of ruling African regimes.

In Kenya, more than 200 statal companies have been sold to private investors and the Nairobi Stock Exchange has abolished all restrictions limiting ownership to local institutions and has opened up share ownership to foreign corporations since January 1994.[61] Inequality is rigidly institutionalized. Kinuthia Macharia, a sociologist from Kenya, notes:

> Despite efforts to redistribute land after independence, inequality in land ownership in Kenya has continued to be a major factor in rural (and increasingly urban) poverty. Inequality in land and income distribution is in itself a major contributor to the persisting growth of poverty in Kenya. With a Gini coefficient of 0.72, Kenya has one of the highest rates of inequality in the world (UNICEF/GOK 1990). It has also been established that the richest 20 percent of the population in Kenya earns twenty-five times what the poorest 20 percent earns. The equivalent for Japan, the world's richest country, is four times (UNICEF/GOK).[62]

In Zambia, the International Monetary Fund is advising the government on the best approach to privatizing state industries so as to attract foreign investment and liberalize markets. Zambia has already privatized some 164 state enterprises and has opened a stock exchange, freeing domestic markets. Mozambique has embarked on a program to privatize 700 state companies.[63] Many of these state companies are being sold to rich white corporations from South Africa that are expanding mining and industrial operations in Namibia, Angola, Zambia, the Democratic Republic of the Congo, and Ghana. In Ghana, the Ashanti Gold Fields has had significant shares sold to large transnational corporations.

What has become most disturbing is the fact that white transnational corporations based in South Africa that were previously prohibited from functioning openly in the rest of Africa as part of the African isolation of apartheid are now penetrating the rest of the continent with lightning speed. The wealth of West Africa has become the target of these predatory organizations, as they have practiced for over two centuries in South Africa. Randgold, a subsidiary of Anglo American, has signed an agreement with the Senegalese government on the exploitation of gold resources. The Senegalese government, very obligingly in line with its posture of neocolonial subservience, granted Randgold an exploration license to mine gold in an area bordering with Burkina Faso that has already proven that it has gold reserves. Randgold has already begun mining in Burkina Faso and Mali. It has also been offered mining rights in Côte d'Ivoire and mineral-rich Gabon. Anglo American Corporation is engaged in exploitation of southwestern Mali as a "gold developer" and has started construction of the Sadiola gold mine in conjunction with a French company as part of a financial contract worth $250 million.[64] The full capacity of the mine was estimated in 1997 to be 10–12 tons of gold per year. Anglo American owns 38 percent of the gold mine, Canada's Iamgold owns 38 percent, the World Bank's affiliate, the International Finance Corporation owns 6 percent, and the Mali government has a minority 18 percent share.[65] Anglo has also penetrated Ghana, as part of a joint exploitative venture in gold prospecting in southwestern Ghana. Anglo has 60 percent of the new gold enterprise; Pacific Comox Resources, 30 percent; and the Ghanaian government, 10 percent. Gencor, another white mining company based in South

Africa, has already taken an 81 percent share in Ghana's Billiton-Bogosu Goldfields, buying it from the Royal Netherlands Shell Group, while holding shares in the Obuom Goldfields in the Bekwai, Ashanti, area. Gencor plans on expanding its parasitic extraction policies in many other parts of West Africa.[66] De Beers has been invited by the Angolan government to mine for diamonds in three areas of Angola for the first time since independence from Portugal in 1975.[67] West African Development Corporation, another white South African–based company, has recently won a contract to develop a free-trade zone in the oil-rich area of São Tomé and Príncipé, the archipelago 160 miles off the coast of Gabon.[68] Despite concerns about the disruption of the fragile ecosystem in the region and questions about the economic payoff, the South Africa company is pursuing an ambitious and lucrative financial development venture.

In the newly liberated Democratic Republic of the Congo, though there is widespread talk of liberating the country from the tyrannical dictatorship of Mobutu Sese Seko under the leadership of Laurent Kabila, the new government's collaboration with imperialism is transparent. In April 1997, a month before Kabila's government took power, a $1 billion tender was signed to explore for copper and cobalt resources at Kolwezi in Katanga Province between American Mineral Fields, a company based in Arkansas, and Kabila's government. The mine dumps at Kolwezi possess an estimated wealth of $10 billion.[69] Anglo American, the South African multinational mineral conglomerate, is also involved in wooing Kabila to have a stake in the lucrative mineral wealth of the Congo. In September 1998, Kabila handed over 80 percent of the interest in the copper industry parastatal Gecamines to Ridgepoint, a large company owned by private Congolese and Zimbabwean interests and headed by Zimbabwean businessman Billy Rautenbach. In January 1999, Rautenbach signed an exclusive contract to market cobalt with the London firm Metal Resource Group, which failed to sell the cobalt stocks, hoping for higher prices, resulting in a cashless return and 25,000 workers without salaries.[70] So much for independence and liberation from Western imperialism and economic exploitation that Kabila so fervently claimed prior to removing Mobutu from power.

Meanwhile, the Central African region is engulfed in a bitter war, with the Congo, Namibia, Angola, and Zimbabwe on one side and Rwanda, Uganda, and Burundi on the other, both sides shamelessly supported by the United States with $125 million in weaponry and training over the past five years.[71] The U.S. Bureau of Mines estimates that Southern Africa contains 90 percent of the global deposits of platinum and palladium resources, 85 percent of chrome, 75 percent of manganese deposits, and 50 percent of gold and vanadium supplies.[72] Africans are once more being sacrificed at the altar of the West's insatiable quest for mineral wealth.

It is palpable from these horrifying facts that the "new South Africa" is being used as a launching pad for the vampiristic extraction of minerals and other vital natural resources in the rest of Africa. Already, South Africa's largest insurance conglomerate, Old Mutual, and Standard Bank, South Africa's leading banking and financial services group, have taken over commercial and financial institutions in several African countries. Since 1993, South Africa almost doubled its exports to African countries from $2 billion in 1993 to $3.7 billion in 1996, with trade surpluses tripling from $1 billion to $3 billion.[73] The European settler-colonialist and capitalist system that has enslaved the African people since the discovery of gold and diamonds in the late 1800s in South Africa intends spreading the epidemic of greed, the pulverization of Mother Earth, and the relentless and unconscionable exploitation of African men and women to the rest of Africa. This system is spreading a plague of economic perversity, as lethal as AIDS, which incidentally is spreading rapidly throughout Africa, leading many African theorists to speculate that AIDS is a disease engineered by the West for the eventual extermination of Africa's population.[74]

It is unequivocal that the current South African government is aware of the duplicitous role of these mining conglomerates. The new regime knows full well that the exploitative economic policies of these mammoth corporations practiced under apartheid will be exported to the rest of "independent" Africa. It realizes, too, that the large capitalist conglomerate tycoons are using their now valuable South African passports to enter African countries where abundant mineral wealth prevails to further draw African workers into the downward spiral of capitalist greed and bloodshed. One is forced to conclude that even though the newly elected South African government is aware of these imperialist designs, it is shameless in its disregard—because the new Black and white ruling class benefits from existing global capitalist arrangements in Africa. The neocolonial regimes of North and West Africa unflinchingly collaborate with this injustice and betrayal of the interests of the working class and underclass masses.

Privatization in African countries is the path vigorously pursued by the IMF and the World Bank under the pretext of "economic growth and prosperity." Privatization essentially benefits big business, and the policy of privatizing major sectors of Africa's economies implies the opening of Africa's industrial wealth to ownership by large foreign transnational corporations that obviously possess substantial amounts of capital to purchase all state enterprises being sold. The argument that these Bretton Woods institutions have made that national governments in Africa can make vast sums of money following privatization and invest these monies in infrastructural development is specious.[75] Privatization, by virtue of its integral connection to the capitalist system, means that these African economies will be further drawn into the spiral of international capital, which will result in the further siphoning of Africa's mineral wealth and natural resources to Western capitalist conglomerates.

In essence, the dictates of these capitalist institutions ensure that African economies will be permanently foreign owned and capitalist controlled, further impoverishing the working-class populations of the continent. The establishment of a stock exchange in Zambia and having an expanding one in Ghana does not in any way promise economic prosperity for the workers. It does ensure, however, that the capitalist elites will make handsome profits and that foreign investors will extract Africa's wealth to the further economic and material detriment of Africa's capital-poor working-class majority. Ghana, which was hailed as the "darling" of the World Bank and IMF since accepting the Structural Adjustment Program in 1983, ironically borrowed 600 billion Ghanaian cedis to finance a 400-billion-cedi deficit, using the remaining 200 billion cedis to pay the interest on its external debt.[76] Ghana is borrowing more money to pay for the interest on monies borrowed earlier, keeping the country in perpetual debt and economic servitude to the West.

Samir Amin, the African political economist, explains that the results of Western economic intervention via the rhetoric of "liberalization;" "deregulation," and "privatization" in underdeveloped countries are uniformly disempowering for formerly colonized nations, a situation that applies directly to Africa:

> Crisis management by national governments proceeds by policies of deregulation designed both to weaken the rigidities of trade unionism and dismantle and liberalize prices and wages; reduce public expenditure (principally subsidies and social services); and privatize and liberalize external transactions. The recipe is the same for all governments and its justification is based on the same vague and excessive dogmatisms: liberalization frees potential initiatives stifled by interventionism and puts the engine of economic growth back on the rails; those who liberalize fastest and most completely will become more competitive in open world markets. But as Marx and Keynes understood so clearly, such liberalization will ensnare the economy into deflationist spirals of stagnation, unmanageable at the international level, multiplying conflicts which cannot be mediated, against the empty promise of future "healthy" development.[77]

Amin avers that the West's emphasis on "democracy" goes along with economic liberalization, obscuring the reality of entering commercial and capital monopolies that are at a premium advantage in the program on liberalization:

> On what basis, with what criteria can these policies be judged or evaluated? Nobody knows. At the same time, the legitimation of choice is reinforced by political and ideological propositions which are as vague, and false, as those advanced concerning economic mechanisms. Economic liberalization is presented as synonymous with political democracy and all critiques of these policies are held to be democratically inadmissible. The merits of economic liberalism are praised in the name of "transparency," the state being considered, by definition, as the locus of opacity (ignoring the fact that the democratic state provides the best conditions for transparency) while in fact—the very real—opacity of private business protected by "business confidentiality" escapes even a passing mention. The social and economic realities of oligopolies, the privileged relations of the private with the public sector, and corruption, are not the object of scientific analysis. Rarely have we witnessed an ideological discourse as extreme as any dogmatic fundamentalism, repeated incessantly by the media and the dominant discourse, as if it were based in established evidence.[78]

In neighboring Botswana, Namibia, Swaziland, and Lesotho, the economic hegemony of European capitalism functioning via South Africa is patently clear. Botswana, for instance, which depends on diamond exports for almost three-fourths of its revenue, has all of its diamonds mined by the Central Selling Organization, a subsidiary of De Beers, which in turn is part of the absent Anglo American Corporation.[79] Anglo American, together with American Metal Climax (which owns the copper-nickel mine at Selebi-Phikwe), owns the colliery at Morupule. Its influence permeates the industrial and commercial landscape of the country in the meat industry and in hotels, construction, banking, insurance, and freight and forwarding.

In Lesotho, South Africa controls all the principal natural assets, diamonds, and water, via Anglo American Corporation and its subsidiary, De Beers Lesotho Mining Company (Pty.) Ltd. Lesotho's scarce water supplies are diverted to support large white-owned industries in South Africa. In Swaziland, where sugar is the main revenue earning crop, twenty-five companies and 263 holders controlled the economy in 1989. The British own most of the sugar corporations through companies such as the Swaziland Sugar Milling Company and the Commonwealth Development Corporation's Vuvuland Irrigated Farms Scheme. South African corporations like Gencor control the asbestos and coal production of Swaziland.[80] Cotton and fruit production are all owned and administered by South African corporations.

In Namibia, economic control by corporations based in South Africa in the areas of mining, commerce, banking, fishing, and agricultural production is also evident. The Namibian economy is still dominated by large white transnational conglomerates, with the most important diamond mine at Oranjemund controlled by Consolidated Diamond Mines, which is owned by De Beers, a subsidiary of Anglo American based in South Africa. In 1991, Consolidated Diamond Mines made a profit of N\$115 million (about \$23 million) from this mine in Namibia.[81] De Beers has rights to mineral and mining rights in Namibia for the next twenty-five years.[82] Rio Tinto PLC, based in the United Kingdom, owns the largest uranium world mine at Rossing, holding rights to lucrative sales deals to Japan, Taiwan, and the European Community. Tsumeb Corporation, a subsidiary of Goldfields of South Africa, owns four base-metal mines and a large copper smelter and lead refinery. Imcor Zinc is another important zinc mining operation owned by South Africa's ISCOR corporation. De Beers controls the largest diamond and uranium mines at Swakopmund. In terms of fishing, although Namibia has one of the most productive fishing grounds in the world, the tenth largest, almost all of it is controlled by South African fishing companies based in Luderitz and Walvis Bay. The total value of fish caught in 1990 was close to N\$1,500 million (about \$400 million), yet much of this revenue ended up in the hands of the giant commercial fishing conglomerates.[83]

In the Zimbabwean economy, as in the case of South Africa, the wealth is owned by large European commercial and mining conglomerates, many of which are based in South Africa and some of which are in partnership with the Zimbabwean government, and a burgeoning Black business elite. African workers are used as cheap labor to produce lucrative profits for these large corporations. According to the Zimbabwe Congress of Trade Unions, 4 percent of the population (mostly white) owned 90 percent of the wealth of Zimbabwe in 1990.[84] The number of parastatal companies grew from twenty-four in 1980 to forty-four in 1993, many of them including members of the Black business elite, as is the case in present-day South Africa.[85]

Zimbabwe attained its independence from British colonialism in 1980 after a protracted and heroic guerrilla struggle waged by ZANU, supposedly for reclamation of the land, during which time guerrilla fighters and their supporters called themselves "children of the soil."[86] Yet repossession and redistribution of the land and the wealth to the indigenous working-class community continues to be a distant dream for the dispossessed in the 1990s.[87] The logic of the Zimbabwean government following its ascent to power in 1980 was to cultivate Black agrarian capitalism by improving credit and market access for the top 20 percent of small-scale commercial farmers, so that they would eventually constitute about 30 percent of the large farm sector, thus progressively undermining the monopoly of white commercial farmers. Sam Moyo, from the Institute for Development Studies, noted that such a program was seriously flawed because it would take a decade to materialize and cost hundreds of millions of dollars, in a cash-strapped economy.[88] The government then abandoned the idea and, following the dictates of the Economic Structural Adjustment Program, declared that it planned to lease "large state leasehold farms formerly held by whites to capitalist Black farmers."[89] President Robert Mugabe was fully aware that these measures would undermine the program for reallocation of lands to the landless rural poor but proceeded with the program anyway. Sam Moyo explains the land-reform strategy and effects of the government program:

> During the 1990s, the state increased the rate and pace at which it pursued the allocation of large state lease-hold farms to black capitalist farmers, including a few blacks in strategic positions within government and the ruling party. The resulting imbalances in gains apparent from these government reform measures have tended to steer public opinion towards a perception that the land and economic reforms as well as the economic nationalism espoused, which are not necessarily based on popular demands, may generate wider public resistance to economic reform, particularly to the present land redistribution program.[90] .

The government had promised to provide land for over 100,000 landless families during the course of the 1980s. By 1989, only 51,000 Black families were resettled on 2.6 million acres of land.[91] By 1999, more than 70,000 families had been resettled under the government's land-reform program.[92] Although the program fell short of the government's goal of resettling 162,000 families, the quality of life of those who were resettled, to the Zimbabwean government's credit, significantly improved. Each family received 0.4 hectares of land for residential occupation, 5.0 hectares for agriculture, and the right to use variable portions of land for communal grazing. Ninety percent of those resettled now own at least ten head of cattle and have received potable water supplies and upgraded facilities for agricultural improvements, schools, and clinics. Resettlement area farmers received more than 6.8 times more from crop sales than farmers in communal areas.[93]

In late 1997 in Zimbabwe, Mugabe declared that commercial land belonging to 1,772 white farmers, constituting 12 million acres of land, would be seized for reclamation by landless Black peasants.[94] Zimbabwe's 4,000 white farmers own one-third of all prize agricultural land in the country. However, in January 1998, the government revised its earlier position, stating that it would repossess fewer farms because it did not have adequate financing to compensate the white farmers. The European Community had offered to lend the government monies for the program, but only on the condition that far fewer farmers were affected, watering down the entire resettlement program.[95] White farmers offered to sell some of their farms to the state as a compromise, to avoid confiscation of white-owned land and farms.[96] In February 1998, white farmers retaliated against the Zimbabwean government's plan on land distribution by withholding payments on a cumulative bank debt of $100 million. They also went on to sell stocks on the Zimbabwe Stock Exchange, which resulted in the loss of some $700 million, precipitating a sharp drop in the value of the Zimbabwe dollar and its devaluation by 50 percent in two months. Finally, an IMF team in Zimbabwe threatened the government with no further loan monies until there was a firm commitment "to compensate white farmers 'fully and fairly' for lands expropriated."[97] Earlier, in 1991, the white-run Commercial Farmers Union had challenged the government's land policy and planned acquisitions of white farms, arguing that the agricultural sector was crucial to the success of the Structural Adjustment Policy as expected by the IMF and the World Bank.[98] These acts of the local white farming establishment and the IMF lend credence to the contention that international capital is in fact *white, racist, anti-Black, anti-working class, sexist, and never color-blind.* It is imperative that all of Black working-class organizations' energies and strategies be devoted to breaking the back of the economic chokehold of white racist capital, the latter classically illustrated in the recent situation of Zimbabwe. Until this objective materializes, Black workers globally can never rest. The U.S. State Department has followed suit in its racist position vis-à-vis the Zimbabwean government's land reform program and refused to support it on the grounds that it was "'racially' based."[99]

White commercial farmers continue to insist that they maintain a position of privilege in being accorded the right to persist in owning the largest and most productive arable land in Zimbabwe, on the absurd grounds that their farms contribute significantly to the agricultural gross domestic product and foreign revenue. They have balked at the seriousness of land reform, dismissing the relevance of Black farming productivity, and have gone to the extent of even taunting government officials by insisting that the farmlands of government ministers be redistributed first.[100] This racist intransigence by white farmers is one very distinct indication of the privilege demanded by whites in Southern Africa, to the point that they believe it their right to adjudicate land policy, an essential task of the state.

The Zimbabwean government, it should be noted, has engaged in extraordinary measures to promote a climate of racial conciliation by privileging white commercial farmers and providing them "with preferential access to economic incentives such as foreign currency for their machinery and other imports requirements."[101] It has treated white commercial farmers with kid gloves and has neglected the Black rural poor in the process. The Zimbabwean government need neither vacillate nor apologize for its land-reform policy, so long as the land and agricultural needs of the Black rural poor are addressed, Sam Moyo argues.[102] The acquisition and possession of land by the rural poor provides an opportunity for this marginalized sector to establish collateral, gain experience in intensifying agricultural production, and most important, in sustaining the rural economy. The land redistribution among the rural poor is an imperative for Zimbabwe's economic survival, given the fact that the country does not have the resources to rapidly industrialize, as with most situations in the underdeveloped world. Sam Moyo contends:

> Access to land must be creatively planned to see Zimbabwe through the current economic transition while expanding output and the income base in the rural areas. In the final analysis, historic grievances over land dispossession and the present inequity in land holding are real political sentiments fuelled by the absence of alternative economic opportunities, given the slow industrialisation in Zimbabwe, as in other parts of the developing world. This is a political problem which government must face with courage rather than with doubts emanating from the pretentious "de-politicisation of the land issue" by politically motivated opponents of land redistribution.[103]

The Zimbabwean government has been ambiguous regarding its claims of addressing land reform. It has politicized the issue in every regard, from its erstwhile position of attempting to appease white farmers on the issue of land reform as part of its policy on reconciliation to threatening farmers with confiscation of farms as part of its land-reform policy. It has generally catered to the development of a Black capitalist class even in its land-reform program, forgoing any sense of egalitarian distribution and creating 800 large-scale Black capitalist farmers and 12,000 small-scale Black capitalist farmers by 1994.[104] A rural Black bourgeoisie has been fashioned, violating the very objectives of land reform and elimination of inequality. The Zimbabwean government has not stood firmly on the principle of accountability to the rural landless poor, who are in desperate need of arable land and who put the government into office in the first place, with up to 50 percent being "deemed near landless or land-hungry."[105] Those in the rural peasant majority have been excluded from participation in the shaping of the land-reform process and feel slighted because their traditional leaders are being excluded from consultation, even though land issues effect rural women and men the most. Sam Moyo decries the continued exploitation of the land-reform issue by the landed elite, at the expense of the landless poor:

> In this situation, the elite are using the poor, economic nationalism and crude power bases as instruments to enhance their accumulation agenda through control over land. . . .
> So far, black landed elites have ignored the land policy demands of the urban poor, the agricultural graduates association, women's groups and various peasant groupings. The state continues to police the numerous "squatters" and "poachers" in general defence of the land tenure rights of the state, large-scale white and black land owners.[106]

Lawrence Shuma, former lecturer in the faculty of law at the University of Zimbabwe and a program legal counsel at the International Development Institute, contends that "administrative authoritarianism continues to characterise its [the state's] relations with the peasantry."[107]

The Zimbabwean government, like those of Mozambique and South Africa, has emphasized modern urban industrialization, to the neglect and detriment of the rural poor. In Mozambique's case, this was a devastating economic blunder on the part of the Frelimo government following independence in 1974.[108] The tendency of African governments, including that of Zimbabwe, to portray economic advancement by extravagant urban construction projects is a delusory tactic that overlooks the fact that Africa is not the West and neither has the resources nor the infrastructure to industrialize itself like the West. An editorial in *Zimbabwe News,* the official organ of the ZANU Patriotic Front, describes this contradictory situation precisely:

> Look at the amount of building construction going on in Harare, Bulawayo, Gweru, and Mutare. Dwelling houses, office blocs, and factories are springing up in many parts of the city, and in large numbers. There is a significant amount of growth taking place. In the area of mining, the indigenous and small-scale miners have greatly increased their share of mineral production. In fact, their production of chrome and gold has exceeded that of the large companies and corporations.[109]

In addition to the above assertion being untrue, such sentiments indicate the distortion of priorities within ruling circles in Zimbabwe.

The citadels of commercial and financial enterprise in Zimbabwe, such as the Confederation of Zimbabwe Industries and the Zimbabwe National Chamber of Commerce, were essentially constituted and headed by whites in the 1980s, despite government prodding toward an integrated white-African organization.[110] In the late 1980s, the Confederation of Zimbabwe Industries expanded its membership to include some prominent African businessmen, but this was short-lived. In 1991, the African members constituted their own organization, the Indigenous Business Development Centre.[111] The Chamber of Mines remained exclusively white throughout the 1980s. It is extremely clear that the white business establishment, accustomed to receiving privileges in former colonial Rhodesia, desired the continuance of being able to make economic decisions independent of the African presence.

During the late 1980s, of the eight largest companies listed on the Zimbabwe Stock Exchange, four companies, Delta, Hippo Valley, Zimbabwe Alloys, and National Foods, were all owned by Anglo American Corporation.[112] Another South African–based corporation, Barlow, owns Hunyani and Plate Glass, together with Lonhro owning David Whitehead and T. A. Holdings, which is Zimbabwean owned. Anglo has purchased larger sections of shares in Lonhro, owns dozens of subsidiary companies in Zimbabwe, and is involved in the finance, transport, property, agriculture, and mining sectors. Old Mutual, the largest insurance company in South Africa, which was founded in England and is worth $7 billion, is also the largest life insurance company in Zimbabwe.[113] The ZANU Patriotic Front has enriched itself by establishing a financial empire worth $62 million under the holding M and S Syndicates, financed by donors and friends from the international community on behalf of the people of Zimbabwe.[114] As Ibbo Mandaza reminds us, many lessons for South African liberation can be learned from the chronology of the Zimbabwe anticolonial struggle, particularly with the incipient stages of the struggle beginning in appeals to the system of Western European liberal democracy.[115]

Receiving no positive response to appeals for freedom from colonial rulers, the Zimbabweans embarked on a national armed struggle culminating in the Lancaster House Agreement between the Zimbabwean liberation movement and the Ian Smith colonial regime in 1979. There, the imperialists foisted their designs on the outcome of the protracted country's struggle for liberation, for instance, by precluding land redistribution for at least ten years after independence and proscribing expropriation of private property except under conditions "in the interests of defense, public safety, public order, public morality, public health and town and country planning," and "only under-utilized land could be compulsorily acquired for settlement of land for agricultural purposes."[116] Appealing to the African masses in Zimbabwe (who subsequently became immersed in the freedom struggle) on the basis of anticolonial sentiment, the Zimbabwean liberation movement was successful in precipitating the downfall of the Smith regime and ascending to the seat of power in 1980 through the majority Zimbabwe National Union party and the Zimbabwe African People's Union party in a Patriotic Front. However, the struggle remained essentially nationalist in content, because its leadership was unwilling to confront the tremendous class contradictions that had emerged during the Smith era. The socialist rhetoric that was espoused by the government shortly after independence remained just that and essentially translated into state capitalism.[117] Radical talk of redistribution of resources within the country was progressively deemphasized. Ibbo Mandaza argues:

> Yet there is no historical evidence to suggest, as others have been keen to extrapolate from this momentous process of the liberation struggle, that this armed struggle encompassed within it even the idea of a socialist revolution. It is true that it had by the late seventies transcended in some respects the earlier strategy and tactics of the early 1960s which saw violence as merely the means to pressure Britain into an intervention that would bring independence sooner rather than later. But the armed struggle, was at best, viewed as a means to dismantle the white settler colonial system and replace it with an African government and at worst, as a pronounced way of pressurizing the imperialists into convening a conference that would bring about an African government in Zimbabwe.[118]

The white-led economy remained virtually intact after independence, with minor modifications at places with the addition of a tiny class of Black politicians, bureaucrats, technocrats, and professionals, prompting the head of the Zimbabwe Stock Exchange, W.A.F. Burdett-Coutts, to confidently declare in 1986:

> Recent Company reports have not only shown improved earnings but chairman's comments have in the main greater confidence than last year. I do believe that it is a credit to our government under leadership of our Prime Minister, Cde. Robert Mugabe, that after six years of independence, the economy of this country should be in relatively such

good shape, that the stock exchange, although still smaller than I would like, is now soundly based and structured and that Black and white are working harmoniously together, for the future advancement of Zimbabwe. It demonstrates that, despite the rhetoric, the policies followed by our government have been mainly *pragmatic*. The increased agricultural production in the communal lands is particularly noteworthy. (italics mine)[119]

Note the ironic references to "Cde," or Comrade, Mugabe by Burdett-Coutts and the paternalistic statement that the economy is still sound, after six years of Black rule! The "pragmatic" approach underscored by Burdett-Coutts and the "vibrant" economy echoes similar sentiments made by the former head of South Africa's Reserve Bank, C. L. Stals, when he talked about the Western imperialist world looking nervously at the South African economy following the termination of apartheid and the inception of Black majority rule, subsequently coining the expression "Afro-pessimism." A classic case of déjà vu?

The point about Blacks and whites "working harmoniously together" in Burdett-Coutts's remarks is not entirely accurate, considering that during Zimbabwe's 1985 elections, most whites averred that they still favored former leader Ian Smith's party because of his protectionist policy toward whites, leaving many Black people surprised and disappointed. The conciliatory posture toward the white settlers that the new Zimbabwean government took shortly after assuming office in April 1980 was best illustrated by then prime minister Robert Mugabe's words,

> We will ensure that there is a place for everyone in this country. We want to ensure a sense of security for both winners and losers. There would be no sweeping nationalization; the pensions and jobs of civil servants were guaranteed; farmers would keep their farms. Zimbabwe would be non-aligned; Let us forgive and forget. Let us join hands in a new amity.[120]

This assurance remained unconvincing to whites. Mugabe's promises sounded strikingly similar to President Nelson Mandela's admonitions to whites that they were critically needed in South Africa following the demise of apartheid in 1994. Mandela declared on the first day after his release from prison in February 1990: "Whites are fellow South Africans and we want them to feel safe. We appreciate the contribution that they have made to the development of this country."[121]

Mandela went on to express the view that the demands for structural guarantees and fears of Black rule by whites needed to be addressed and promised that Black rule would not lead to "Black domination."[122] These statements ring hollow when we realize that we have no evidence of any Black government in Africa or anywhere in the world, historically or contemporaneously, that has dominated whites. It is interesting to note that these conciliatory gestures were akin to views expressed by Jomo Kenyatta in 1963, when he addressed the bastion of the white settler community, Nakuru, requesting them "to stay and farm in the country" and urged them to "learn to forgive one another."[123] Of course, as Chango Machya w'Obanda points out, Kenyatta had acquired a huge farm in Nakuru and had absolutely no intention of expropriating lands that white settlers had stolen from Africans. Similarly, in Namibia, the rhetoric of redistribution of white settler–occupied land among Africans shortly after independence in 1990 was gradually squelched after President Sam Nujoma was courted by wealthy white farmers and even offered farmland for hunting, leading to him being treated as a benign and "avuncular figure."[124]

It is further ironic that amid promises of security for both "winners and losers" in Zimbabwe, most Blacks continue to live in economic and social insecurity, while most whites (the so-called losers) continue to enjoy economic security under Black majority rule, as is the case in contemporary South Africa. In fact, whites are returning to Zimbabwe in droves, following their mass exodus in 1980, when they feared that their positions of white social and economic privilege would come under attack by a Black government. The views of a returnee encapsulate the general sentiment of whites on this issue and demonstrate that deep-seated anti-African racial prejudice and the expectations of privilege for whites had never dissipated:

A lot of people had that feeling of fate about this country. . . . But a lot of people are surprised. I think they've managed to maintain most of the standards—medically, educationally, the police . . . I didn't think of England as my country. I came back two years ago . . . I will say there are a lot of things wrong with this country, but by God, that has run a pretty good show. We have a house, a swimming pool, a tennis court, three servants. We live very well. Someone is cooking dinner tonight. Someone is watering the garden tomorrow. Someone is looking after the tennis court. And they don't bother us and we were on the other side of the fence. We're not living in Rhodesia. We're living in Zimbabwe.[125]

The tragedy of this disposition of persistent anti-African racism is that the struggle to rebuild a society free of racist ideology becomes more compounded. Equally disconcerting is the lack of courage demonstrated by the Black leadership of countries like Zimbabwe and South Africa to candidly raise such critical questions so that the pathology of such attitudes can be eviscerated from the national socioeconomic landscape. The level of confidence and optimism that the Zimbabwean masses had in the Mugabe regime throughout the 1980s has steadily eroded to the point that in December 1997, March 1998, and June 1998, mass demonstrations by workers and students intensified, exerting pressures for President Robert Mugabe to resign. The Zimbabwe Congress of Trade Unions organized a massive stay-away in March 1998, protesting against a 2.5 percent development levy and a 17 percent general sales tax, resulting in police confronting thousands of demonstrators.[126] The students denounced the corruption, and nepotism of the government and an alleged student leader of the demonstrations, Learnmore Jongwe, blatantly exposed the government's fraudulent policies on "indigenization" and "empowerment" when he declared: "Indigenisation [a system of empowering blacks] has been privatised, monopolised, and politicised [to the extent that] huge government contracts have been won by people from the ruling province in this country owing to coincidence or design."[127]

Zimbabwe's government, which began on a platform of supposed radical transformation and socialist justice, has progressively teetered off-course, resulting in widescale discontent on the part of the working class and, for the first time, a growing repudiation of the historic Robert Mugabe as a "dictator." Mugabe, in turn, has retorted angrily to calls for his retirement, contending that he is the one who will determine when that would be.[128] While the economy teeters on the verge of collapse and hundreds of thousands of unemployed young people despairingly hope for a miraculous transformation, Mugabe has defiantly directed his vitriol against the IMF and the World Bank, Britain and the United States, and the white commercial farmer community, accusing them of the deliberate destabilization of Zimbabwe. Author and political commentator Chenjerai Hove was caustic in his assessment of Mugabe's adamance and denial of his political incompetence when he charged that "there is sickening ritual on the part of our leaders to substitute rhetoric for planning, to scapegoat and to shoot from the hip in the hope that the smoke will clear and sound will cover up the reality."[129]

Although one has to concede that the imperialist powers and their financial surrogates, the Bretton Woods institutions, have devastated Zimbabwe and the rest of Africa financially, through their noose-shaped, strangulation-effect policies, the Zimbabwean government has also made critical mistakes and has deviated from its ostensible original mission of emancipating the poor from the ranks of poverty and providing decent education, health care, and employment. It has allowed the IMF and World Bank to enter the country and exert leverage in the vital areas of the economy and land reform, knowing from experience in the early 1980s that collaborating with these institutions would result in impoverishment and disempowerment of the working classes. Contradictorily, it has squandered hundreds of millions of dollars earmarked for rehabilitation and the resettlement of former guerrilla fighters and the purchase of white commercial farms for Black resettlement. The country's top-ranking officials have bought some 500 white commercial farms scheduled for distribution among landless peasants, for their private use.[130]

The Zimbabwean government continues to spend vast amounts of money on military hardware, up to 1 billion Zimbabwean dollars (about $25 million) on weapons and aircraft, and Z400 million ($10 million) on riot equipment, all at a time when funds are being severely drained in areas of education, health care, and employment creation.[131] Over 200,000 female secondary school students were forced to leave school in late 1992 because they could not pay their school fees; charging students fees had been introduced as part of government policy to meet Structural Adjustment Policy dictates by the IMF and World Bank.[132] The rolls of the unemployed have grown since then, reaching 1.3 million youth in 1992.[133] The government of Zimbabwe has confused priorities since its sanguine inception in 1980 and now finds it impossible to acknowledge its fatal flaws.

In the face of the political events described above in many parts of Africa, one is compelled to raise the question as to whether Africans are living in more authentically democratic societies today. Since Africans in these nations have been granted the democratic right to vote for a political party of their choice, does that make them free? The workers are not better off with incomes per capita at $741 per year in moderate income countries and $239 for the poorest countries and poverty or near-poverty rates at 70 percent.[134] There have been few schools built, there are fewer clinics constructed, and few people are being permitted the right and resources to build decent homes, nor are they granted the fundamental right of earning just and livable wages, following the elections in these various African states. Africans cannot be considered politically free so long as the working classes are denied the freedom to economic justice because of these imperialist, capitalist, and neocolonial systems.

The neocolonial reformers and their puppet masters in Western Europe, North America, and Japan then remind us that change cannot come overnight. The Black working-class masses, however, pointedly ask: How long is overnight going to be? We on the continent have become so gullible and malleable that we have begun to mouth the pious and empty sanctimonious rhetoric of our former slave masters in our yearning for freedom. Even though we know that we are not free because the majority of our people live in overcrowded slums and ghettos in Katatura, Nairobi, Soweto, Dakar, Lagos, or Cairo; or for that matter, in Brixton, London, or in Roxbury, outside Boston in the United States; even though we are deprived of adequate food, let alone permitted to read the writing on a ballot box, we come to shamelessly aver that we are now free in this era, which many academics describe as "postcolonialism." Many of us are not willing to concede more accurately that we are now living in the phase of neocolonialism, that we live under a more sophisticated and modern colonialism, the objective material conditions of which many of the working-class masses realize and experience daily. Neocolonialism is

> a continuation of the colonial economic heritage which represents a compromise between the indigenous and alien economic interests. The primary aim of neocolonialism is to maintain the former colony as a dependency, as a controlled source of raw materials as well as market for investment and the sale of goods manufactured both in the metropolitan countries and overseas by local subsidiaries of foreign firms.[135]

The result of the skewed and perverted political economies in Africa has been a distorted social culture and existence for the majority of African people, pauperization of the working classes, and the further marginalization and impoverishment of rural and urban unemployed men, women, and youth, while bolstering the wealth of the tiny ruling bourgeoisie and its elite allies. This represents an overall condition of underdevelopment and untrammeled exploitation. This situation of cyclical poverty, illiteracy, ill health, and economic hardship has induced a sense of demoralization, frustration, and powerlessness unprecedented in Africa's history. Chango Machyo w'Obanda summarizes this predicament of impoverishment and pauperization aptly:

> Deepening underdevelopment and impoverization of the rural areas leads to migration, especially of the youths, into the urban areas thereby compounding the problem of over-

crowding and slum development in the towns where conditions of living are most appalling.

Other social evils the people of Africa are faced with because of the adverse effect of SAPs [structural adjustment policies] include an increasing rate of crime—lawlessness, robberies, thuggery, etc.—both in the urban and rural areas; prostitution and other forms of social deviance despite the scourge of AIDS; increasing school drop-outs as many parents fail to raise escalating school fees, and malnutrition both of the children and adults as people cannot afford to feed themselves well. People cannot afford decent clothing—even relatively better-paid civil servants and teachers are forced to resort to buying second-hand clothing from Europe and the U.S. which has an adverse effect on our textile industry. Nor can people afford decent housing. Family instability, drug abuse and alcoholism, juvenile delinquency, child abuse, unmarried spinsters as head of households, uncared-for widows, orphans and the aged are all with us today (Leakey, 1961, p. 20). This is what "modernization" really means—social rejection! Then there is the increasing mortality rate as many find it too expensive to go to hospitals which have introduced cost-sharing schemes; many women cannot go to hospital for delivery because they cannot afford the fees. It should also be noted that many workers and peasants cannot afford to send their children to school, while higher education is more and more becoming a privilege of only the rich and well-to-do families.

The fact is that African youths are so demoralized, frustrated and helpless due to the working of an unjust inhuman system . . .

. . . Culturally, the African of today has lost a sense of direction. The African, due to colonial and neo-colonial mentality has lost the core of moral, ethical and aesthetic values that could serve as his or her lodestar to life . . . What people seek after is to imitate others as a monkey does a human! This is defended in terms of "international standards," "fashion," or "modernity." African collective consciousness and social commitment are steadily disappearing, giving way to the worst kind of crude individualism of human-eat-human.[136]

Even though the above excerpt may be exaggerated in some instances because it certainly does not obtain in every part of Africa, Chango Machyo w'Obanda does capture the gravity of Africa's pain under the scourge of neocolonialism and capitalism. It is certainly the capitalist West that thrives on the continued extraction of resources from the underdeveloped world, functioning as "predators" as Chinweizu asserts, particularly in view of the fact that in 1997 and 1998, Africa paid out more than $1 billion to the IMF, even while African indebtedness to Western banks and financial institutions increased by 3 percent to $226 billion.[137] Every month, from the beginning of 1982 through 1991, Africa has been forced to pay $1 billion in debt repayments, a total of $120 billion.[138] These payments were made to ostensibly reduce Africa's debt levels; yet Africa's debt increased by more than 113 percent from $77 billion in 1982 to $164 billion in 1992.[139] The net result is that Africa is a debtor continent and owes some $300 billion to Western financial institutions. The shocking by-product of this transfer of payments is the death of 21 million people since the beginning of structural adjustment in the late 1980s and the denial of basic education to 90 million African women and girls by the year 2000. Malnutrition has affected 40 percent of African children in the late 1990s, up from 25 percent in 1985, and in most countries involved with the IMF and the World Bank, urban unemployment has reached 60 percent and real wages have dipped 30 percent, up to 90 percent in extreme cases.[140] This contemporary genocide is little known in the West, even though Western institutions and governments are responsible for this form of abject economic slavery. Concomitantly, African neocolonial governments function as co-conspirators when they agree to collaborate with the policies of the IMF and World Bank, fully cognizant of the devastation of their poorest populations. Their sycophantic character compels these regimes to cater to the needs of the wealthy financial powers located in the West and to their indigenous commercial elites, who derivatively benefit from such extractionist policies, brutalizing the African poor in the process.

Adebayo Adedeji has calculated that the debt figure is essentially exaggerated because it ignores the fact that $70 billion is attributable to high interest and losses as a result of lowering of exchange rates in relation to the U.S. dollar, that $54 billion was lost in export earnings as a consequence of commodity prices dropping, and that $30 billion left the continent in flight capital between 1974 and 1985.[141] The debt stranglehold is part of the silent "Third World War" that the West has waged and continues to wage on the underdeveloped world, what Susan George has termed the Financial Low-Intensity Conflict.[142] However, Africa must refuse to pay this "debt" to the West since it is the West that owes Africa trillions in reparations for reaping economic benefits from five centuries of slavery and colonization and, in effect, causing genocide.[143] The decision to refuse to be held in perpetual servitude and indebtedness to the West can only be taken by a liberated, socialist, united, and integrated Africa.

It is indeed incredible that right-wing and some liberal academics still laud the success of capitalism in Africa, knowing the scenario described above. It is obscene and unconscionable that anyone can still heap praises on capitalism's policies and presence on the African continent. Africa's underdevelopment, pauperization, and misery must be permanently arrested and sent to the gallows without delay. The real criminals benefiting from Africa's exploitation need to be seized and put away for life: the IMF, the World Bank, and capitalist agencies and corporations. To cherish the hope of the IMF and the World Bank to "reform" and reconstitute themselves in favor of genuine and equitable economic assistance to the underdeveloped world, as Robert Naiman and Neil Watkins do in their informative preamble article, is to naively assume that Western imperialism will undo itself and work against its own exploitative interest.[144] It is tantamount to asking the vulture to change its predatory nature. It is precisely for this reason of breaking away from the yoke of dependency that Africa desperately needs a cultural revolution that will in turn set into motion a socialistic political economic revolution, the makings of which will be described in Chapter 6, where the potential Azanian revolutionary model is discussed, with derivative implications for the rest of Africa.

Summary and Conclusion

Even though many have been led to believe that most of Africa became independent from the European colonial powers in the 1960s, it is evident that dependency is the accurate way to describe most countries in Africa, with the possible exception of Libya. Although not all nations have been discussed in detail here, similar situations obtain in the rest of northern, eastern, western, and southern Africa, where transnational corporations extract vast amounts of minerals and natural resources and export manufactured products to these countries, which end up paying for them through loans at exorbitant rates of interest set by the World Bank and the International Monetary Fund.

In the next chapter, Pan-African responses to the problematic of neocolonialism in Africa will be discussed.

Notes

1. Ihechukwu Chinweizu, *The West and the Rest of Us: White Predators, Black Slavers, and the African Elite* (New York: Vintage Books, 1975), 340.

2. Kwame Nkrumah, *Africa Must Unite* (New York: Frederick Praeger, 1963), xii.

3. Samir Amin, *The Class Struggle in Africa* (Reprint, Cambridge: Africa Research Group, 1964), 26, 38–39. Amin does an excellent job in this article in analyzing the colonial subjection of Africa and its subsequent domination by neocolonialism, particularly in North and West Africa.

4. Wayne Nafziger, *Inequality in Africa: Political Elites, Proletariat, Peasants, and the Poor* (Cambridge: Cambridge University Press, 1988), 37.

5. Thomas J. Biersteker, *Distortion or Development? Contending Perspectives on the Multinational Corporation* (Cambridge: MIT Press, 1978), 6.

6. *Weekly Mail*, October 9, 1995.

7. *Business Day* (Johannesburg), October 4, 1995.

8. Kwame Nkrumah, *Africa Must Unite* (New York: Praeger, 1963), xvii.

9. See, for instance, Ali Mazrui's *The Africans: A Triple Heritage* (Boston: Little Brown, 1986), 179–180, and Kwame Nkrumah's *Neo-Colonialism: The Last Stage of Imperialism* (New York: International Publishers, 1966), for an illumination of the illicit and arbitrary partitioning of Africa that totally disregarded indigenous ethnic and social affiliations and merely focused on colonial Europe's commercial needs.

10. Moses Nwoye, "Analysis of Problems and Prospects of Kwame Nkrumah's Pan Africanism in Africa, 1945–1970," Ph.D. diss., Howard University, 1974, 110.

11. The segmentation of "sub-Saharan" Africa has no functional point of reference for Africa—the Sahara Desert is a geographical product and has no bearing on the continent needing to be divided, just as the Sonora Desert has no basis in dividing the state of Arizona in the United States! This imposed division is predicated on insidious attempts to separate "Black Africa" from supposedly "white Africa," the latter being associated with North Africa. It is important to understand that the roots of all Africans are indigenous and Black, be they Tuareg in North Africa, or Khoi Khoi in Azania. "Black Africa" is just as much part of the racist myth of Africa as "white Africa" is.

12. Notwithstanding classical Marxist theoreticians' attempts to perceive a class analysis without seeing the confluence of the categories of race and class, it must be stated that racism

is the prevailing and dominant ideology by which capitalism survives. See George Frederickson, *White Supremacy* (New York: Oxford University Press, 1981), and Stephen Cell, *The Highest Stage of White Supremacy* (Cambridge: Cambridge University Press, 1984).

13. See the incisive speech of Malcolm X, reflecting on the ineffectiveness of the ballot system in the U.S. democratic system because it did not substantively empower oppressed Black people, in "The Ballot or the Bullet," in *Malcolm X Speaks* (New York: Grove Press, 1965).

14. *People's Weekly World*, February 3, 1996.

15. National Public Radio report, October 31, 1995.

16. Videotaped interview by the author with Mrs. Titi, July 11, 1997, assisted by Rev. Xolisi Mfazwe, a cleric with the Ethiopian Orthodox Church.

17. Adebayo Adedeji, "Popular Participation, Democracy, and Development," in *Nigeria: Renewal from the Roots: The Struggle for Democratic Development* (London and Atlantic Highlands, N.J.: Zed Books, 1997), 14.

18. Joseph Hanlon, *Peace Without Profit: How the IMF Blocks Building in Mozambique* (Dublin: The International African Institute, in association with Irish Mozambique Solidarity, Oxford: James Currey, and Portsmouth, N.H.: Heinemann, 1996), xv. Hanlon furnishes a substantive description of the role of the West in strangling Mozambique, particularly detailing the callous roles of the IMF and the World Bank, with very moving accounts of familial and societal devastation of a country struggling to be free from colonialism and neocolonialism since 1974.

19. Ibid.

20. Ibid., 24.

21. B. O. Nwabueze, "Constitutionalism in the Emergent States," in Henry Bretton, ed., *Power and Stability in Nigeria, 1962* (London and Enugu, Nigeria: C. Hurst, in association with Nwanife Publishers, 1973), 119.

22. Ibid., 111.

23. Cited by P.N.C. Okigbo, "Economic Implications of the 1979 Federal Constitution," in *The Nigerian Economy: A Political Economy Approach* (London: The Nigerian Economic Society and Lagos: Longman, 1986), 21.

24. Joseph Umoren, *Democracy and Ethnic Diversity in Nigeria* (Lanham, Md.: University Press of America, 1996), 98–102.

25. *West Africa*, February 1–14, 1999.

26. Abubakar Momoh, "The Political Economy of Transition to Civilian Rule," in Said Adejumobi and Abubakar Momoh, eds., *The Political Economy of Nigeria Under Military Rule (1984–1993)* (Harare, Zimbabwe: SAPES Books, 1995), 21.

27. Daniel Fogel, *Africa in Struggle: National Liberation and Proletarian Revolution* (San Francisco: Ism Press, 1986), ix–x.

28. Momoh, "The Political Economy of Transition to Civilian Rule," 18. This book is an informative source on the contradictions of the successive military regimes in Nigeria and the manner in which the Bretton Woods institutions and structural adjustment policies were used to enrich the ruling military and the tiny business elite groups, all in connivance with international capital.

29. Ibid., 19.

30. Gesiye Angaye, "Petroleum and the Political Economy of Nigeria," in *The Nigerian Economy: A Political Economy Approach* (London: The Nigerian Economic Society, and Lagos: Longman, 1986), 65.

31. *West Africa*, February 1–14, 1999.

32. *New African*, January 1999.

33. See Michael Fleshman's article "Obasanjo Takes Conrol," published by the Africa Fund, New York, June 17, 1999, and listed on the Black Radical Congress listserv.

34. Ebenezer Babatope, *Nigeria: Towards the "Revolution": Essays on the 2nd Republic* (Ibadan, Nigeria: The Skeletal Publishing Company, 1981), 77.

35. See, for instance, Eliphas G. Mukonoweshuro's *Colonialism, Class Formation, and Underdevelopment in Sierra Leone* (Lanham, Md.: University Press of America, 1993), 48–49, for an expatiation on the backdrop of economic neocolonialism in Sierra Leone.

36. The destructive and predatory policies of the IMF and the World Bank are documented in John Week's *Development Strategy and the Economy of Sierra Leone* (New York: St. Martin's Press, 1992).

37. *Chicago Tribune*, July 8, 1999.

38. Nnamdi Azikiwe, *Liberia in World Politics* (Westport, Conn.: Negro Universities Press, 1934), 151.

39. Stephen S. Hlophe, *Class, Ethnicity, and Politics in Liberia: A Class Analysis of Power Struggles in the Tubman and Tolbert Administrations from 1944–1975* (Washington, D.C.: University Press of America, 1979), 71–72. Hlophe furnishes a substantive historical analysis of the confluence of state formation and class configurations in Liberia, dating back to the American Colonization Society "founding" Liberia in 1821 and subsequently according Liberia state sovereignty in 1847. Contrary to Eurocentric views paraded in the media, Liberia was not founded by freed African "slaves" but by European colonists from the United States. See the chapter "The Idea of Liberia, in Amos Sawyer, *The Emergence of Autocracy in Liberia* (San Francisco: Institute for Contemporary Studies, 1992), where Sawyer makes it abundantly clear that Robert Finley, a Presbyterian clergyman, together with William Thornton of the Quakers and John Caldwell of the American Bible Society (all God-fearing Christians!) were the ones who conceived the idea of colonizing Liberia, thus ridding the United States of freed Africans while maintaining the viability and profitability of slavery. Gus Liebenow also discusses the neocolonial character of the modern Liberian state in chap. 2 of *Liberia: The Quest for Democracy* (Bloomington: Indiana University Press, 1987).

40. Adedeji, *Popular Participation, Democracy, and Development in Nigeria*, 14.

41. *West Africa*, February 1–14, 1999.

42. Cited by Robert Naiman and Neil Watkins, *A Survey of the Impacts of IMF Structural Adjustment in Africa: Growth, Social Spending, and Debt Relief* (Washington, D.C.: Preamble Center, April 1999). This statistic is based on a World Bank Report, *Poverty in Côte d'Ivoire: A Framework for Action, 1997*, which was cited in *External Review*, 102.

43. Ibid., 14.

44. Ibid.

45. Ibid.

46. *New African*, May 1999.

47. *West Africa*, February 1–14, 1999.

48. *New African*, June 1999.

49. Ibid.

50. *Chicago Tribune*, July 10, 1999.

51. Mohammed Ali, *Ethnicity, Politics, and Society in Northeast Africa: Conflict and Social Change* (Lanham, Md.: University Press of America, 1996), 97.

52. Mohammed Hassen, "Eritrean Independence and Democracy in the Horn of Africa," in Amare Tekle, ed., *Eritrea and Ethiopia: From Conflict to Cooperation* (Lawrence, N.J.: Red Sea Press, 1994), 87–88. For a more detailed elaboration on the politics of the Horn of Africa, see Peter Woodward, *Sudan 1898–1989, The Unstable State* (Boulder: Lynne Rienner, 1990);

David Laitin and Said Samator, *Somalia: Nation in Search of a State* (Boulder: Westview Press, 1987); Mansour Khalid, *Nimeiri and the Revolution of Dis-May* (London: KPI Limited, 1985); and I. M. Lewis, *A Modern History of Somalia: Nation and State in the Horn of Africa* (Boulder: Westview Press, 1988).

53. Ali, *Ethnicity, Politics, and Society in Northeast Africa*, 97.

54. Antoine Lema, *Africa Divided: The Creation of "Ethnic Groups"* (Lund, Sweden: Lund University Press, 1993), 46.

55. *World Press Review*, March 1998.

56. Lema, *Africa Divided*, 43–44.

57. Archie Mafeje, *The Theory and Ethnography of African Social Formations* (London: CODESRIA, 1991), 14.

58. Ibid., 15.

59. Kofi Hadjor, *On Transforming Africa: Discourse with African Leaders* (Trenton, N.J.: Africa World Press, and London: Third World Communications, 1987), 59.

60. Ibid., 65.

61. *Weekly Review* (Nairobi), January 6, 1995.

62. Kinuthia Macharia, *Social and Political Dynamics of the Informal Economy in African Cities: Nairobi and Harare* (Lanham, Md.: University Press of America, 1997), 169.

63. *Finance Africa* (Johannesburg), November–December 1997, 19.

64. *African Business*, February 1995.

65. *New African*, April 1995.

66. Ibid.

67. *West Africa*, April 26, 1996.

68. *New African*, May 1999.

69. *Mail and Guardian* (Johannesburg), January 9–15, 1998.

70. *New African*, May 1999.

71. Clarence Lusane, "U.S. Arms Both Sides in Congo," *San Jose Mercury News,* February 20, 2000.

72. *World Press Review*, January 2000.

73. *Finance Africa* (Johannesburg), November–December 1997.

74. *World Press Review*, July 1999. See also the cover story of the *New African*, April 1998, which describes the racism associated with AIDS in Africa and the manner in which the West has used and continues to use Africa as an experimental ground for testing vaccines against the AIDS virus.

75. The World Bank and the International Monetary Fund were founded at an international conference in Bretton Woods, New Hampshire in 1944, with the objective of manipulating commerce, finance, and trade so that the recession of the 1930s would not be repeated. These organizations were also intended to generate financial support for the industrialization of Europe and Asia following World War II. See Hanlon's *Peace Without Profit*, 24.

76. *Finance Africa* (Johannesburg), November–December 1997.

77. Samir Amin, *Capitalism in the Age of Globalization* (London and Atlantic Highlands, N.J.: Zed Books, 1997), 96.

78. Ibid., 96–97.

79. Margaret Lee, *SADCC: The Political Economy of Development in Southern Africa* (Nashville, Tenn.: Winston Derek, 1989), 74.

80. Ibid., 75–76.

81. This information was supplied to the author by the Embassy of Namibia, Washington, D.C., in March 1995.

82. *Business Day* (Johannesburg), November 25, 1995.

83. Namibian Embassy information, March 1995.

84. Dan Mokonyane, *The Big Sell Out by the Communist Party of South Africa and the ANC* (London: Nakong Ya Rena, 1994), 48.

85. Elton Gray Razemba, *The Political Economy of Zimbabwe: Impact of Structural Adjustment Programmes, 1980–1993* (Harare, Zimbabwe: Aroclar, 1993), 143.

86. Lawrence Shuma, *A Matter of (In)Justice: Law, State, and the Agrarian Question in Zimbabwe* (Harare, Zimbabwe: SAPES Books, 1997), 1. Shuma's book is extremely helpful in understanding the legal context whereby the indigenous landless rural peasantry has by and large been deprived of land reclamation in the government's supposed land reform program.

87. Sam Moyo, *The Land Question in Zimbabwe* (Harare, Zimbabwe: SAPES Books, 1995), 8.

88. Ibid., 7.

89. Ibid.

90. Ibid.

91. Peter Moll, *The Great Economic Debate* (Johannesburg: Skotaville, 1991), 139.

92. Bill Kinsey, "Land Reform, Growth, and Equity: Emerging Evidence from Zimbabwe's Resettlement Programme," *Journal of Southern African Studies*, Vol. 25, No. 2, June 1999, 173.

93. Ibid., 181, 183.

94. *New York Times*, November 20, 1997.

95. National Public Radio News, Washington, D.C., January 14, 1998.

96. "White Farmers Offer Compromise in Zimbabwe," *Arizona Republic*, January 25, 1998.

97. *People's Weekly World*, Febuary 14, 1998.

98. Shuma, *A Matter of (In)Justice*, 128.

99. Moyo (*The Land Question in Zimbabwe*, 10) provides a delineated and comprehensive analysis of the process of land reform in Zimbabwe, identifying problems with policies and strategies, and proposing concrete alternatives for successful land redistribution among the landless peasantry.

100. Ibid., 9.

101. Ibid., 12.

102. Ibid., 18.

103. Ibid., 19.

104. Ibid., 280.

105. Ibid., 139.

106. Ibid., 289.

107. Shuma, *A Matter of (In)Justice*, 79.

108. Hanlon, *Peace Without Profit*, 13.

109. *Zimbabwe News*, official organ of ZANU PF, December 1998.

110. Ibbo Mandaza, "The Post White Settler-Colonial Situation," in Ibbo Mandaza, ed., *Zimbabwe: The Political Economy of Transition, 1980–1986* (Dakar, Senegal: CODESRIA Book Series, 1986), 63.

111. Tor Skalnes, *The Politics of Economic Reform in Zimbabwe: Continuity and Change in Development* (London: Macmillan, and New York: St. Martin's Press, 1995), 101.

112. Joseph Hanlon, *Beggar Your Neighbors: Apartheid Power in Southern Africa* (London: Catholic Institute for International Relations, in collaboration with London: James Currey and Bloomington: Indiana University Press, 1987), 201.

113. Ibid., 203.

114. *New African*, May 1998.

115. See Mandaza's classic article, "The Post White Settler-Colonial Situation."

116. Shuma, *A Matter of (In)Justice*, 39.

117. Ibid., 54.

118. Cited by Mandaza, "The Post White Settler-Colonial Situation," 29.

119. Ibid., 60–61.

120. Ibid., 42.

121. Eileen Riley, *Major Political Events in South Africa, 1948–1990* (New York: Oxford Facts on File, 1991), 216.

122. Ibid.

123. Chango Machyo w'Obanda, "Conditions of Africans at Home" in *Pan Africanism: Politics, Economy, and Social Change in the Twenty-First Century*, ed. Tajudeen Abdul-Raheem (New York: New York University Press, 1996), 39.

124. Archie Mafeje, *The National Question in Southern African Settler Societies* (Harare, Zimbabwe: SAPES, 1997), 12. Nujoma's courting by white farmers was conveyed to the author in interviews in Namibia in July 1992.

125. Cited by Mandaza, "The Post White Settler-Colonial Situation," 58.

126. *New African*, May 1998.

127. "Pressure for Mugabe to Quit," *Weekly Review* (Nairobi), June 5, 1998.

128. From the *Zimbabwe Chronicle*, May 13, 1999. Cited in *Zimbabwe Press Mirror* (Bonn), May 16, 1999.

129. *Financial Gazette* (Harare), June 3, 1999.

130. Skalnes, *Politics of Economic Reform in Zimbabwe*, 161.

131. *Herald* (Harare), May 13, 1999. Cited in the *Zimbabwe Press Mirror* (Bonn), June 16, 1999.

132. *Herald* (Harare), November 16, 1992.

133. Razemba, *Political Economy of Zimbabwe*, 97.

134. W'Obanda, "Conditions of Africans at Home in Pan Africanism, 51.

135. Cited by Edwin Madunagu, *Nigeria: The Economy and the People: The Political Economy of State Robbery and Its Popular-Democratic Negation* (London: New Beacon Books, 1983, 1984), 8.

136. W'Obanda, "Conditions of Africans at Home in Pan Africanism, 52–53.

137. Cited by Naiman and Watkins, "A Survey of the Impacts of IMF Structural Adjustment in Africa."

138. Susan George, "Uses and Abuses of African Debt," in Adebayo Adedeji, ed., *Africa Within the World: Beyond Dispossession and Dependence* (London and Atlantic Highlands, N.J.: Zed Books, in association with the African Centre for Development and Strategic Studies [ACDESS], Ijebu-Ode, Nigeria, 1993), 60.

139. Ibid., 63.

140. Ibid.

141. Adebayo Adedeji, 1989 Foundation Lecture, *Africa in the 1990s: A Decade for Socio-Economic Recovery and Transformation or Another Lost Decade?* (Lagos: The Nigerian Institute for International Affairs, 1991), 22.

142. Susan George, *A Fate Worse Than Debt* (New York: Penguin, and London and New York: Grove Press, 1988), 5. Also cited by Adedeji, *Africa in the 1990s*, 19–20.

143. The following organizations have launched intense campaigns to advance the cause of reparations for the enslavement of Africans and the accompanying devastation of Africa: The National Coalition of Blacks for Reparations in America (NCOBRA), the African Repara-

tions Movement (ARM), and the African National Reparations Organization (ANRO). The Black Radical Congress, founded two years ago, has also made the question of reparations a priority in its political advocacy.

144. Naiman and Watkins, "A Survey of the Impacts of IMF Structural Adjustment in Africa," 25.

5

Pan-Africanism and the Struggle Against Colonialism and Neocolonialism

Since 1995 marked the fiftieth anniversary of the Pan-African Congress in Manchester and much of my analysis of the current situation in South Africa stems from a Pan-Africanist reading of African history, it is necessary to briefly revisit this significant meeting of 1945. Based on this excursion, the discussion will examine the prospects for the intensification of working-class Pan-Africanist struggle in South Africa particularly and within the African continent generally. This visitation is also of import in learning from the errors of preceding Pan-African formations and in understanding tactical weaknesses and building on concrete strengths, with the purpose of advancing the continental struggle against imperialism, neocolonialism, and capitalism.

The assertion of a Pan-Africanist identity is not a novel one. Noted historical Black thinkers and leaders from the Caribbean, Africa, and the United States such as George Padmore, Kwame Nkrumah, W.E.B. Du Bois, and Marcus Garvey were all vigorous Pan-Africanist advocates, the concept of Pan-Africanism originating with Henry Sylvester-Williams, a lawyer born in the West Indies who organized the first African conference in London in 1900.[1] Unfortunately, not many women were featured as prominent leaders within this historical movement, owing to the patriarchal contexts within which these figures functioned, even though recent womanist scholarship has demonstrated that Amy Jacques-Garvey, Garvey's last wife, did in fact exert considerable influence on the development of Pan-Africanist and Black nationalist thought.[2] A very positive development issuing from the Seventh Pan-African Congress in Kampala, Uganda, in April 1994, was the formation of the Pan-African Women's Liberation Organization, which had its first meeting in Harare, Zimbabwe, in August 1995, and has begun to launch chapters throughout Africa and the African Diaspora.[3]

One of the cardinal premises of Pan-Africanist philosophy is the mutualistic, reciprocal, and inclusive concern for Africa's descendants globally, an "African nationalism to replace the tribalism of the past: a concept of African loyalty wider than the 'nation' to transcend tribal and territorial boundaries."[4] George Padmore, a leading Pan-Africanist theoretician of the 1950s, viewed Pan-Africanism as offering an "ideological alternative to Communism on the one side and Tribalism on the other" and embracing "the federation of regional self-governing countries and their ultimate amalgamation into a United States of Africa."[5] "An injury to an African anywhere, is an injury to Africans everywhere" is one aphorism that derives from such an understanding, signifying the extended family concept implicit within historical African sociocultures, translated into political and economic proportions.[6]

Historical Pan-Africanist Struggle and South Africa: The Pan-African Congress in Manchester, 1945

One of the features of the Pan-African meeting in Manchester that distinguished it from previous such meetings in 1900 and 1927 was that it attracted high-profile figures who had become titans within the global Pan-Africanist movement, such as George Padmore, Kwame Nkrumah, Jomo Kenyatta, Peter Abrahams, and C.L.R. James.[7] Even more important was the fact that for the first time, over 200 African worker, trade union, and peasant association representatives from the African continent attended, unlike the previous Pan-African gatherings, where relatively few grassroots working-class people from the continent were prominently represented. There was a refocusing on the African continent and its struggles for independence and liberation from European colonialism, underscoring the centrality of Africa in the global Pan-Africanist movement and radicalizing the erstwhile bourgeois character of preceding Pan-African congresses.[8] Africans in the Diaspora, particularly in the Caribbean, the United States, and Europe, were reminded that the victory of Africa over colonialism and imperialism was a precondition for the emancipation of Black people everywhere.[9] Describing the aura of the Fifth Pan-African Congress in Manchester in 1945, W.E.B. Du Bois, considered the "father of Pan-Africanism" and the organizer of the first Pan-African Congress in London in 1900, asserted:

> The 200 delegates of the Fifth Pan African Congress believe in Peace. How could they do otherwise when for centuries they have been victims of violence and slavery? Yet if the world is still determined to rule mankind by force, then Africans as a last resort may have to appeal to force, in order to achieve freedom, even if force destroys them and the world . . . We are determined to be free; we want education, the right to earn a decent living; the right to express our thoughts and emotions and to adopt and create forms of beauty. Without all this, we die even if we live. We demand for Black Africa autonomy and independence so far and no further than it is possible in this "One World" for groups and peoples to rule themselves subject to inevitable World Unity and Federation.[10]

Kwame Nkrumah, the first president of independent Ghana, a Pan-Africanist pioneer, and one of the axiological figures at the 1945 congress, echoed the distinctive tone of the gathering in contradistinction to previous Pan-African gatherings and declared forthrightly:

> The goal is the liquidation of colonialism and imperialism from the continent of Africa. Previous Pan African Congresses had laid emphasis on agitation for amelioration of colonial conditions. They called for reforms and pressed for nothing more than a voice by colonial people in their own government.
>
> The Fifth Pan African Congress struck a new note. Those of us from Africa, more numerous at this assembly than at the earlier ones, had decided that reformism offered at best a delaying strategy. At worst, it could be met just as sharply as outright demand for complete and absolute independence . . .
>
> . . . we shot into limbo the gradualist aspirations of our African middle classes and intellectuals and expressed the solid down-to-earth will of our workers, trade unionists, farmers and peasants who were decisively represented at Manchester, for independence.[11]

The affirmation of the Fifth Pan-African Congress was for all Africans to govern themselves, and the clarion call for working-class leadership in anticolonial struggles was echoed for "workers and farmers to organize effectively—colonial workers must be in the front of the battle against imperialism—your weapons—the Strike and the Boycott—are invincible."[12]

The bold call for self-determination and the outright rejection of imperialism and capitalism and the advocacy of working-class leadership of the African struggle symbolized new approaches of activists in the Pan-Africanist movement as organizational mechanisms were formed to implement resolutions from the Fifth Congress into concerted action. Kwame Nkrumah identified the enemy of Africans as "imperialism." Africans had had enough of passive appeals to European colonial powers for their freedom. The voices of appeal had loomed into demands for liberation and independence, utilizing force if necessary. Chango Machya described the mood of the historic gathering:

> Africans were unwilling to starve any longer while doing the world's drudgery, in order to support by our poverty and ignorance a false aristocracy and a discredited imperialism. Monopoly of capital and the role of private wealth and industry for private profits were condemned. . . . After the conference, Pan Africanism entered a new phase of Positive Action. But the effect of this new attitude depended upon the degree to which African peoples were organized. Organization was held as the key for freedom. The Fifth Congress called upon the colonial people to close their ranks and form a united front between the intellectuals, the workers, and the farmers in the struggles against colonialism. From the Congress various delegates began to work for the implementation of the Congress declarations. Kwame Nkrumah took the initiative to organize West Africa by forming what was known as the West African Secretariat in London. Its aim was to

struggle for territorial self-government and "to promote the concept of a West African
Federation as an indispensable lever for carrying forward the Pan African vision of an ultimate *United States of Africa.*"(emphasis mine)[13]

There was also a strong sense of the need for the elimination of the artificial political boundaries that were imposed by European colonialism on Africa, the resolution
to work for African continental unity, and the recognition that class contradictions
were real manifestations on the continent as a result of colonialism and capitalism.
Elenga M'buyinga, a leader of the Cameroon People's Union (UPC) and an incisive
critic of the corrupt neocolonial nature of the OAU, explained:

> The Fifth Congress was a turning point for Pan Africanism. Class struggle was recog-
> nized, in principle, at least, as the driving force of history, including African history (al-
> beit from 1945 onwards); in other words, "Marxist socialism was adopted as a philoso-
> phy." The Fifth Congress was equally explicit on the question of African unity. One
> resolution in particular stressed that the "artificial divisions and territorial boundaries
> created by the imperialist powers are deliberate steps to obstruct the political unity of
> the West African peoples." It was only after the Fifth Congress that Pan Africanist ideas
> began to find widespread roots in Africa itself, usually in close association with the
> struggle for independence. The link between Africa's independence and its unity was
> thus made clearly apparent.[14]

Of particular importance was the dynamic character of the Fifth Congress, which
emphasized action, liberation, and movement toward unification of the continent
predicated on various anticolonial and anti-imperialist struggles. Kwame Nkrumah
described this critical purposefulness: national independence leading to African
unity. This limited objective was combined with a wider perspective. Instead of a
rather nebulous movement concerned vaguely with Black nationalism, the Pan-
African movement had become an expression of true working-class nationalism.[15]

Specifically relevant in the context of this discourse were the resolutions adopted
at the congress on the situation of white minority rule in South Africa. At a session
chaired by W.E.B. Du Bois, Peter Abrahams and Marko Hlubi from the Pan-African
movement in South Africa described the horrors of the colonial oppression of Black
people and the unmitigated economic exploitation of Black labor by Europeans.
Humiliating decrees such as the Pass Laws and the gross disparities in educational
opportunities for Europeans and Black people were recounted. The reverberating
theme was "Down with Imperialism."

Solidarity statements on South Africa were echoed by African representatives from
around the world. In response to this situation of virulent racism and repression, the
Fifth Congress passed a resolution condemning the violation of human rights in
South Africa by the European colonial minority, called for a franchise for all people,
freedom of movement and occupation, equal rights for all devoid of discrimination,
and reform of the legal and civil system as well as land policy and pledged "to work
unceasingly with and on behalf of its non-European brothers and sisters in South

Africa until they achieve the status of freedom and dignity."[16] The congress made it clear that it regarded "the struggle of our brothers and sisters in South Africa as an integral part of our common struggle for national liberation throughout Africa."[17] The slogan at the end of the resolution was *Mayibuye! Afrika!* (Let Africa Return!). This theme of Pan-Africanist struggle in South Africa at the Fifth Congress at Manchester in 1945 is instructive for critical appraisal of recent developments in the country.

The Obstacle to Pan-African Working-Class Unity: Neocolonialism

The system of Western European rapacity, enslavement, and exploitation extant in colonialism continues today in the system of neocolonialism. Neocolonialism is an extension of the power of Western European capitalism and imperialism into Africa, this time facilitated by indigenous African ruling tyrants who are beholden to their former colonial European masters and assure the continued extraction of mineral wealth and other natural resources for utilization by Western European industrial capitalism.

Achieving "independence" has meant little to the vast majority of African people because the same structures of colonialism are manifest and perpetuated in the current nation-state configuration of Africa, under which each government continues to legitimate its own existence. Those who hold the reins of power in almost every situation seek first and foremost to enhance their ruling-class interest. The system of cheap African labor for those large European corporations and conglomerates that strip Africa of its resources and impoverish its people is still maintained. The African nation-states established by colonialism have no basis in the independent evolution and interests of indigenous African cultures and societies but rather derive their roots from European political designs in Africa, reproducing the violence and alienation, repression and impoverishment of their colonial founders.[18] As Basil Davidson opines: "The colonial state was illegitimate because it was a conquest state . . . It may easily be seen, today, exactly in the light of its 'totalitarianism in government,' that the post-colonial nation state, as the inheritor of the colonial state, has proved to be a disaster."[19] For the sake of staving off imperialism, these boundaries must therefore be abolished.

The Organization of African Unity, which was formed in 1963 as a body to promote inter-African unity, is itself a legacy of the colonial period, because it assumes the moral and historical legitimacy of Africa's existing boundaries, reflected in the clause that prohibits intervention of any one state in the "internal affairs" of any other state. Such legacies perpetuate colonial divisions of Africa and a weakened continent that make it easy prey for acquisitive and predatory European powers.

The principal solution and weapon of defense against this colonial legacy is the establishment of a Pan-African working-class-based form of government that discards the strictures and divisions of the colonial boundaries that have perforated Africa. Kwame Nkrumah, the leader of Ghana, the first African country to receive its independence from European colonial rule in 1957, worked and lived for the realization of the dream of Pan-African unity, proclaiming that "Ghana's independence would be meaningful only within the larger context of liberation and independence for all the peoples of Africa."[20] He was overthrown in a coup in 1966, orchestrated by the forces of Western imperialism, just as he sought common ground with other Pan-Africanist-leaning leaders such as Julius Nyerere in Tanzania and Sekou Toure of Guinea. He founded the All African People's Revolutionary Party and the West African Secretariat in London, which sought to unify progressive anti-imperialist forces all over the continent and within the African Diaspora in Europe, the Americas, and the Caribbean.

The imperialists realized that a united Africa would be a stronger Africa and thus sought to destabilize the movements afoot for the consolidation of African peoples and organizations. They penetrated organizations like the Organization of African Unity through surrogate neocolonial regimes in the Congo (Mobutu Sese Seko, who overthrew Patrice Lumumba through help from the CIA in 1961), Côte d'Ivoire (Felix Boigny), and Senegal (Leopold Senghor), which opted to continue their participation in the confederation of Francophone colonies, as opposed to agitating for the independence of African countries and a unitary African continental confederation. Regarding the fierce battles fought for the independence of the Congo from Belgian colonialism and the perfidious role played by Western imperialism in propping up Mobutu Sese Seko, Nkrumah was patently clear in his advocacy of unified liberation struggle:

> If we allow the independence of the Congo to be compromised in any way by the imperialist and capitalist forces, we shall expose the sovereignty and independence of all Africa to grave risk. The struggle of the Congo is therefore our struggle. It is incumbent on us to take our stand by our brothers in the Congo in the full knowledge that only Africa can fight for its destiny.[21]

Kwame Nkrumah and Sekou Toure emerged as the strongest spokespersons in the aftermath of the Fifth Pan-African Congress held in Manchester, England, in 1945. Many organizations representing various anticolonial movements and worker unions attended. The result was the Bamako Conference in 1946, which echoed clarion calls for true African independence from Europe. This movement was opposed by Houphouet Boigny of the Côte d'Ivoire, who claimed to be Pan-Africanist and for African unity, while being against authentic independence from the European powers. The first Conference of African People was held in Accra, December 5–13, 1958, where Algeria and the Cameroon, which were engaged in a massive war against the French led by the National Liberation Front and the Cameroon People's Union, respectively, were accorded strong support. At this meeting, it was stressed that "the ultimate goal of Pan Africanism was federation of the entire continent," with Nkrumah lobbying for a program of unity via a federalist model.[22] Regional groups were proposed as the stepping stones toward gradual continent-wide unity. It was the December gathering that prompted the formation of the All-African People's Organization, led by Nkrumah, which functioned as an open forum to discuss pressing problems on the continent and promised support to groups fighting colonialism.[23]

In Monrovia, Liberia, August 4–8, 1959, the spirit of compromise changed the anticolonial liberationist fervor, with the Provisional Government of the Algerian Republic going unrecognized and the UPC of the Cameroon, an anti-imperialist organization, being asked to find a truce in its struggle against the French. In January 1960, the Second African People's Conference met in Tunis. A strong anti-imperialist resolve emerged at the meeting, with Patrice Lumumba being elected to the Leadership Committee. A political resolution condemning Boigny's *Franco African Communauté* was passed and the formation of a volunteer brigade to assist the Algerian anticolonial struggle was proposed. The Third Conference of African States met in Addis Ababa in June 1960, and then in Cairo in March 1961, where it approved "the use of force to eliminate imperialism" regarding anticolonial struggles being waged in Algeria, Cameroon, South Africa, and the Congo. It was at this conference, followed by the one held in Casablanca January 3–6, 1961, that strong sentiment was expressed calling for the "complete abolition of all the frontiers arbitrarily imposed by imperialism." The Casablanca Group formed a charter and maintained consistent support of all movements struggling against imperialism in Africa.

Interestingly, five months later, May 8–12, 1961, another conference was held in Monrovia, attended by puppet leaders from the Cameroon like Ahidjo and others from Senegal, Côte d'Ivoire, and other French colonies, with the exception of Guinea. The important outcome of this meeting was that it asserted, in opposition to the Casablanca Charter of Pan-African Unity, the principle of "non-interference in the internal affairs of other states." Even Nigeria's Azikiwe was represented at this conference of surrogate regimes, promoted by Western imperialist powers. The stage had been set for the dissolution of the more radical caucuses that had met in Casablanca and Monrovia.[24]

This conference presaged the formation of the Organization of African Unity formed in Addis Ababa in 1963, so that the OAU eventually became a watered-down imperialist-orchestrated organization that adhered to keeping neocolonial African regimes in power. The very purpose of forming the OAU, which was originally to unite African forces across the continent, was thus defeated by imperialism's external political pressure and the collusion of their surrogate regimes in Africa.[25] It is this legacy of sham independence (which in actuality is an incremental dependence on Western imperialism) that resulted in African underdevelopment, manifest in illiteracy, economic exploitation, homelessness, and the sociocultural disintegration of the lives of the African masses.

Revolutionary Pan-Africanism:
A Radical Response to Black Oppression

Pan-Africanism was a direct and trenchant response to the fragmentation of Africans globally and signified an urgent call to all of Africa's descendants that the subjugation of Africans by colonialism and capitalism necessitated forging a united Black front to arrest the course of underdevelopment. One of the premises of Pan-Africanist philosophy is the mutualistic, commensalistic, and all-inclusive concern for African descendants globally. In the closing address of the Third Annual Conference of the American Society of African Culture, Jaja A. Wuchuku, then Speaker of the Nigerian House of Assembly, echoed this view of Pan-African interconnectedness by stating:

> The interests of twenty million Negroes are entwined with those of two hundred million kith and kin in Africa. The American Negroes should appreciate that as long as our continent exists and thrives, as long as there are African states respected in the world community of nations, they themselves will have a full growth—which is what they require. Their full contribution will be appreciated, their role as human beings will be greater. That's all we want.
>
> African unity and Pan-Africanism: your African is an African, no matter where he or she is found.[26]

Within this configuration, the struggles of Africans in the United States against American capitalist civilization and cultural imperialism becomes an intrinsic issue on the agenda of Africans struggling in South Africa against white colonial rule and racial capitalism. Concomitantly, the struggles of the African and colored working classes of the Caribbean, the Pacific, and Asia against capitalism, imperialism, and European cultural hegemony become inextricably connected with the battles being waged by Black working classes in South Africa and the United States. In the particular contexts of the United States and South Africa, it is becoming more apparent that the African peoples of these nations cannot realize the goals of liberation and self-determination in isolation from one another, since both peoples are afflicted by a particular common racism endemic to these situations.

Black people in the United States, specifically the Black working class and underclass, are victimized by American racial capitalism, a mammoth system that terrorizes the oppressed peoples of the world and that cannot be toppled by Africans in the United States going at it alone. The independence and liberation of South Africa from colonial capitalism would be a material bolstering of oppressed Africans living in the United States in which the last remaining scourge to Black humanity, *apartheid*, is finally destroyed, so that Black people everywhere could at least hold their heads high in the aftermath of the victory of Black humanity over white racist and capitalist inhumanity in South Africa.

Radical socioeconomic and historical analysis similarly reveals that working-class and underclass Black people of South Africa cannot defeat the monster of apartheid capitalism alone, since the apartheid system is a composite transmogrification that was created, nourished, and maintained by the edifice of Western European capitalist civilization. The analysis conducted in the preceding chapters underscores the fact that Black people in South Africa are not involved in a struggle only against the white minority ruling South Africa but also against the entire Western capitalist world. George W. Shepherd Jr., a scholar at the Graduate School of International Studies Center on International Race Relations at the University of Denver, concurs with this view, contending that South Africa is an integral cog within the Atlantic imperialistic system, which is

> a concept drawn from regional and dependency theory. It is used to indicate the interdependence of the Western industrialized core powers and the dependency of South Africa in that system . . .
>
> . . . South Africa is seen as a semi-peripheral power of the system which has a high degree of dependency upon the core powers (the United States, the United Kingdom, Germany and France) while providing essential services to them. South Africa also acts as a subsidiary of Atlantic capitalist powers to penetrate economically the rest of the Southern African periphery. The dependency relationship of South Africa upon Western powers is defined in numerous ways, but particularly in cultural, military, technological and corporate terms. Much confusion has been caused by the failure to perceive South Africa's role as essentially an extension of the Atlantic world in Africa.[27]

The organic nature of racist exploitation by the forces of colonialism and capitalism of African peoples in the United States and South Africa behooves an organic response on the part of both colonized communities: protracted struggle and mutual support of each community's effort to liberate itself.

The Pan-Africanism that I advocate in this study is one involving the *people*, specifically the rural peasantry and urban working and underclasses, with special accent on women, and not the ruling elites who have historically utilized the concept of Pan-Africanism for elitist purposes such as Senghor of Senegal and even Kwame Nkrumah of Ghana. In Nkrumah's case, though he was the African leader most committed to the unification of Africa under the banner of a United States of Africa, he was still not able to extricate himself from a bourgeois-based Pan-African movement that rejected the leadership of the movement by the working classes. Nkrumah correctly viewed industrialization of the continent as the key to Pan-African economic success and proposed economic and political integration of the continent, yet he failed to see that economic and political integration among ruling-class elites would only benefit the bourgeoisie in Africa.[28] The solution of Pan-African unity needs to be qualified by a working-class revolutionary character where the means of production is eventually owned, harnessed, and developed by the talented and skilled African working classes for the benefit of the African working classes. Commenting on the ideological confusion surrounding the Sixth Pan-Africanist Congress in Dar es Salaam, Tanzania, Ofuatey Kodjoe, a Pan-Africanist scholar, cautions against this banding of the concept of Pan-Africanism to justify oppressive and alienating ends:

> It does not require any great astuteness to see that this pronouncement [a declaration defining Pan-Africanism as class struggle] amounts to no more than paying lip service and using rhetoric in order to confound Pan-Africanists, and defuse their revolutionary

commitment. After all, how can we accept the definition of Pan-Africanism as a class struggle from the representatives of *petit bourgeois* regimes which exploit the African masses; how are we to interpret an appeal to aid liberation movements; and how seriously are we supposed to take the suggestion that neo-colonialism be eliminated from representatives of states whose policies nurture neo-colonialism?[29]

It is also an acknowledged fact that one of the leading architects of Pan-Africanist philosophy, Kwame Nkrumah, notwithstanding the global legacy of African unity that he left, ironically failed to "comprehend the centrality of democracy to the dynamics of national liberation."[30] The case of Ghana's downfall in 1966 through a military coup was an unfortunate result of the failure of the Pan-Africanist titan to establish a mass peasant grassroots base, including working women, as Cuba successfully did in the 1960s, for instance, within his own nation.[31] Pan-Africanism without democratic roots in the masses, like any undemocratic movement, is doomed to failure. The lesson of Ghana illuminates the prerequisite that Pan-Africanism and African nationalism must be undergirded by participatory democracy in which urban working-class and rural peasant women and men are actively involved at all levels of the liberation struggle. Pan-Africanism must be a working-class and people-based movement that intrinsically involves women, as opposed to elitist male Africanist leaders merely in conversation with each other.[32] It is precisely for this reason that Elenga M'buyinga, the vice president of the National Revolutionary Council of Manidem—the Manifesto for Democracy in Cameroon, distinguished between "revolutionary Pan-Africanism," which "expresses the interests of the working classes and the poor peasantry, allied to the other exploited classes of Africa," from "Pan-African demagogy," which "expresses the interests of the ruling African bourgeoisie, their allies and their masters—the imperialists and world reaction."[33]

It was Marcus Garvey who emphasized the urgency of Pan-Africanist unity and consolidation in the United States in the early 1900s. His founding of the Universal Negro Improvement Association was unique among nationalist movements in the United States in that it embraced working-class and underclass Africans numbering up to 4 million in 1920, the largest gathering of Africans into a single organization in American history.[34] Garvey sought to convince the African masses in the United States of the imperative of African unity in the struggle for emancipation and the dignity of Black people globally. He was instrumental in empowering Africans in America with the concept of Blackness and Africanity: "Apart from rehabilitating the colour 'Black' he shook the Black masses of the Diaspora into an awareness of their African origin. Without setting foot on African soil he created for the first time a real feeling of international solidarity among Africans and persons of African stock."[35]

One of Garvey's limitations was that he did not understand the need for socialist principles within the Universal Negro Improvement Association, a critical oversight that weakened the level of potential radical political resistance of the organization to the edifice of capitalism. Garvey sought the construction of an "African empire" on the scale of that of Britain and the United States, without repudiating the fundamentally exploitative and patriarchal imperialist system.[36]

Concurrently, Pan-African radicals must be careful in extricating themselves from the vitiation of movements with which capitalist ideology indoctrinates all: that a movement is dependent on a single leader or a couple of leaders who come to assume deity status.[37] We must work diligently and tirelessly toward making all radical movements for change, including the Pan-Africanist movement, mass-based, grass-roots, and never centered around the individual personality of some charismatic cultic figure. If our movements are contingent upon the charisma of a single person, then they are inevitably weak and powerless, because imperialism will assassinate the lead figure, historically witnessed in the tragic instances of Malcolm X, Martin Luther King Jr., Patrice Lumumba, Amilcar Cabral, Steve Biko, Samora Machel, Maurice Bishop, and the like.

The Role of Revolutionary Ideology

The analysis of one of the greatest theoreticians of decolonization in Africa, Frantz Fanon, a psychiatrist from Martinique, is pertinent in this regard when discussing the foundational transformation of contemporary South African society mediated by Black working-class culture. Fanon reflected upon his experiences in the anticolonial struggle in Algeria in the 1940s and 1950s. His insightful and penetrating analysis of Algerian colonialism is instructive for the context of uprooting capitalism and neo-colonialism in South Africa. Fanon is illuminative in helping liberation activists to comprehend the magnitude of colonial repression and capitalist exploitation, their essentially conflictual character vis-à-vis the Black working class and the course for the positive abolition of these twin evils. He averred: "Decolonization is the meeting of forces opposed to each other by their very nature, which in fact owe their originality to that sort of substratification which results from and is nourished by the situation in the colonies."[38]

Radical Black working-class culture purveys the ingredients for resisting colonial repression and capitalist exploitation. Don Mattera, an Azanian poet, artist and writer, opines: "Culture is a way of becoming,"[39] a view that underscores the notion that radical Black working-class culture inculcates the vision of what Black people can *really be* as decolonized and liberated people in Africa.[40] Although European colonial bourgeois culture has functioned to legitimate the oppressive, exploitative, and corrupt behavior of apartheid society, Black working-class cultures of resistance have provided the "point of critique and creativity"[41] of this oppressive system and wield the power to fundamentally transform these structures through the nexuses of solidarity.[42]

In the South African context, a revolutionary ideology needs to firmly take hold within the Black working class, particularly the rural peasantry, the largest segment of the Black population. This ideology needs to be African centered, working-class based, and female affirming. The inculcation of this revolutionary ideology needs first and foremost to view the South African/Azanian situation as one integral to that of African history and not as an extension of colonial Europe, as is so often the case presently.

One of the most pressing problems facing all Black people in South Africa is the psychological devastation of Black minds, whereby three and a half centuries of colonization, enslavement, exploitation, and servitude have forced us to internalize our conditions of subjugation. We have come to believe that somehow Africa has always been a "backward" and oppressed continent and that we are destined to suffer as *Black* people, almost adhering to the biblical Hamitic myth, which states that the descendants of Ham were cursed by God to be slaves, concretized in recent historical events like slavery and colonialism.

The military and material conquest of Azania by Europeans has convinced us that whites are masters of the world and will continue to define Africa's future for generations to come. Black South Africans generally view themselves as a conquered people, powerless to stave off the mighty West and therefore basically surviving to adjust to this seemingly perpetual historical reality. We have instilled a defeatist attitude on account of this European conquest of Africa in general, and Azania in particular. We have ingested our inferiorization as Africans based on the distorted inverse of European superiority and supremacy. The result is a conscious and unconscious aping and emulation of European behavior and culture that makes us believe that what is best for Europe is best for Africa and that Western European cosmologies will move Africa in the direction of "advancement," "development," and "progress." In the South African context, this predicament of African inferiority and European supremacy is acutely compounded.

There are three axiological reasons for this prevailing sense of inferiority among Black working-class people:

A. Black workers, specifically African workers, are the most exploited sector of the Black population and are most susceptible to ideological manipulation, which relegates Africans to epitomes of inferiority and backwardness by European colonial imperialism and capitalism because of historical and contemporary material deprivation. They are made to normatively accept the role of whites as *baas* (master or boss) and to deferentially respect all people with wealth or money as powerful, viewing themselves as powerless in the face of overarching economic and social structures and institutions that reinforce Black inferiority and white supremacy. Since being full-blooded indigenous African is considered to be the lowest order of humankind by the colonial-apartheid system, thereby producing an anti-African worker casteism, Black workers are constantly subject to the scourge of being "African," "worker," and "poor," tolls of triple jeopardy in the world.

B. Black workers have the least access to educational resources and thus know little of the far-reaching African history that preceded European colonialism in Africa, causing them generally to accept the Eurocentric view of African history that avers that Africans are a "primitive" people who are indebted to white men for introducing civilization to Africa.

C. Black workers constitute the sector that is most dependent on the economic machinations of capitalism, since they often live from day to day or week to

week, surviving on progressive paychecks for food and shelter, with no fall-
back or provisions of savings or other possessions like middle-class Blacks.

This situation of ongoing oppression must be forcefully arrested. The ideological
enslavement of all Black people, particularly Black workers, must be earnestly addressed by all liberation movements in the post-apartheid era. Ultimately, one realizes that material liberation for all Black people, especially Black workers, requires
the preceding actualization of psychological liberation, underscoring the truism by
Steve Biko, founder of the Black Consciousness Movement in Azania, that "the most
potent weapon in the hands of the oppressor is the mind of the oppressed."[43] Carter
G. Woodson, the pioneer of the reeducation movement for all people of African descent in the United States in the early part of the century, reiterated this important
point regarding the educational and socialization process in the United States vis-à-
vis Black people:

> No systematic effort toward change has been possible, for taught the same economics,
> history, philosophy, literature and religion which have established the present code of
> morals, the Negro's mind has been brought under the control of his oppressor. The
> problem of holding the Negro down, therefore is easily solved. When you control a
> man's [woman's] thinking you do not have to worry about his [her] actions. You do not
> have to tell him [her] to stand here or go yonder. He [she] will find his [her] "proper
> place" and will stay in it. You do not need to send him [her] to the back door. He [she]
> will go without being told. In fact, if there is no back door, he [she] will cut one for his
> [her] special benefit. His [her] education makes it necessary.[44]

A decisive element of this reeducation process is the need to address the question of
the stigmatization associated with Africanity. Under apartheid-colonialism, racial
stratification was perversely construed from ten different racial categories from "pure"
white at 00 to "pure" African, defined as Bantu, the latter being forced to carry a passbook with 7 digits, identifying them as the lowest on the racial pyramid of classification. Being Black or a person classified as "non-white" was discriminating and damag-
ing enough; being African (full-blooded and untainted by any racial admixtures) was
devastating in terms of social, economic, and political marginalization. Individuals
who were classified as "Colored" (people of mixed African and European or Asian
blood) were viewed as being more socially acceptable than Africans, though essentially inferior to whites. Many people of Indian descent, particularly those who were
descended from indentured laborers, though often darker in complexion (because of
their Black-skinned South Indian heritage) than Africans, were viewed as "better"
than Africans because Asia was perceived to be a continent that though backward in
contradistinction to Europe, was relatively "more civilized" than Africa.

Ironically, the racist propagators of the "African-uncivilized" myth failed in their
contradictory and warped ideology to acknowledge that India was at one time considered part of Ethiopia and populated by Dravidian Blacks, whose descent was
African.[45] The principal reason for according these "Colored" groups some level of

mobility beyond that of Africans, of course, was to effect the traditional colonial policy of "divide and conquer" and to create buffer "Colored minority" communities among the Black oppressed that would feel beholden to whites for receiving "preferential treatment" and would repudiate identification with Africans, the majority group. During the 1994 elections, this racist tactic was successfully deployed by the National Party (F. W. De Klerk's party) in the Western Cape, where "Coloreds" were urged to vote for the National Party so as to prevent the possibility of being "swallowed" by the African majority. The antiquated apartheid slogan *swart gevaar* (Black danger or threat) was evoked by National Party leaders. This strategy was essential to the maintenance of white supremacy and the inferiorization of all Blacks, with the target of the racist vitriol being "full-blooded" Africans.

It is for this reason that the philosophy of Black Consciousness needs to be reimplanted within the fabric of Black social existence in South Africa, so that "Coloreds" and "Indians" will come to viscerally embrace the Black identity by conjoining fully with the indigenous African population. The perfidious character of whites determining the criteria of "Blackness" is still patent in the post-apartheid era, as a result of the persistence of white power following the crushing of the Black Consciousness Movement in the late 1970s and early 1980s, to the advantage of white supremacy and continued fragmentation of the Black community. This colonial design must be occluded at all costs so that a solid bulwark of Black unity can be established, one that is unabashedly pro-African and proudly Black at its core.

Today, the legacy of racist prejudice and stigma of anti-African caste-ism continues, even in post-apartheid South Africa. The most classic illustration is reflected in the fact that it is still socially acceptable to specifically hire African women as maids and nannies, a practice followed even by many "Coloreds," "Indians," and upper-middle-class Africans. This pathology of the exploitation of African women must be uprooted, eviscerated, and destroyed. It must not be permitted to persist under any new dispensation of post-apartheid, even under the supposed philanthropic pretext of "providing employment for desperately poor Africans." Paradoxically, this was the very practice of European colonials that led to the institutionalization of African female servitude in South Africa and that is now being observed by many "liberated" and economically well-to-do Blacks, under the pretext of "providing employment" to poor Black women.

Many middle-class Blacks from the United States who have either migrated to or are working in South Africa as part of the ostensible attempt to reconstruct the Black economy following apartheid's demise also participate in the same shameless neocolonial behavior. If these financially successful Blacks would like to address the needs of poor Black women, one wonders why they do not contribute their resources to educational and other economic improvement programs for low-classed Black women, who are forced to work for the wealthy and middle classes because they are unable to find gainful and productive employment in South Africa's capitalist class–tiered economy. On a related issue of the intimidation of poor African women, it is disconcerting to note that some Black women from the United States are involved in administering the Lindela Accommodation Center in Randfontein on the West Rand for people arrested as "illegal immigrants," as part of the Dyambu Trust created by some high-ranking women from the African National Congress such as Adelaide Tambo and Deputy Home Affairs Minister Lindi Sisulu.[46]

The anti-African aura that pervades South Africa in this post-apartheid epoch needs to be addressed at every level by the liberation movements in the country, certainly by those that are still committed to the cause of Black working-class and national liberation, such as the Azanian People's Organization, the recently formed Socialist Party of Azania, a splinter AZAPO grouping, and the Pan-Africanist Congress of Azania. A pro-African revolutionary ideology needs to take hold, whereby all people living in South Africa will come to understand the depth of African history and evolution.

Black people need to comprehend the fact that Africa was the continent of the cradle of human civilization, responsible for pioneering discoveries in science, medicine, mathematics, agriculture, astronomy, architecture, religion, and the like, the oldest being in Nubia and Egypt, which possessed formidable classic civilizations for three thousand years prior to Aristotle, Plato, and Hippocrates and the subsequent birth of Christ. Although there were social contradictions and conflicts between the ruling classes and the laboring classes in early African civilizations as prototypical of *all* societies, Africa nevertheless contributed foundationally to the shaping of human civilizations at a time when Europe was steeped in epistemological ignorance and social chaos and unheard of in global commerce and trade. It is important to emphasize in this pro-African revolutionary reeducation that Europe was the last continent in the world to be civilized in the classical sense of human civilization. Western Europe received much of its impetus for technological invention from Africa and Asia, since it possessed little autochthonous technical and artistic innovation in its early history. The Parthenon in Greece, for instance, is minuscule when compared to the early temples of Luxor and Karnak in Egypt and Nubia, and it was built 2,000–3,000 years after the latter. Christine Qunta, a womanist activist, makes this point lucid when she writes:

> One of the more significant contributions that white scholars and interest groups have made to the illogical notion of European superiority is to blot out the historical achievements of Black people—These Africans—who invented writing, who were the first brick and stone masons and whose mathematical genius and architectural skills enabled them to construct marvels such as the Great Pyramid of Giza (for over four thousand years the world's tallest building) and the city of Great Zimbabwe . . . are today portrayed as a people with no history of civilization before European conquest.
>
> The Black writer and researcher should not fall into this racist trap, but should recognize, as Dabi Nkululeko points out (later in this book), "that one of the major fronts in the war against colonialism is writing our own history." This fundamental principle is often overlooked, including [by] feminists.[47]

Such an exposé of African history is not to romanticize or reify Africa's past; it is, however, important to underscore the point that great scientific and technological achievements were made by Africans at a point when Europe was overcome with epistemological ignorance.

Proceeding to highlight the interwovenness of the structures of racism and sexism in the configuration of the oppression of Africans and the fact that both evils must be opposed in any revolutionary transformation, Qunta argues:

> African women must speak for themselves. They should also decide for themselves who they are, where they are going, what obstacles face them, and how to remove these. Acknowledging of this will signal European women's willingness to critically inspect the roles their own societies have played *and are playing* in the oppression of African people in general and African women in particular; this will allow them to play a supporting role where necessary. (italics mine)[48]

It is imperative that an African-centered educational system replace the existing Eurocentric educational curriculum in South Africa's schools, colleges, and universities. A thorough re-Africanization of South African society must be instituted so that the legacy of its perfidious and distorted pro-European colonial character is uprooted in all spheres. A rewriting of South Africa's and Africa's history must immediately occur, so that European colonial and capitalist aggression in Africa can be seen for what it really is: the cultural, social, and economic destruction of Africa and the intended genocide of Africa's peoples. The truth of African history must be told in diffuse ways at all educational, social, and cultural levels so that all Black working-class people particularly come to embrace African identity with solid positiveness and pride, as opposed to perpetually aspiring toward Western European modes of development and achievement. South Africa is not an extension of Europe but, like Egypt, is an integral part of the African continent. It cannot be viewed as a reflection of European settler-colonial culture as is currently the case in the economic and social aura of the country's cities and towns. Frene Ginwala, a speaker of the South African Legislative Assembly, reiterates this point, though the new government apparently has difficulty recognizing the magnitude of this problem of anti-Africanity and pro-Europeanization: "One of the fundamental changes facing South Africa is its emergence as a democratic country, an African nation with an African culture on an African continent; not as a European nation with a European culture artificially located on the African continent."[49]

Neither the government nor all of the people of South Africa can afford to obscure the truth about Western Europe's genocidal history and presence in Africa, for the sake of healthy race relations in the country. Such oversight will permit the cancer of white supremacy to persist and the perversity of European violence against Africans for five centuries to be suppressed and go untold. So long as such untruths are not confronted on a fundamental level, the entire South African nation will live a historical lie, building a new society on immoral and rotten foundations, resulting in the negation of African history and the perpetuation of anti-African exploitation. The ideological distortion that made Africans feel ashamed of being African now needs to be revolutionized by the assertion and teaching that *the shame lies not in*

Africa or in being African but in the European colonization and plunder of Africa and its subsequent exploitation by Western capitalism.

The revolutionary ideology that is being proposed here must be African-centered and uproot anti-African racist ideology, but it must also attack the very nature of the educational system in the South African context. It was often argued during the days of formal apartheid that all that Black people needed to have for the effective governance of a democratic and free South Africa was education. Based on the existing post-apartheid situation, it is evident that education as implied under the present dispensation is defective and seriously ineffective. A pivotal ingredient for the launching of the Azanian working-class cultural revolution is a comprehensive program characterized by the process of consciousness-raising or what the Brazilian educator Paulo Freire has termed "conscientization."[50] *Conscientization* is crucial because it does not merely subscribe to the notion of education as an antidote to the conditions of illiteracy and impoverishment, social degradation and personal humiliation among the oppressed, but it also raises questions about the content of education. It is distinctive from the education that is generally received in capitalist and Eurocentric societies.

First, conscientization starts off with the concrete experiences of the working classes and employs a dialogical approach in the very application of the educational process by affirming their culture and talents in the resolution to the problematic of oppression. Second, it is not an education for *consumption,* a feature of educational curricula in capitalist societies, such as in South Africa and the United States, for instance, where the dominant educational philosophy and pedagogical methodologies employed subscribe to a promarket capitalist ideology. Students, workers, youth, and the elderly are all schooled toward upholding the capitalist system and indoctrinated by its tenets into believing that it is the most intellectually sophisticated and successful system in the world. The educational system is also geared toward legitimating and accepting the existing economic arrangements under capitalism that favor the wealthy and educated and marginalize the poor and less educated. Finally, conscientization is a protracted process rooted in a philosophy that makes education a vocationally liberating experience, one in which oppressed people are made conscious of the histories and structures of oppression in their society and in the world, then schooled into formulating approaches that are designed to challenge and overthrow oppression. Paulo Freire explains this liberatory educational approach in the context of cultural action and cultural revolution:

> Conscientization is not a magical charm for revolutionaries, but a basic dimension of their reflective action. If men [and women] were not "conscious bodies," capable of acting and perceiving, of knowing and re-creating, if they were not conscious of themselves and the world, the idea of conscientization would make no sense—but then neither would the idea of revolution. Authentic revolutions are undertaken in order to liberate men [and women], precisely because men [and women] know themselves to be oppressed, and be conscious of the oppressive reality in which they exist. But since, as we have seen, men's [and women's] consciousness is conditioned to reality, conscientization

is first of all the effort to enlighten men [and women] about obstacles preventing them from a clear perception of reality. In this role, conscientization effects the ejection of cultural myths that confuse the people's awareness and make them ambiguous beings.[51]

It is through the *conscientization* of the Black rural peasantry and urban working-class women, men, and youth that a radically new cultural and political outlook can be cultivated, one that will embrace an indigenous African cosmology that refuses to tolerate any form of degradation of working-class African people, vehemently attacks stigmatization notions associated with Africanity, and eschews every aspiration toward Western European capitalist ideals.

Black Revolution and the Environment

The current and recent historical situation in South Africa shows that Western progress and industrialism have been held as the socioeconomic paradigm for the future, without explaining the correlative features of persistent exploitation of Black workers, the devastation of Mother Earth and natural resources like wood, agricultural land, and forests, the contamination of the country's rivers and air as a result of dumping the end products of mass industrial production, the emissions of carbon dioxide from vehicle exhausts, and the commodification of all human and natural relations within the ecological system. Apartheid, in addition to being a violently racist system, was also fundamentally inefficient and wasteful in the utilization of natural and material resources. Typical of Western European industrial capitalist culture, which uses energy, power, and land in proportions of excessive consumption as opposed to concern for conservation and preservation, apartheid has also encouraged this kind of wastefulness and uneven energy usage in South Africa. For instance, even though the Electricity Commission of South Africa produces 60 percent of Africa's electricity, 70 percent of the South African population (almost all Black) in the late 1980s and early 1990s did not benefit from this electrical network, leaving about 12 million people forced to use fuelwood as a primary means of energy.[52] Whites consume an average of ten times more water than the urban Black population in South Africa. Soil erosion in Kwazulu-Natal Province has increased tremendously since the advent of Europeans, particularly affecting areas like the Tugela Basin.[53] The result of high population densities and intensive agricultural cultivation on small land areas has naturally resulted in massive soil erosion.

The growth of "informal dwellings," often created by Black people from the Bantustans who have been forced to seek employment for survival in the cities, has also contributed to Black misery, indignity, and humiliation. In such areas, people live in shacks, with no clean water and little or no sanitation facilities, and thus become victims of both poverty and disease, which leads to an environmental hazard through unhealthy living conditions, all attributable to the insouciance, brutality, and racism of a colonial system called apartheid. In Durban, on the southeastern coast of South Africa, over 60 percent of the population—some 1.8 million people—are compelled to live amid such environmental violence.[54] The effects of coal burning in Black township communities like Soweto, with a population of about 3 million people, has not only been environmentally destructive because of sulfur oxide entering the atmosphere but, worse still, has caused serious health problems for the residents of the townships. Children especially suffer from fatal diseases like lung cancer, abestosis, and mesothelioma.[55] The roofing of houses with asbestos materials in Black townships is yet another health hazard that Black people continue to suffer.

It is clear in many of the facts listed above that environmental devastation in South Africa is fundamentally the result of European colonialism's model of economic development. It is exacerbated by years of political apartheid, wreaking havoc on the country's ecosystem, air, water, land, plant, and animal life and, most important, causing immeasurable suffering for the majority of South Africa's people.

In any discourse on a post-apartheid society, the issue of environmental reconstruction and reparation must be primary. It is at this juncture that the principles of indigenous African religioculture can be didactic for developing a consciousness and disposition that emphasizes preservation of natural resources and the protection of Mother Earth. The Green Movement and other European-based environmental organizations need to realize that it is the indigenous colored peoples of the world like the American Indians and Africans who were the first conservationists. Much of the religiocultural values of these indigenous peoples embrace a cosmology in which all life is sacred, recognizes that Mother Earth is a Living Spirit and a supreme manifestation of the Creator and affirms that all of nature is organically interdependent and precious within the Creator's complex design of the universe. Mother Earth herself is a living being, where the ancestral spirits reside. The reason that forced removals are such an abomination to Black people is because they not only signify a physical dislocation but also represent a *spiritual abortion*, by attempting to rupture the ties that Black people have with their ancestral spirits through living on lands where their ancestors have been buried.

Given the destructive side effects of Western European industrial-technological development on global ecosystems, it ought not be a romantic and farfetched notion to suggest that indigenous African spirituality needs to be earnestly considered in determining the modus operandi of the economy and models of sustainable development. A question that warrants raising but that may initially sound injudicious, given the current dependence of the South African economy on mining, mineral exploitation, and gas drilling, is whether we ought not explore alternative forms of sustainable, cost-efficient, and environmentally sensitive modes of economic development and energy utilization since Mother Earth continues to be plundered, ravaged, and wounded by mining drills, which has been occurring for over a century in South Africa. Black people have lived on lands for millennia without becoming obsessed with the idea of bleeding Mother Earth in order to unearth gold, diamonds, and precious metals for the *sole* purpose of selling them for profit, while impoverishing others. Ten thousand miners are either injured, disabled, or killed in South Africa's underground mines each year, as they pay the ultimate price for maintaining an economy that is predicated on mineral exploitation and human devaluation.

The model of Western modernity has been held in front of the eyes of all African people like a carrot dangled before a horse, with the objective of seducing Africans, including African workers, into accepting Western cultural consumption and imperialism as the yardstick for human advancement and success. The price paid for this obsession with Western progress is poignantly evident in the deaths of close to 2 million people in Asia as a result of drinking and washing in waters polluted by the dumping of industrial wastes, breathing toxic gases from the proliferation of automobiles and industrial vehicles, and the inhalation of deadly fumes from forest fires recently burning out of control in places like Malaysia and Indonesia, because poor people are desperately seeking land for cultivation.[56]

Western industrialization has left death and devastation of Mother Nature in its wake, precipitating conditions of global warming that have induced progressively lower average rainfall over the past decade in the Northern and Southern Hemispheres and air and water contamination that severely hampers the quality of life for billions of the world's people. The recent natural disasters of flooding and droughts associated with the El Niño and El Niña phenomena, with unprecedented floods, tornadoes, and storms all over the world, were exacerbated by the warming of the oceans, heavily attributable to Western industrialization.

Clearly, the model of Western progress as embraced by the new South African government needs to be challenged, given the increased mortality rates among East Asia's industrialized powers, which have also followed blindly in the path of Western industrialist and capitalist development without considering the heavy toll on the quality of human life and the liquidation of natural ecosystems. For instance, sulfur emissions from the burning of coal in China cause acid rain that results in $5 billion worth of damage to agricultural crops annually. The quality of life wrought by Western industrialization leaves much to be desired, and the new South African government needs to be confronted on this fundamental "progress" issue, given the conflicting success stories of Western industrialism in Asia. The process of *conscientization* as envisaged in the cultural revolutionary transformation proposed here would inculcate a qualitatively and radically different value system, one that is spiritually and culturally indigenous in its essence, in that it is proworker, anti-individualistic and collectively defined, socialistic in its core, and embracing of female leadership and direction.

Revolutionary Transformation and Indigenous African Spirituality

Indigenous spirituality is foundational for the execution of the cultural revolution necessitated in Azania's transformation. The spirituality suggested here is not a chimerical and opiate ideology as prototypical of segments of African Christianity, which defends Western traditions in Africa and espouses pacifism and nonviolence in response to capitalist and militarist oppressor systems. It is not the absurd ostensible Christian philosophy as advocated by the Truth and Reconciliation Commission formed in South Africa a few years ago.

The Truth and Reconciliation system is headed by former Anglican archbishop Desmond Tutu and was created by the new government as part of an effort to supposedly facilitate reconciliation between former apartheid oppressors and those who are victims of apartheid violence. The commission is not only tainted by a "Christian" aura because it is chaired by a clergyperson but also because of its emphasis on reconciliation and harmony between oppressor and oppressed, in adherence to essential "Christian principles."

The tragic and ridiculous irony of the commission, whose work is hailed by the new government as a precondition for the establishment of racial peace in South Africa, is that the opponents of apartheid are viewed on the same par as those who were perpetrators of apartheid. Many Black apartheid resisters from the liberation movements, who may have caused casualties or fatalities in their opposition to the apartheid system, are now called upon to apologize for their "crimes" and request amnesty from the Truth and Reconciliation Commission. Apartheid killers are placed on the same level as guerrilla fighters who courageously fought and sacrificed families and security to advance the liberation struggle. This preposterous scenario has been lauded by many academics and liberal observers in the West, including people like Donald Woods, an associate of Steve Biko, who expressed his amazement that Black people in South Africa were still so forgiving following centuries of colonial violence and oppression.[57]

Again, the Christian ideology has been subtly inserted into the political culture of post-apartheid South Africa, essentially led by white liberals and their Black supporters, in the movement toward reconciliation between Blacks and whites in South Africa. One wonders why it is that when it comes to reconciliation between white oppressors and the Black oppressed, the price is so cheap: mere apologies and applications for amnesty. One also ponders why this process of reconciliation is so devoid of substance that it entails no sacrifice on the part of whites or reparations to Blacks for past acts of dispossession and victimization by colonial violence.

The Truth and Reconciliation Commission signifies a ploy by white oppressors to continue to hold the reins of economic and material power in South Africa, under the pretext of racial reconciliation and harmony and in accordance with the historical principle of reconciliation among "Christians." This folly underscores the manner in which Black people have been duped by Western Christianity and forced to accept humiliation, injustice, and exploitation, all under the auspices of South Africa being a "Christian" country.

It is instructive to note that Jewish victims of the Holocaust in Nazi Germany who escaped the gas chambers are never placed in the position of having to forgive and forget the horrific crimes perpetrated against them, and to this day, four decades after the tragic Holocaust occurred, Nazi criminals are still sought, apprehended from places as far away as Argentina, tried by Israeli courts, and sentenced to long prison terms. Those former Nazi guards and officials, believed to be responsible for atrocities in the concentration camps, are made to pay for their heinous deeds. Kader Asmal, South Africa's minister of education, like many victims of apartheid, has described it as a form of genocide, and in this sense, comparable to Nazism, even though some critics charge that this is "conflating it [apartheid] with the Holocaust."[58] The standards employed in meting out justice to victims of Nazism and genocide ought to also apply to Black people. The move to absolve the perpetrators of apartheid crimes is racist and devoid of any sense of authentic morality, and it violates the tenets of Christian ethical behavior because it refuses to recognize the *genocidal* character of apartheid. These exculpatory actions need to be vigorously challenged in international forums.

Many in the West have been reflecting euphorically on the activities of the Truth and Reconciliation Commission in South Africa, hailing it as Hillary Clinton, First Lady of the United States, put it after a visit to the country in 1997, as being "rooted in the Easter spirit of forgiveness and reconciliation."[59] This entire issue of "forgiveness and reconciliation" needs to be critically scrutinized, particularly when one considers, for instance, the fact that the murderers of Steve Biko came forward on February 6, 1997, and confessed that they participated in Biko's assassination. Not only did they beat him savagely, it was revealed, but they may have also poisoned him. Ntsiki Biko, the widow of the Black Consciousness Movement leader, was understandably devastated when she learned that those who had killed her husband on September 12, 1977, were planning to request amnesty for their heinous crimes. She, together with the Azanian People's Organization, has legally challenged the credibility of the Truth and Reconciliation Commission and its ability to arbitrate the principles of justice, given the provisions that allow for criminals who committed wide-scale murders of anti-apartheid activists to merely publicly confess and consequently become eligible for governmental pardon.

The Truth and Reconciliation Commission heard testimonies from hundreds of Black people who had been victimized by apartheid police and security force members. The commission appeared to dwell more on some vacuous form of "reconciliation" than on any emphasis on "truth." It appeared that it desired to hear the emotional ventilation of Black people's grief, horrific tales of torture, rape, murder, assassination, and brutality, then determining that a resolution to these painful accounts of victimization could be accomplished by offering indemnity to those guilty of apartheid crimes so long as they confessed their guilt and were remorseful. Some families were offered forms of monetary compensation.

The absurdity of this administration of justice meted out to Black people in South Africa was highlighted when Ntsiki Biko retorted in disbelief when Desmond Tutu asked her and other victims of apartheid violence to forgive those who had perpetrated atrocities against their families: "There is a lot of talk about reconciliation, but I don't know who is supposed to be reconciled with whom. What I want is the proper course of justice."[60]

Ntsiki Biko expresses the frustration of many Black people in South Africa who feel that Black life has been horribly cheapened. The maximum that victims' families can receive as compensation from the state is R15,000 ($2,500).[61] In many instances, the compensation monies are essentially used by destitute families to provide decent burials for victims of apartheid violence, as in the case of Angeline Nonkonela, who died in township unrest in 1992, for instance.[62]

Why must Black people be urged to forgive and forget the tragedy of the loss of the lives of their loved ones, when the Jewish victims of the Nazi holocaust during World War II are never asked to do likewise? Is Black life worth less than white life? The constitutional legitimacy of the Truth and Reconciliation Commission is currently being challenged by the families of those assassinated for their resistance to apartheid: Steve Biko, Victoria Mxenge, and the Ribeiros. One can understand the intense anger of Ntsiki Biko and hundreds of Black mothers and fathers, sisters, and brothers in South Africa, who saw loved ones disappear without trace or discovered charred remains of their burned bodies, for their resistance to apartheid oppression. There is no more pride in stating that "[t]he generosity of spirit displayed by many—most—of the victims has yet to be matched by a similar spirit on the part of the perpetrators," as Kader Asmal, Louise Asmal, and Ronald Suresh Roberts have asserted.[63] Black people are unable to demonstrate generosity with their lives after 300 years of colonialism—because they have nothing more to give.

The question that warrants raising is the double standard employed in the administration of justice for Black people. In the West, all people who are convicted of committing crimes are punished with the full weight of legislative law. Pardon is certainly denied to those who perpetrate capital crimes as part of the West's general observation of justice, albeit flawed in many instances. However, when it comes to Black people in South Africa who have been victims of the horrific crimes of brutality and murder by apartheid's police, justice is conveniently abrogated. Instead, Black families are asked to forgive those who stamped out the lives of their loved ones and to forget the sadistic apartheid past, as part of the exigency to construct a new nonracial and peaceful society. Black people are asked to forgive the killers of their family members and accept the prospect that these criminals go scot-free. Black people are being asked to sacrifice their dignity once again, as they did under apartheid, this time to turn a blind eye to the need for justice even though they have historically been victims of injustice.

Now that Black people in South Africa have some latitude to make the culprits who committed crimes of inhumanity against Black opponents of the oppressive apartheid system accountable, the Truth and Reconciliation Commission has stepped in to arbitrate its form of justice. The commission has been hailed by many whites as the epitome of the conciliatory spirit of humanity. Somehow it appears from such sentiments that many whites feel that Black people must be forgiving and conciliatory when it comes to injuries and atrocities against Blacks, in order to "keep the peace." Yet if most of the victims of apartheid crimes had been white and the perpetrators Black, white families today would hardly abdicate their demand for justice and permit the offenders to go free. The white world was quite unwilling to grant clemency to O. J. Simpson in the United States following the guilty verdict against him in the recent wrongful death civil trial, even after a jury found him "not guilty" in a criminal trial.

The problem with the notion of "truth and reconciliation" as propounded by the authors of the commission with the same name, including Archbishop Desmond Tutu, is that it does not embody *justice* and subscribes to a convoluted and perverse sense of justice and truth. The commission is the result of the negotiated settlement to the end of white minority rule in South Africa, where the African National Congress and the then South African ruling Nationalist Party injudiciously agreed to absolve those who had committed apartheid crimes, as well as those who may have been convicted because of attempting to destroy apartheid. The moral logic deployed to place the victims of apartheid on the same par with those who were instruments of apartheid is the product of opiate Christian ideology.

A further problem with the Truth and Reconciliation Commission is that it sets a racist standard for Black humanity, assuming that somehow Black people are immune to the pathology of injustice and can be made to feel content with hearing the truth, without subsequently anticipating justice as many other peoples do. For five centuries of colonization and slavery, Black people have been forced to sacrifice their cultures, lives, and dignities for the material enrichment of the Western world. Liseka Mda, a Black professional in South Africa, concurs with the assessment that the Truth and Reconciliation Commission is a sham. She avers:

> Three hundred and forty-four years after the first white settlers landed in South Africa, it is the indigenous population that wallows in poverty, completely destroyed as a people in life, limb and mind by a colossal crime against humanity. And to add insult to injury, this genocide is being peddled as a minor mistake, for which no one can be blamed. Not really.
>   That is the truth.
>   It is the truth that we find ourselves in this position because sworn enemies sat around the table and knocked together a few conditions for doing business together.[64]

Liseka Mda's angry convictions are shared by many Black South Africans. For all who observe the changing context of Africa, it will be mindful to realize that there can never be veritable truth or reconciliation without justice, not in South Africa, nor anywhere else in the world.

It is crucial that an indigenous African working-class-based spirituality must pervade the South African landscape and erase the devastating ideological effects of four centuries of Western European colonialism and capitalism in the country. Black people have been brainwashed and whitewashed by the manipulation of the Christian ideology posing as "religion," resulting in a form of Black "mentacide" where Black people view the well-being of their white oppressors as being more important for a viable future than their own welfare and tacitly accept the disparagement of "Africanity." The religious roots of Western Christianity and its legacy of pacifism and coexistence with structures of oppression, specifically colonialism and capitalism, must be uprooted and supplanted by a revolutionary spirituality that views all modals of oppression as demonic and defends the right of all creation, particularly working-class people, to be free by "any means necessary."

The conceptualization of divinity here is framed in the mode of a Creator whose creative power is observed in the acts of resistance of African ancestors who waged wars to stave off colonial invaders and in a society where sacred ancestral and other spiritual ceremonies function to instill a sense of power among working-class African men and women to the point that fear of terror and death is exorcised by the spirit of resistance that emanates from the Creator Spirit, the Ultimate Being of Liberation. Revolutionary African working-class spirituality also places a check on brash self-seeking egotistical inclinations, a pursuit that has enveloped many independence leaders during the colonial era and in this neocolonialist period, such figures as Nelson Mandela of South Africa, the late Sekou Toure of Guinea, and, most recently, Laurent Kabila of the Congo. It is moored in a cosmology that affirms the materialist proportions of reality while rejecting the materialistic preoccupations of capitalist civilization and culture. It is thus simple in that it is not ostentatious, and it is compassionate in that it elevates the interests of the neediest sector of society, Black workers, as the subject of primordial concern. It views materiality as utilitarian, for the enhanced well-being of humanity and all creation and repudiates the notion of materialism as an end in itself, as in capitalist culture. It does not perceive commodity or materialistic accumulation as the raison d'être of life, as does capitalist religion. Neither is it individualistically defined, as under capitalist rubrics, but always collectively characterized, viewing communal interest as intrinsic with and coterminal with that of the individual. It does not view money as an *idol,* as is evident in capitalist society, but understands its mundane utility and its limitations in providing ultimate personal and social fulfillment.

Perhaps the assertions about this indigenous African spirituality and its resilience and tenacity are best borne out by the fact that most peasant folks on the African continent, particularly those in the rural areas, survive on the meager resources that they possess, yet always collectively distribute even these paltry possessions among the neediest. Most African peasant women and men do not have capital financing or vast sums of money; they survive and struggle to survive through an indigenous religioculture that embraces simplicity, moral responsibility, respect for the individual and for the natural world; that educates the young through the mediums of age-old ancestral wisdom; and that prepares children for survival for future generations. Human nature is not innately selfish, as capitalism would have us believe, based on Adam Smith's notion of individual selfishness being at the heart of human culture. On the contrary, Africans, like other indigenous colored peoples, epitomize the collective ethic, predicated on an ancestral spirituality that shares, does not hoard for self, and respects creation as opposed to commodifying it for extractionist purposes. Malidoma Soma, an African spiritual leader from Burkina Faso, renders an illuminating version of this powerful spiritual world in his classic, *Of Water and the Spirit,*[65] where he biographically describes the role that his indigenous Dagara ancestors and spiritual leaders play in personal transformation and social harmonization.

It is this very spirituality that needs to become the basis for revolutionary warfare that is geared toward the repossession and reclamation of the land of Azania, which was confiscated and is now held by the European settlers in the country. Land signifies the basis of indigenous peoples' existence, in South Africa, Africa, North America, and across the indigenous world. It denotes sacred space, where the ancestors are buried, and unites all living things with Mother Earth. It connotes a living Spirit, with whom all life must walk in harmony. The Sisala community of southwestern

Ghana believe in the divinity of both the Sky and Earth, perceiving that Sky is more intolerable of human violations and that Earth, "though powerful in its own right, is more understanding of human frailties and less dreadful."[66] Mother Earth assumes such supreme significance in Sisala religioculture that an earth priest functions as a mediator between the people and Mother Earth.

Symbolism of the Earth and Sky are two basic determinants for fertility in agriculture. Edward Tengan notes:

> It is thus symptomatic that the highest authority in the village, the *totina*, be vested with the authority of overseeing that such conditions (both moral and ecological) prevail within the village so it would not impair the relationship between the Earth and the Sky. Thus, harmony within the village at all levels, which both generates and is generated by good rains from the sky above and a good yield from the earth below, is ultimately the fruit of a healthy communication between the Earth and the Sky.[67]

Land is not merely a material possession that can be alienated by conquest as was the case under European colonialism, hence the resistance of indigenous peoples in Africa, the Americas, the Middle East, Asia, and the Pacific to European annexing and expropriation of lands. Land reform does not need to be defined between "moral-emotional" (emphasizing a historical injustice or "rational pragmatic" (emphasizing economic and market forces) as Themba Sono suggests.[68] The land-reform and redistribution program so vitally necessary in South Africa's colonial context is a moral and spiritual vocation, because returning lands will reconnect all Africans to their historical ancestors. It is a rational and sensible economic policy because for the first time since the entrenchment of colonialism, it will provide indigenous people with the economic wherewithal to establish viable and self-sustaining agrarian-based economies.

Neither is land a piece of real estate, as European capitalism unabashedly assumes; in the African and indigenous worldviews, land is a spiritual entity that must be preserved and kept in harmonious accord with the spirit and ancestral world. The Gikuyu understanding of the need for reestablishing an African religiocultural cosmology in the launching of political and economic struggle, as was evident in the Mau Mau uprising in Kenya (yet which has unfortunately been gradually undermined by neocolonial elitists), may also have currency in South Africa's liberation struggle. Jomo Kenyatta described this concept of religiocultural and socioeconomic fusion:

> Land tenure is the most important factor in the social, political, religious, and economic life of the tribe. As agriculturalists, the Gikuyu depend entirely on land. It supplies them with the material needs of life, through which spiritual and mental contentment is received. Communion with the ancestral spirits is perpetuated through contact with the soil in which the ancestors of the tribe lie buried. The Gikuyu consider the earth as the "mother" of the tribe, for the reason that the mother bears her burden for about eight or nine moons while the child is in her womb, and then for a short period of suckling. But

it is the soil that feeds the child through lifetime; and again after death it is the soil that nurses the spirits of the dead after eternity. Thus the earth is the most sacred thing above all that dwell in or on it. Among the Gikuyu the soil is especially honored, and an ever-lasting oath is to swear by the earth *(koirugo)*.[69]

The Bantustan reservation system, which forcibly dislocated and relocated Black people from traditional ancestral lands since 1913, just like the reservation system that confines Native peoples in North America, was not merely a social and personal affront and physical injury; it marked a spiritual decapitation of sacred ties to the land of the ancestors. Any movement for change in South Africa must take this historical issue into account and ensure return of confiscated lands to the indigenous people as a sine qua non for the establishment of veritable Black nationhood.

Following the paradigm of *chimurenga* (revolutionary struggle) in Zimbabwe, Black working women, men, and youth, need to invoke their ancestral spirits in the summoning to assume the cudgels of struggle as demanded by the liberation movement. In Zimbabwe, guerrilla fighters appealed to spirit mediums to sustain them in the fierce bush war waged against the Smith regime, considering shortages of food and aggressive military incursions by the Rhodesian armed forces. Thousands lost their lives and were injured in the fourteen-year independence war. Yet the indigenous people of Zimbabwe rallied around the cry of return of ancestral lands and defeat of the enemy settlers, and they prevailed.[70] The Zimbabwean fighters and their allies believed that their ancestral spirits would only rest when their land was free.

In Azania, the whites are entrenched in the land and view it as rightfully theirs. They are intransigent as settler-colonialists and have to be forced into giving back what they have plundered and pillaged. The struggle to repossess and return these stolen lands will be a long and arduous one, possibly involving wide-scale guerrilla warfare and escalated armed struggle by an organized and *conscientized* Black rural peasantry and urban working class, a struggle that could be drawn out for decades. The fermentation of revolutionary religiocultural consciousness that rebukes every oppressive design and structure, particularly land dispossession by white settlers, is a process that will take years and is something for which Black working-class people are currently unprepared.

Ultimately, it is the mobilization of hundreds of thousands of Black working-class cadres from the cities, towns, and villages, organized by revolutionary organizations like the Socialist Party of Azania and the Azanian Congress of Trade Unions (something that is occurring presently, often involving frustrated guerrilla fighters from the ANC's military wing, Umkhonto We Sizwe) that will spark the movement toward land reclamation.[71] Fearlessness of death is inspired by the revolutionary spirit of the ancestors, like Bambhata, Shaka, and Cetshwayo, who gave their lives fighting for preservation of Azanian lands against white invaders.

For Black Christians, it is the revolutionary life of Jesus, the Ancestor, par excellence, that serves to capture the principles of fearlessness, courage, and determination to wage armed struggle for the return of Africa's lands. White settler-colonialists and capitalists and their Black surrogates will not surrender without a protracted and massive "fight to the death." The forces of oppression will be forced to surrender to the forces of liberation, principally because the oppressed are fighting for the realization of the truth, and history is on the side of the oppressed, a point that may take half a century to establish but which ineluctably will occur. The Azanian working class has to accept this eventuality of revolutionary warfare and upheaval, inasmuch as such processes are so painful and traumatic. *The only real solution to the bestiality of neocolonialist capitalist oppression in South Africa is also the most difficult one: revolution.* The days of a deferential "yes-*baas*/no-*baas*" culture will come to pass, and the mental and psychological enslavement that Black workers suffer under the yoke of white power will be broken. Its rupture will only materialize by cultural and armed revolution, and Black workers need to accept and prepare for this eventuality. At that stage of realization, whenever that does occur in the near or distant future, indigenous African and revolutionary socialist principles will paradoxically coincide in that both necessitate preparation for and involvement in this lifelong struggle for the return of Africa's stolen soil and the restoration of Azania to its rightful owners.

There can be no authentic material reconciliation between Blacks and whites in South Africa/Azania until veritable justice is *practiced* in that country. Authentic justice entails that the land stolen by white settlers be returned to the African people unconditionally and then redistributed according to nonracial principles. Whites in South Africa/Azania, or in any part of Africa for that matter, can no longer shield themselves behind the cloak of privilege and continue to monopolize ownership of South Africa's land, mines, and natural resources, while expecting that this position of domination and immoral prerogative of ownership will go unchallenged. If people of European descent, or Asian descent, for that matter, intend to live in peace in Africa, they must be willing to live under authentic African rule and expect to be accorded egalitarian treatment on the same par with indigenous Africans and not anticipate privileged socioeconomic status because of their heritage, descent, and culture. They must accept the cultures of African people as definitive of life of the people on the continent, just as Africans who immigrate to Europe are expected to accept and live according to the dictates of European rule.

There is no place for racial privilege on any grounds in Africa, or anywhere else in the world. It is only when these conditions exist that whites in Africa will come to be accepted as authentic residents of Africa and nonracialism will be realized. After all, "race" is a social construction of European colonial anthropology and has no scientific basis on grounds of biology or genetics. Racism does exist as a result of the ideology of racism propagated by the European colonizers who perpetrated genocide of indigenous peoples of the Western Hemisphere and of Africans through the horrific slave trade. Race must be deconstructed by all academics and activists of goodwill so that we can all realize that there is only *one race—the human race—*and our phenotypical differences are incidental to our definitions as human beings, just as the color of a person's eyes has absolutely nothing to do with his or her intelligence or abilities.[72] Ultimately, a revolutionized and liberated Africa will be characterized by a raceless society.

Language Policy and Education in Post-Apartheid South Africa

The question of language in educational policy, as rudimentary as it may seem, is nevertheless an extremely controversial issue in post-apartheid South Africa, particularly in the area of education. Given South Africa's entrenched colonial history and the neocolonialist orientation of the country's present leaders, this issue portends to become one of the most explosive in transformation discourse because it will determine the future cultural direction of the nation, as was the case in the rest of Africa. David Birmingham, a British historian, confirms the centrality of language in the educational process and the manner in which it was used to subvert Africa's indigenous cultural base when he writes:

> One of the penetrating cultural transformations brought to Africa was the acceptance of colonial languages as the languages of administration and justice in most of Africa's successor states. In a few emerging republics the ruling classes even preferred French, Portuguese or English in their social and political discourse and allowed their children to grow up ignorant of traditional vernaculars. Education, especially higher education, remained predominantly in the European mould, with European textbooks and teachers.[73]

Post-apartheid South Africa appears headed in this direction of cultural catastrophe, as extant in most of postindependent Africa, hence the focus on this subject. Principally, the South African cultural and socioeducational conflict in this regard centers around whether English and Afrikaans will continue to be hegemonic in scope as under apartheid, with indigenous languages continuing to function in a background capacity confined to common social usage by Africans, or whether indigenous languages will become institutionalized within educational, political, and economic circles, so that indigenous language materials will be developed in a range of spheres to establish a truly multilingual culture, rooted in the African ethos. Nhlanhla Maake, a South African linguistic scholar, underscores the magnitude of this historical legacy of cultural and linguistic hegemony in the context of literature:

> The comparison of the qualitative growth of the literature written in African languages with English and Afrikaans literature is an often reiterated tune, with subtle but somehow overt references to its inferiority, if not implying an inability of the speakers of these languages to produce fine literature.
>
> Such comparisons deliberately or inadvertently lose sight of the various factors which were responsible for shaping such a context. The social, cultural, political and economic conditions for Afrikaans and English literature have always been and *still are* different from those which the African language literatures developed. The latter literatures have been under siege since their birth, and their state and status in the corpus of creative writing is a testimony to the consequence of the well-meaning but stultifying control and censorship of the missionaries, and also censorship by the South African publishing houses in collusion with government policy. (italics mine)[74]

What becomes patently clear in the illumination of the linguistic and cultural ethos of Azania is that the tenacity of European colonialism is formidable, to the point of now attempting to convince us that although African languages are necessary within the new political dispensation, we need to accept the normalcy of the supremacy of both English primarily and Afrikaans secondarily. Such proponents, including some so-called linguistic experts, claim that South Africa is entering a new era of the industrial and technological boom and, in order to be economically competitive, must retain its European languages as the national languages, particularly English. Abram Mawasha argues:

> Education and training as we understand them today mean largely Western-style education. The language of this education is English. These two factors define an educated Black South African as one based in Africa, but educationally Westernized and with English as an important means of expressing such education. This could well confuse and confound the idea of being educated with whiteman-ness.[75]

Nonetheless, there is already the talk about "education going to the dogs" and widespread fear of progressive lowering of standards with any suggested move toward the Africanization of education and society in Azania. This type of racist ideology is akin to the point made by Nahishakiye Leocadie, a linguistics researcher from Burundi, who explained that the government of Burundi was opposed to or reluctant to embrace a policy that expanded the use of the indigenous language, Rundi, in schools as a means of replacing the colonial French language because such practices would set the country back and lower academic standards.[76] It is a stark and disturbing fact that in post-apartheid South Africa, as in "independent" Burundi, the inveterate distorted view that equates excellence and intelligence with Europe and European people and backwardness and fatuity with Africa and Africans still prevails on subtle and overt levels, severely infecting our educational systems in the process.

From the vantage point of clear-thinking African linguistic and educational experts, though, this ideology of white supremacy and standard of human normatization in Africa must be broken, and Azania could very well be the place where this cultural-linguistic conflict and contradiction will resolve itself. In an address at Rhodes University in Grahamstown in 1990, P. T. Mtuze, a linguistics scholar at the University of South Africa, stated:

> No, we can no longer allow circumstances, under the cloak of pragmatism or any kind of justification, to recolonize our minds. African languages, as some academics have contended vociferously, and so rightly, must be promoted at all costs to play their role in the new South Africa. They represent the spirit of Africa, the culture of the African soil and the hope of a re-born South Africa, if this circle from freedom through serfdom and back to freedom is to be complete. They are our last shield against cultural conquest . . .
>
> The struggle is not only political. It has never been only political. Nor has it been purely linguistic. But it has always been underpinned by cultural considerations and

complicated by psychological factors. A new South Africa must be seen to be part of the continent of Africa—free from all forms of colonialism, antiquated or modern, overt or covert.[77]

Mtuze underscores the rationalization for the infusion of African linguistics teaching, training, and use in the task of reestablishing the African identity and restoration of self-esteem of a people who have been stripped of their dignity by the ferocity and rapaciousness of European settler-colonialism and capitalism:

> Local languages, unlike the *lingua franca*, protect the weak from the strong; they give the mother-tongue speaker her first relationships to mother earth and the world; like the vegetation and topography, they are peculiar to the place. They give a person identity and a proper pride in her herself and her heritage, on a scale small enough to comprehend. They confer a diversity and an independence of thought and vision in the unwielding conglomerate states . . . That is why they must be promoted and encouraged . . . so that children in schools will have pride and not shame in the culture of their own family and community. And it is in the University that the creative forces are generalized and fostered to ensure that the schools will indeed promote that pride.[78]

Neville Alexander, an important leader and articulator of Black perspectives in the discussion of linguistic policy in post-apartheid South Africa through organizations like the National Language Project (since 1985), also reiterates the need for the serious reassertion and adoption of indigenous African languages in the evolution of a new sociocultural and economic dispensation in South Africa. He views linguistic transformation as axiological in the defeat of ethnicization and fragmentation as sown by the seeds of European colonialism and as a means of breaking the ideological economic stranglehold of the white ruling class. He opines: "Clearly the oppressed people have to forge weapons out of the same materials (language, nation and culture that were used by the colonizers) so that they can defend themselves and break the domination of the ruling group."[79]

Alexander suggests that our very notions of ethnicity, language, and culture need to be redefined in the discourse on language policy since so much of our thought process has been vitiated by the perversity and designs of European missionaries and colonizers: "Many of the ethnic divisions that are today a concrete reality did not exist, even in a conceptual form before the end of the nineteenth century."[80]

As an example, he points out that the division between the Ronga and the Gwamka among the Tonga-speaking people was a result of the differences between the Spelonken and Coastal branches of the Swiss mission and that the Tsonga is a "classic instance of ethnic differences whose roots may be traced to an obscure linguistic debate between two Swiss missionaries."[81] He insightfully reminds us that African history has been written by the European conquerors who rule and that languages are always products of such conflict between the powerful and the powerless: "The fate of languages is decided in the course of class struggles in which the actual linguistic elements are seldom pertinent at all."[82]

Based on this analysis, Alexander declares the objective of language policy in the post-apartheid South Africa to be the following:

> Our main goal in the sphere of language policy in this period up to a liberated post-apartheid South Africa/Azania must be to facilitate communication between the different language groups that comprise the population of South Africa, in order to counteract the isolating effects of Verwoedian apartheid language policy. At the same time, this means that we have to encourage multilingualism among our people. At the very least, people ought to know their home language and English, but the ideal situation would be one in which every person in South Africa would be able to speak fluently his or her home language, English, and one or more other regionally important languages.[83]

The apprehension about ethnic domination by one linguistic community over another in an authentically liberated Azania needs to be addressed forthrightly for its racist and pejorative projections. Nhlanhla Maake incisively contends in this regard:

> Contrary to the views of some outsiders, who can neither understand nor speak any South African Bantu language, it is evident that the Nguni languages, either in the form of Zulu or Xhosa, are becoming more widespread and generally acceptable as the main vehicle of communication among Africans, without any implication of political domination by those who speak those languages as their first language. This is evident to anyone who can speak South African Bantu languages and has traveled in South Africa. It is particularly true in the Southern Transvaal, where as a result of more than one generation of inter-marriage among African mother tongue language groups, it is rare to find a monolingual umZulu, umXhosa, moSotho, etc.[84]

It needs to be emphasized that the subject of linguistic transition from European languages to African lingua francas in post-apartheid South Africa does not merely suggest a mere infusion of "African wine into European wineskins." On the contrary, it implies a revamping of the entire colonial-capitalist system and the system's use of language in general, from one of serving the white ruling class and the tiny Black elite to one that empowers working-class Black men and women. The Soweto uprising of 1976 in which thousands of students marched against the imposition of Afrikaans as the principal language of instruction in African schools signified both a rejection of Afrikaans, symbol of white supremacy and settler-colonialism in Southern Africa, and the Eurocentric educational system that relegated Africans to a position of subservience and inferiority. The voices of resistance and critique echoed by Soweto evoked a different basis to language itself, where language would be dynamically used in a manner that facilitated the socioeconomic development of Black working-class communities and did not reproduce the pathologies of a racist and exploitative capitalist society.

The task of the Black working class that speaks a variety of Nguni and Sotho languages in South Africa as a part of the wide-scale proletarianization of Black people under capitalism is thus the massive overhauling of the country's sociocultural, political, and economic system. This revamping would relocate South African society, shifting it from a condition of Anglo-Dutch settler-colonialism and monopoly capitalism to one of African socialistic working-class egalitarianism, predicated on the independent evolution of the African sociocultural ethos.

South Africans/Azanians need to initiate a cultural transformation that is Africanist in character and working-class based and biased. In this process, indigenous African languages would become reestablished and reinstituted in a large-scale educational program of the post-apartheid government. At this stage in South Africa's evolution and transformation, it is necessary to consolidate internal African forces and dynamics, and indigenous African languages and cultures need to be enforced at all educational, social, and political levels. We need to sever the tentacles of Western "aid" that hold us in perpetual economic servitude to global capitalism and expedite the process of reorienting the post-apartheid government introspectively toward Africa, so that it will institute policies like creating an integrated Southern African economic community with a singular currency, instead of the recent South African–European Community Trade Agreement that undermines the economies of Southern Africa.[85] C. M. Rubagumya, a Tanzanian linguistics scholar, argues persuasively for us to reject this pattern of linguistic colonialism, which has implications for the manner and direction in which African societies and economies evolve:

> The "long term" benefit that Africans are supposed to get from English and French are questionable. The unequal power relationship derived from the imposition of "English" on "developing countries" is well demonstrated by Tollefson (1991). In his critique of the concept of English as a tool of technological development for "Third World" countries, Tollefson makes two important points. First, the spread of English helps to maintain the existing unequal power relations between the rich and poor countries of the world. This is because in order to get access to scientific and technological information available in English, "developing" countries depend on developed countries for support institutions and facilities such as research centers and computer hardware and software. This "aid" is on many occasions used for ideological purposes by rich countries to influence events in the poor countries.[86]

We in South Africa and Africa must cultivate our indigenous cultural and linguistic resources through all educational institutions and mediums and painfully realize that participation in the global economy requires selling Black people to the capitalist system that terrorizes the world, particularly people of color. We need to inculcate a sense of national self-appreciation and indigenous cultural embracement rooted in a working-class foundation in South Africa and the rest of Africa first and foremost, as a precondition for engaging with the world outside Africa.

Both the Chinese and the Japanese have done tremendously well in this area, without surrendering pride in their languages, cultures, and heritages. In fact, they have also been successful in some economic areas, depending on the criteria that one employs. China continues to engage in global commerce and geopolitics. It uses Chinese as its lingua franca and has developed the Chinese language in every sphere of science and technology. When it enters the arena of international relations, it resorts to interpreters. Africa can do the same, notwithstanding our linguistic diversity. Multilingualism in Africa should not be used as the principal reason for adopting European languages (essentially the language of our colonizers) as the lingua franca on the continent. Many advocates claim that the reason for maintaining the supremacy of European languages in Africa is to advance economically and technically. In actuality, much of what we have seen of the expanded use of these languages is the enrichment of whites and the paltry class of African elites who use these languages as educational weapons to keep the African working classes subservient and ignorant, at the expense of the downgrading of indigenous African languages. Speaking from the Nigerian context, Olasope Oyelaran cogently asserts in this regard:

> It is my considered opinion that the vast majority of Nigerians are excluded in the primary sense from participating in the productive life of the nation. A minority—just because they are the minority can read or write the English language—see themselves as the anointed heirs to the imperial masters whose purpose they (the minority) . . . served effectively, in order to strip Nigerians of their "meanings, values, and activity" in those undertakings geared to the promotion of the Nigerian's own perceived well-being. This act of exclusion serves to liquidate the nation's well-being in order to enrich the imperial metropolis.[87]

We can no longer ignore the destructive and debilitating effects of colonial languages in our deliberation of contemporary language policies in Africa. The contention for the primacy of indigenous African languages does not eliminate the need for offerings of European languages since European languages are still the global lingua franca. Neither does it imply a disregard or rejection of those communities that use European languages as a primary language. Nor does it suggest a "totalitarian" imposition of African languages. However, it is important for all African students to learn to speak and read these languages as *secondary* languages, so that Africans are able to understand and communicate with developments on the global stage. In essence, the bilingualism of Africans normatively speaking English and an indigenous language (which is currently characteristic of those in urban middle-class settings, predominantly) becomes a province of operation for the masses. The significant difference, however, is that Africans will, under these circumstances, use African languages as the primary means of educational, political, and economic communication, with European languages relegated to the arena of international relations, as opposed to the current situation that generally reflects the inverse.

Culturally and cosmologically, Africans and other indigenous peoples have always embraced diversity as healthy for their societies. Cultural difference does not by virtue of its character imply chauvinism, hierarchization, discrimination, or disparagement of one tradition over another, as one observes regularly within European colonial culture; rather, it suggests strength through variety. It is precisely for this reason that over 1,000 languages are still spoken throughout Africa today, after millennia of intercultural interaction that encourages diversity and cultural pluralism.

This debate is necessary and healthy so that the emergent policy of the post-apartheid government realizes the magnitude of the task of decolonization and deconstruction, coupled with the need to address the needs of the struggling Black working class.

Summary and Conclusion

Pan-Africanism is the necessary ideological framework to advance the Azanian and continental African revolutionary struggles, but it must refuse to repeat the mistakes of Pan-Africanist history while building on its strengths. In the final chapter, based on the critical observations made herein concerning neocolonialism in South Africa and the rest of Africa, the strengths and limitations of the Black working-class movement will be examined for their potential for revolutionary transformation, with principal emphasis on a sector that has historically been disregarded by many radical theoreticians of struggle: indigenous women.

Notes

1. Abdul Said, *The African Phenomenon* (Boston: Allyn and Bacon, 1968), 110. Said provides a good exposition on the development of Pan-Africanist ideology and the struggles for African unity in the 1960s.

2. Amy Jacques-Garvey, the second wife of Marcus Garvey, was one of the noted women leaders in the Back to Africa Movement led by Marcus Garvey in the 1920s. The point about her influence that has hitherto gone unrecognized is made in an article by Karen Adler entitled "Always Leading Our Men in Service and Sacrifice," in Esther Ngan-Ling Chow, Doris Wilkinson, and Maxine Baca Zinn, eds., *Race, Class, and Gender: Common Bonds, Different Voices* (Thousand Oaks, Calif.: Sage Publications, 1996).

3. See the remarks of Makini Roy-Campbell, chairperson of the African Women Liberation Organization, and Tajujdeen Abdul Raheem, general secretary of the Global Pan African Movement, in *National Seminar Series: Women Organizing for the Future*, Ranche House College, August 7–9, 1995, Harare, Zimbabwe, discussing the historic founding of the new organization.

4. This is part of a summarization of one of Pan-Africanism's distinctive political objectives, described in Colin Legum's *Pan Africanism: A Short Political Guide*, rev. ed. (New York: Frederick A. Praeger, 1965), 38. The reason the word "tribe" and the notion of "tribalism" is rejected as an arrogant and racist assumption is that the word "tribe" was conceived by European colonial anthropology in its attempt to categorize and control the world of indigenous peoples by disparagingly segmenting the world's majority into supposedly "uncivilized" tribes in contradistinction to superior "civilized" Western European nations. This fallacious mythology is destructive of all indigenous peoples and needs to be eviscerated from the entrails of all literature.

5. George Padmore, *Pan Africanism or Communism? The Coming Struggle for Africa* (London: Dennis Dobson, 1956), 379.

6. This notion of a common African heritage and identity of all people of African descent around the globe, particularly in Africa, the United States, and the Caribbean, functions as the basis for the waging of political struggles for self-determination within the ambit of the

Black world. Regardless of where they live, Blacks are united by virtue of historical oppression and the desire to be liberated. These ideas are expressed through the proliferation of writers, thinkers, and activists in the Black world in Fred Horde and Jonathan Lee, eds., *I Am Because We Are: Essential Readings in African Philosophy* (Amherst: University of Massachusetts Press, 1995).

7. Ron Walters, *Pan Africanism in the African Diaspora* (Detroit: Wayne State University Press, 1993), 95.

8. Jon Woronoff, *Organizing African Unity* (Metuchen, N.J.: Scarecrow Press, 1970), 23.

9. See, for instance, John Markakis, "Pan Africanism: The Idea and the Movement," Ph.D. diss., Columbia University, 1965, 168.

10. Cited in John Hendrik Clarke, ed., *Pan Africanism and the Liberation of Southern Africa: A Tribute to W.E.B. Du Bois* (New York: African Heritage Studies Association, 1978), 15–16.

11. *History of the Pan African Congress* (London: Hammersmith Bookshop, 1963), v.

12. Ibid., 6.

13. B. Chango Machyo, *Pan African Makerere Adult Study Center Pamphlets, No. 3,* Makerere University College, Kampala, Uganda, 1968, 10–11.

14. Elenga M'buyinga, *Pan Africanism or Neo-Colonialism: The Bankruptcy of the O.A.U.* (London: Zed Press, 1982), 34.

15. Kwame Nkrumah, *Africa Must Unite* (New York: International Publishers, 1963), 135.

16. *History of the Pan African Congress,* 58.

17. Ibid.

18. See, for instance, the excellent article by Michael Crowder, "Whose Dream Was It Anyway?" in Timothy Welliver, ed., *African Nationalism and Independence,* Vol. 3 (New York and London: Garland Publishing, 1993), which argues that any assessment of the postindependence period in Africa must be placed squarely on the shoulders of European colonizers, who trained and shaped Africa's leaders in this era, and that the "criteria by which Africa is being judged are Eurocentric ones" (p. 24).

19. Basil Davidson, "For a Politics of Restitution," in Adebayo Adedeji, ed., *Africa Within the World: Beyond Dispossession and Dependence* (London and Atlantic Highlands, N.J.: Zed Books, in association with the African Centre for Development and Strategic Studies [ACDESS], 1993), 24–25.

20. Alex Quaison-Sackey, *Africa Unbound: Reflections of an African Statesman* (New York: Frederick Praeger, 1963), 62. This text provides some important detail in understanding Kwame Nkrumah's persistence in the struggle for African continental unity and illuminates the history of the Pan African Movement in the context of international relations and global geopolitics.

21. Kwame Nkrumah, "The Congo Struggle," in *I Speak of Freedom: A Statement of African Ideology* (New York: Frederick Praeger, 1961), 257.

22. Said, *The African Phenomenon,* 114.

23. Ibid., 115.

24. Thomas Hovet, "African Politics in the United Nations," in Herbert Spiro, ed., *Africa: The Primacy of Politics* (New York: Random House, 1966), 122.

25. Elenga M'buyinga furnishes a comprehensive and cogent treatment of the theme of neocolonialist roots of the Organization of African Unity, rendering it essentially ineffective in realizing Pan-African unity and independence, in *Pan Africanism or Neo-Colonialism.*

26. Jaja A. Wuchuku, "The Relation of AMSAC and the American Negro to Africa and Pan-Africanism," in American Society on African Culture, ed., *Pan Africanism Reconsidered* (Berkeley and Los Angeles: University of California Press, 1962), 376.

27. George W. Shepherd Jr., *Anti-Apartheid: Transnational Conflict and Western Policy in the Liberation of South Africa* (Westport, Conn., and London: Greenwood Press, 1977), 11.

28. See, for instance, Kwame Nkrumah's important text *Africa Must Unite* (New York: Frederick Praeger, 1963), where Nkrumah engages in an excellent historical analysis of European colonial political economies in Africa yet proposes the industrialization solution and African unification, without considering that many of the Pan-African leaders like Sekou Toure, Modibo Keita, and Haile Selassie were all preoccupied with self-preservation. These leaders thus failed to establish working-class revolutionary governments, needed as a precondition for African industrialization, which ought to be geared toward the benefit of the working-class and not the privileged few.

29. W. Ofuatey-Kodjoe, "Pan-Africanism in Crisis: The Need for a Redefinition," in *Pan Africanism: New Directions in Strategy* (Lanham, Md.: University Press of America, 1986) 5.

30. Manning Marable, *African and Caribbean Politics* (London: Verso Press, 1987), 149. Marable cites C.L.R. James's criticism of Nkrumah, who although a dynamic Pan-Africanist advocate and ideologue, ignored the working-class people of his native Ghana, precipitating a military coup that was unopposed by most of the Ghanaian people, from whom Nkrumah was quite detached. Marable's book is informative in providing a critique of nationalist movements, which, in the name of national independence and revolutionary struggle, suppress the working classes and deny grassroots democratic participation.

31. It is tragic that such a talented, charismatic, and knowledgeable leader as Kwame Nkrumah, leader of the first country in Africa to be freed from the shackles of European colonialism in 1957, came to such a fate. However, his literary works still have relevance, especially *Consciencism: Philosophy and Ideology for Decolonization* (New York: Monthly Review Press, 1970), which describes the epistemological basis for the decolonization of African nations and the need for intellectual insight in the utility of material force. *Revolutionary Warfare* (New York: International Publishers, 1970) describes the methodology and tactics of armed struggle, written for nations struggling against colonial powers.

32. Both Marable and Ofuatey-Kodjoe delineate this critique of Pan-Africanism and Africanism instructively, a factor about which all Black/African liberation advocates and social critics need to be cognizant. It underscores our central contention that rigorous social and economic analysis and criticism is vital to revolutionary liberation deliberations.

33. M'buyinga, *Pan-Africanism or Neo-Colonialism*, 177. M'Buyinga is provocative yet perspicacious in his critique of the OAU and the role of subterfuge that Pan-Africanism plays in concealing the neocolonial dictators of Africa, who participate in the exploitation of Africa's workers and peasantry, servicing the greed of Western capital. His analysis of the moribund state of the Organization of African Unity and his proposal that "the African working class and poor peasantry take hold of the leadership of the revolutionary movement in each country and in the continent as a whole" (p. 5) is worth noting.

34. Cited by Alphonso Pinkney, *Red, Black, and Green: Black Nationalism in the United States* (Cambridge: Cambridge University Press, 1976), 44. Pinkney explains that although there may be some dispute about the number of Black people who were members of the Universal Negro Improvement Association by scholars, it is obvious that the organization perceived millions in its membership, inclusive of those who were not formal "card-carrying" members but supporters.

35. P. Olisanwuche Esedebe, *Pan Africanism: The Idea and Movement, 1776–1963* (Enugu, Nigeria: Fourth Dimension Publishing, 1980), 79. Esedebe's historical delineation of the evolution of the Pan African Movement is instructive, especially the role that Black clergypersons such as Edward Blyden and James Johnson played in this Black nationalist development.

36. See, for instance, Marcus Garvey's *Philosophy and Opinions of Marcus Garvey,* ed. Amy Jacques-Garvey, Vol. 1 (New York: Universal Publishing, 1923–1925), 68–72.

37. See Wilson Jeremiah Wilson's work, *The Golden Age of Black Nationalism: 1850–1925,* especially chap. 10 (Hamden, Conn.: Archon Books, 1978), where Moses describes these nationalist movements as "bourgeois" expressions, questioning whether even Garvey's ideas were radically new in comparison to the historical Black nationalist tradition represented by W.E.B. Du Bois, Henry Macneil Turner, Martin Delany, Alexander Crummell, and the Ethiopian movement. Rodney Carlisle's book *The Roots of Black Nationalism* (Port Washington, N.Y.: National University Publications, Kennikat Press, 1975), particularly chaps. 12 and 13, provides a reflective summary of these nationalist movements.

38. Frantz Fanon, *The Wretched of the Earth* (New York: Grove Press, 1963), 36–37. For an insightful treatment of this theme of decolonization and the need to address the psychological effects of colonialism on the mind of Africans, see his other seminal work *Towards the African Revolution* (New York: Monthly Review Press, 1967), especially chap. 7, "Decolonization and Independence." Albert Memmi's writings are equally didactic, particularly *The Colonizer and the Colonized* (Boston: Beacon Press, 1965).

39. From a videotape of the *Cultural Boycott,* BBC, 1989.

40. The point of defiance of colonial-state culture is also emphasized by Achmat Dangor from Azania in *New African* (Durban), July 24, 1989.

41. J. Holland and P. Henriot, *Social Analysis: Linking Faith and Justice* (New York: Orbis Books, in collaboration with the Center for Concern, 1983), xiii.

42. Buti Thlagale, "Towards a Black Theology of Labor," Seminar on Black Theology entitled "Revisiting Black Theology," 1983, 29.

43. Steve Biko, *I Write What I Like* (London: Bowerdean Publishing, 1996).

44. Carter G. Woodson, *The Miseducation of the Negro* (Trenton, N.J.: Africa World Press, 1990), xiii.

45. See, for instance, John Jackson's *Introduction to African Civilizations* (New York: Citadel Press, 1970), and W.E.B. Du Bois, *The World and Africa* (New York: International Publishers, 1965), for an illumination of this historical background. Runoko Rashid has also written extensively on this subject in *Kushite case-studies* (Compton, CA: Compton Community College, 1983).

46. See *Mail and Guardian,* August 1–7, 1997. This was confirmed in an interview with Strini and Asha Moodley, leaders and activists in the Black Consciousness Movement, January 2, 1998.

47. Christine Qunta, ed., *Women in Southern Africa* (London and New York: Allison and Busby, in association with Skotaville Publishers, 1987), 13. For an elucidation of the African origins of Greek civilization and Afro-Asiatic influences on classical Greek civilization, see Cheikh Anta Diop, *The Origin of Civilization: Myth or Reality?* (New York: Lawrence Hill, 1974); Martin Bernal, *Black Athena, Volume I: The Fabrication of Ancient Greece 1785–1985* (New Brunswick, N.J.: Rutgers University Press, 1987); *Black Athena, Volume II: The Archaeological and Documentary Evidence* (New Brunswick, N.J.: Rutgers University Press, 1996); and Theophile Obenga, *Ancient Egypt and Black Africa: A Student's Handbook for the Study of Ancient Egypt in Philosophy, Linguistics, and Gender Relations* (London: Karnak House, 1992).

48. Qunta, *Women in Southern Africa,* 13.

49. Frene Ginwala, "The Women Are the Elephant: The Need to Address Gender Oppression," in Susan Bazilli, ed., *Putting Women on the Agenda* (Johannesburg: Ravan Press, 1997), 64.

50. See Paulo Freire's historic work *Pedagogy of the Oppressed* (New York: Continuum, 1970) for a detailed explication of the notion of *conscientization.*

51. Paulo Freire, *The Politics of Education: Culture Power and Liberation* (South Hadley, Mass.: Bergin and Garvey, 1985), 89. Freire discusses the subject of cultural praxis and revolutionary transformation as a result of conscientization of oppressed peoples in the section "Cultural Action and Conscientization."

52. B. Huntley, R. Siegfried, and C. Hunter, *South African Environments into the 21st Century* (Cape Town: Human and Rousseau and Tafelberg Publishers, 1990), 69.

53. Ibid., 38.

54. Terence Wulfsohn and Brocas Walton furnish a detailed description of the agony of Black people living in such deprivation in *Restoring the Land: Environment and Change in Post-Apartheid South Africa*, ed. Mamphela Ramphele, with Chris McDowell (London: Panos Publications, 1991), 106.

55. See Emmanuel Kgomo, "Access to Power: Smoke over Soweto," in Ramphele, *Restoring the Land,* 117–121.

56. *New York Times*, November 28, 1997. A two-section expose on pollution in Asia was featured on the front pages of the *New York Times* on November 28 and November 29, 1997.

57. *National Public Radio*, February 6, 1997.

58. Kader Asmal, Louise Asmal, and Ronald Suresh Roberts, *Reconciliation Through Truth: A Reckoning of Apartheid's Criminal Governance* (Cape Town: David Philip, Oxford: James Currey, New York: St. Martin's Press, 1996), X.

59. *Arizona Daily Star,* March 28, 1997.

60. *World Press Review,* June 1996, 10.

61. This was conveyed in a speech commemorating Steve Biko's Twenty-Second Assassination Anniversary by Lybon Mabasa, president of the recently formed Black Consciousness Movement organization, the Socialist Party of Azania, and a split from the Azanian Peoples' Organization, on September 15, 1998 at the University of Arizona.

62. *Drum,* November 26, 1998.

63. Asmal, Asmal, and Roberts, *Reconciliation Through Truth*, xiv.

64. *Tribute,* December 1996.

65. Malidoma Soma, *Of Water and the Spirit* (New York: Penguin, 1994).

66. Edward Tengan, *The Land as Being and Cosmos: The Institution of the Earth Cult Among the Sisala of Northwestern Ghana* (New York: Peter Lang, 1991), 87.

67. Ibid., 56–57.

68. Themba Sono, *African Perspectives, Selected Works No. 4, The Land Question: Healing the Dispossessed and the Surplus People of Apartheid* (Mmbatho: Institute for African Studies, and Pretoria: Centre for Development Analysis, 1993), 11.

69. Jomo Kenyatta, *Facing Mt. Kenya* (New York: Vintage Books, Random House, 1962), 22.

70. Lawrence Shuma, *A Matter of (In)Justice: Law, State, and the Agrarian Question in Zimbabwe* (Harare, Zimbabwe: SAPES Books, 1997), 1.

71. Patrick Mkhize, general secretary of the Socialist Party of Azania and general secretary of the Azanian Workers Union, based in Kwazulu-Natal, has already spearheaded this initiative, and hundreds of young people are undergoing military training in the region. This fact is known to the authorities and was even publicized on South African public television in mid-1997. The authorities are unable to clamp down on the group because they used wooden weapons and ammunition in military drills, wisely setting no illegal precedent! The author interviewed Mr. Mkhize in Austerville, Kwazulu-Natal, in December 1997.

72. Works such as Frantz Fanon's *Black Skin, White Masks* (New York: Grove Press, 1967); Joel Kovel's *White Racism: A Psychohistory* (New York: Pantheon, 1970); Barry Schwartz and

Robert Dische's *White Racism: Its History, Pathology, and Practice* (New York: Dell Publishing, 1970); and Ruth Frankenberg, ed., *Displacing Whiteness: Essays in Social and Cultural Criticism* (Durham and London: Duke University Press, 1997) are very useful resources in the epistemological deconstruction of white racism.

73. David Birmingham, *The Decolonization of Africa* (London: UCL Press, 1995), 7.

74. Nhlanhla Maake, "A Survey of Trends in the Development of African Language Literatures in South Africa: With Specific Reference to Written Southern Sotho Literature c. 1900–1970," *African Languages and Cultures*, Vol. 5, No. 2, 1992, 158–159.

75. Abram Mawasha, "The Problem of English as a Second Language as a Medium of Education in Black Schools in South Africa," in *Language: Planning and Medium in Education*, papers presented at the Fifth Annual Conference of the South African Applied Linguistics Association, University of Cape Town, October 9–11, 1986, 114.

76. Nahishakiye Leocadie, "Language Planning in Burundi: Policies and Implementation," Ph.D. diss., University of Wisconsin, Madison, 1991, 102.

77. P. T. Mtuze, "The Role of African Languages in a Post-Apartheid South Africa," inaugural lecture delivered at Rhodes University, August 9, 1990, 10.

78. Ibid., 13.

79. Neville Alexander, *Language Policy and National Unity in South Africa/Azania* (Cape Town: Buch Books, 1989), 11.

80. Ibid., 23.

81. Ibid.

82. Ibid., 52.

83. Ibid., 52–53.

84. N. Maake, "Dismantling the Tower of Babel: In Search of a New Language in Post-Apartheid South Africa" in R. Fardo and G. Furniss, eds., *African Language, Development and the State* (New York: Routledge, 1994), 118.

85. *Africa Recovery*, United Nations Department of Public Information, Vol. 13, No. 1, June 1999.

86. C. M. Rubagumya, epilogue to *Teaching and Researching Language in African Classrooms*, ed. Rubagumya (Clevedon, England: Multilingual Matters, 1994), 156.

87. Olasope Oyelaran, "Language, Marginalization, and National Development in Nigeria," in E. N. Emenanyo, ed., *National Languages and National Development in Multilingualism: Minority Languages and Language Policy in Nigeria* (Agba, Nigeria: Central Books, 1970), 24.

6

Black Union Praxis and Worker Culture: Revolutionary Prospects and Limitations

The revolutionary Black working class organizing in South Africa embodies the precept of collective solidarity in its consciousness and praxis, albeit punctuated by the penetration of capitalist culture, which compels some leaders of the unions to surrender revolutionary principles for material gain in white corporate South Africa. People like Cyril Ramaphosa and Jay Naidoo immediately come to mind. In some sense, a reclamation of the principles of revolutionary Black Consciousness that inspired organizational commitment and fearlessness, independent thinking and open defiance, as demonstrated by organizations like the Black Allied Workers Union in the 1960s and 1970s, needs to recur as we begin the new millennium. The symbol of the Black fist breaking from the chains of oppression that was used by organizations like the Black People's Convention in the 1970s needs to be brought back, so that it can function to reconscientize people about the true nature of the condition of Black oppression today and become a catalyst for Black working-class solidarity and cohesion. The struggle for a socialist Azanian republic is a long, protracted, and arduous course that will take several decades to materialize. The colonization of Africa and its subsequent subjugation by capitalist hegemony has lasted almost five centuries; the undoing of these tenacious oppressor systems could take the span of the current new century. ·

The question that many radical critics from the left, particularly the "white left," raise in response to the post-apartheid scenario is: How does one account for the "growing, almost total, hegemony which the ANC has achieved?"[1] The question of the nonrevolutionary character of the Black working class and rural peasantry in

South Africa is being raised by numerous left-wing critics, many of them rooted in the Western radical, Marxist, Marxist-Leninist, or Trotskyite traditions. These critics often fail to understand the particularity of context, specifically that obtaining in Africa, most pointedly, South Africa. They insist on imposing a Eurocentric Marxism-Leninism or a Stalinist or Trotskyite analysis of the Russian Revolution on all other situations in the colonized world, including South Africa.[2] There is no "one-size-fits-all" solution from these European critical traditions, radical as some of them are, for the continent of Africa, or for places outside Europe, for that matter. For instance, Mao Tse-tung clearly understood the primacy of indigenous Chinese socialism and the principal nature of the Chinese social context in the socialist struggle and refused to subject the Chinese movement to rigid Marxian-Leninist principles as Russia had hoped.[3]

Revolutionary Limitations and Possibilities of the Black Working Class

During the 1980s, Black-led trade unions were effective in establishing dynamic nexuses of unity among Black workers, melding critiques of white power with those of capitalist monopolies. C.R.D. Halisi confirms this assertion when he states:

> Recently several black consciousness unions have flourished and formed themselves into their own federation. These unions include the Black Allied Mining and Construction Workers Union, the Insurance and Assurance Workers Union of South Africa, and the Amalgamated Black Workers Union. Black consciousness unions demand black leadership of the union movement and seek to combine black consciousness and trade union objectives.[4]

This front of Black worker unity suffers serious perforation today, as philosophies, ideologies, tactics, personalities, and perceptions of the new ruling class deeply divide Black workers and undermine the thrust toward revolutionary organizing.

The common denominator of revolutionary socialist Black Consciousness needs to be firmly sown in the minds and hearts of the Black rural peasantry and urban working class, in their language and on their cultural terms. Revolutionary groups need to realize that it is imperative that revolutionary principles be maintained and adhered to and that Black peasant and working-class organizations, however variegated, need to unite around fundamental liberation *principles*. Failure to abide by revolutionary principles will result in even radical groups like the Azanian People's Organization being swayed by the seduction of money, media attention, and visibility in the bourgeois communities of South Africa. No revolutionary organization is capable of advancing the interests of Black peasants and urban workers by going along with the neocolonialist capitalist tide that has unfolded in post-apartheid South Africa, under the guise of a "free and democratic country." It is vital to realize that the Azanian revolution will not be fought on television or in the press, reminding us of Gil Scott Heron's classic of yesteryear, which is still relevant: "The Revolution will not be televised!" At any rate, the television medium in South Africa signifies the titivation of the Western mind in Africa and essentially represents white capitalist frivolity, whether it emanates from Europe, the United States, or South Africa.[5]

The global dimension of the Western market economy that has infiltrated and permeated the underdeveloped world, including Africa, has saturated the South African political economy. South Africa's recent signing of a wide-ranging trade pact with the European Union for the next twelve years and its total disregard of the need to create an *integrated Southern African economy* with a *singular currency*, which could then trade with non-African nations, is one clear testimony of this fact. The poor and working classes around the world are being fed the daily dose of capitalist medicine that neutralizes their consciousness and benumbs their proclivity toward challenging the injustice of the global market economy. Many are lulled into believing that "economic growth" and "balanced budgets" will benefit them as well and that the explosion of computer technologies, the Internet, and lightning-speed digital communications will provide the path to economic well-being. This ideology of enslavement and containment needs to be systematically and relentlessly arrested. Revolutionary organizations in Azania need to launch massive literacy and conscientization campaigns in poor urban and rural communities that underscore the fundamentally fraudulent and exploitative character of the ruling elites. The Black poor need to understand that the supposed "booming economy" of the United States in the 1990s, for instance, essentially booms for the wealthy, at the cost of impoverishing the poor. As Diego Ribadeneira, a journalist with the *Boston Globe,* writes:

While these have been flush years for the wealthy, those at the bottom of the financial ladder have not only been left out of the economic surge—they've actually fallen several rungs . . . In the 1970s, the richest 10 percent of the American population controlled 50 percent of the wealth, while the rest of the population divided the remaining 50 percent, according to U.S. government data . . . During the 1990s, the percentage of wealth enjoyed by the richest 10 percent has climbed to 70 percent.[6]

In the United States, child poverty increased by 21 percent between 1979 and 1989, so that today there are more than 15 million children living in poverty out of a poor population of 37 million people, and 9 million children lack basic health care.[7] Globally, the end result of the market boom is the shocking statistic that the combined wealth of the world's 358 billionaires exceeds that of the incomes of people of 45 percent of the world's population.[8] What goes largely unmentioned is that this obscenity of wealth is directly linked to the inhuman debt burdens placed on underdeveloped nations by the Western imperialistic powers, so that African, Asian, and Latin American nations squeeze their economies, slash spending on education and health care, and deny welfare access to the poor, all to pay the towering $300 billion debt to the West. Western capitalism does not understand the principle of Jubilee widely promoted by activists in Europe in June 1999, because capitalism is a system of exploitation that irreversibly thrives on the impoverishing of the workers and the toiling laborers of this world.[9]

In the United States and the rest of the Americas, this level of exploitation and humiliation is compounded because of the mounting genocide that the indigenous people of the Western Hemisphere have experienced and continue to face following 500 years of European settler-colonialism.[10] These occurrences are not inevitable and mystical happenings that transpire by some wave of a magic wand or by virtue of some ideas falling from the sky; they are the products of deliberate and systematic planning by the ruling elites of the world's united imperialistic system.[11]

It is imperative that revolutionary Black organizations in South Africa use the putative "democratic space" created by the post-apartheid dispensation, as Archie Mafeje maintains, to engage in reconscientizing the Black worker and rural poor peasant communities about the way the post-apartheid political economy really functions. Black workers and the poor need to understand the impact of global markets on the evolutionary dynamics of the post-apartheid state and the functionary role that Black elites play within this essentially extractionist system. They need to overcome the aura of wonder and awe at their national leaders such as newly elected President Thabo Mbeki and Deputy President Jacob Zuma and realize that these leaders are not the economic saviors portrayed by the bourgeois media. They need to extinguish the sentimental tingling and quivering that they experience when seeing the country's president has a Black face, difficult as this is considering decades of white rule. Barney Mokgatle, a leader of the Soweto uprising of 1976 and the organizing secretary of the Socialist Party of Azania, described it aptly when reflecting upon Black attitudes in the 1999 elections, "Black people need to think with their heads and not be swayed by what's in their hearts."[12]

One strategy to circumvent the onslaught of bourgeois political formations is to establish a new political party that principally represents the interests of the Black rural peasantry and the Black urban classes in South Africa and that is also socialist and womanist in philosophy, such as a Black Socialist Workers Party of Azania. Perhaps the Socialist Party of Azania would consider such a formation in discussions with the most revolutionary segment of Black workers. It is of vital importance to understand that the majority of Black workers in South Africa, though possibly inclined in the socialist direction, will not vote for socialism because they realize that the ANC is the most powerful political bloc and is bound to win any election. A tendency to want to go with the political winner, notwithstanding reservations about the winner's political efficacy, still prevails among the overwhelming Black voting population. At this stage, it is imperative that the most radical sectors of Black workers unite across all tactical and petty fractious divides so that all the revolutionary segments can be consolidated. This constitutes a small minority of the working class, but it wields the potential of establishing a groundswell of support within the Black peasant and worker community, in the aftermath of the growing disillusion with the cheap and petit-bourgeois politics of the ANC and other minor political parties. The task of building a revolutionary base is arduous, protracted, and time-consuming and will cause much frustration within worker ranks, given the low level of financial resources available. Yet it can be done, and needs to be done, to redeem Azania from a possible genocidal fate, by capital.

However, should the state resort in openly hostile fashion to suppressing radical Black peasant and worker organizing, conscientization, and mass mobilization, another plan ought to be conceived, this one involving an underground strategy. In some sense, the indigenous cultural medium of oral communication (particularly surreptitiously) may need to be adopted as another strategy of revolutionary communication. It may be necessary to inaugurate an underground movement among Black workers so that the field may be prepared and the seeds sown prior to the harvest of revolutionary action. It is clear that capitalist society is not about truth and integrity; in fact, it fears truth and thrives on lies. The revolutionary Black movement in Azania may need to operate significantly underground so that its strategies, actions, and organized resistance are hidden from the bourgeoisie. Such revolutionary "baking" can only occur in the oven of Black determination once the entire Black working class has been *reconscientized*, by understanding the nature and workings of contemporary capitalism, the manifestations of the colonial state, the role of imperialism, the evil of sexism, and the realization of the need for the eradication of all these anomalies in Azania in particular, and Africa in general. It will take some earnest cultural discipline and organization to engage in the surreptitious dimension of the movement, yet as it was advanced in Zimbabwe, it can be done in South Africa, although admittedly the latter situation is much more difficult.

Black rural peasants and urban workers need to know and understand the history of Black working-class resistance, as delineated in Chapter 1. The 1987 mine-workers strike, the longest in South African working-class history, is a case in point. Some of the principal issues that caused difficulties for the National Union of Miners in that strike were the control of the worker hostels, the supply of food, and the intense violence and intimidation by the police and security establishment in the mines. These were sobering concerns for the union, and in the event of future strikes geared toward wresting power from the corporate mining enterprise, the union would need to lay out contingency plans for accommodations, food supplies, and self-defense, so that worker action is not violently and physically disrupted by the powerful economic colonizers of the country. Careful and strategic planning is called for, particularly in establishing concrete chains of solidarity strike actions from sister unions that would cumulatively shut down the wheels of industry and the engine of the capitalist-propelled economy.

The solidarity and effective organization of Black workers implies the potential for the acquisition of power, so long as Black peasants and workers realize that the ultimate objective is not merely higher wages within a capitalist structure but a takeover of farms and industries so that wealth is collectively owned and equitably shared by Black peasants and workers. The ideology of revolutionary Black socialism and self-determination needs to be firmly rooted in the minds and spirits of Black peasants and workers so that they will refuse to be pushed around by white capital or by Black surrogates for white capital, for that matter, embodying a militant African spirit of noncooperation and determination that eschews every fear of the power and threat of death uttered by capital. It is abundantly clear from the thousands of Black workers who were killed and injured in the hundreds of strikes organized that many Black workers are not afraid to lose their lives for the liberation of Black nationhood. What is needed is the adoption of a posture of militant armed self-defense by Black unions that shields Black workers against police violence, which is likely in the anti-worker, antistrike ethos of post-apartheid South Africa.[13]

Determination and perseverance pay off, especially when workers are organized and united. The only defect in the 1987 National Union of Miners strike was the fact that the workers were unarmed and thus vulnerable to armed state attack and that they were not able to take over the mines. In the spate of actions envisaged in the revolutionary praxis of Black miners, the miners would need to be armed to protect themselves against police and military violence, especially since initially, their strikes would be nonviolent, and secondly, their objective would be to assume full control of the mine in question as part of a wider strategy of worker ownership and control of the mines. Such actions may be perceived to be unrealistic and unworkable. Yet they have never been attempted within a *revolutionary* ethos, making the context of action radically different. A classic example of this revolutionary organization is the cadre of fighters led by Patrick Mkhize, general secretary of the Azanian Workers Union in Kwazulu-Natal.[14] Although the South African authorities attempted to charge him with unlawful military organizing, they were prohibited from doing so because the fighters used wooden weapons, not real rifles or live ammunition. Such innovative revolutionary strategies are emerging on the Azanian landscape, even though many are dismissed as insignificant or innocuous by bourgeois critics. Yet when revolutionary Black workers are organized, their being pitted against capitalist oppression is anything but benign.

Indeed, Black workers, the bulk of the labor force in South Africa, possess the potential of bringing the wheels of the South African economy to a grinding halt. No facet of South African civil life could continue without the labor of Black workers, who are critical to: land cultivation; crop planting and harvesting; transportation of essential food for the country's entire population; oil refining for transportation; all basic services such as transportation in the motor, railway, and aviation sectors; food packaging; processing of all agricultural and industrial products; construction; sanitation; motor assemblies; textile production; environmental cleaning and restoration; and all spheres of mining. This dependence of the South African economy on Black labor is an axiological point for radical mobilization, contingent upon the critical political consciousness of Black workers and their ability to organize themselves into a cohesive resistance force that could present a united bulwark of worker activism against the forces of capital. The crucial element that was absent in the 1987 strike was sustainability and contingency plans that would have furnished essential survival ingredients: food and shelter. The strike organizers ought to have anticipated the long-drawn-out character of the strike and organized food cooperatives within the Black community that would help striking workers survive so that they were not forced into starvation.

Such revolutionary organization was extant, for instance, during the Algerian war of independence against French colonialism led by the National Liberation Front and the Cuban anti-Batista, anticapitalist revolution. Although the South African situation is far more complex, modernized, and Westernized than the Algerian or Cuban landscapes, the *principle* of revolutionary organization of the working class is essentially the same. Noncooperation with the economic edifice of capitalism is an arduous task because all physical survival is tied to and dependent on the pole of capitalist material supply: food, energy supplies, and so forth. Capitalism forces Black workers to cooperate with its exploitative nature; failure to do so means certain premature physical death.

The Black working-class culture that one finds throughout the African continent is able to survive with the barest minimum of material ingredients: limited supplies of food, rest, and shelter, as one witnesses in many rural communities in Africa. It is this indigenous culture of tenacity, subsistence, and indomitable persistence that will function as the redeeming instrument of revolutionary Black praxis, in the face of economic constraints and limitations imposed by capitalism. After all, most of the rural working-class folk in African villages do not depend on capital since they have no capital; yet they survive and propel themselves into the future by protecting and advancing the interests of the children and the young through collective sharing of the meager material resources that they possess. The Black peasantry and working class in Azania needs to regain this noble and revolutionary principle in its offensive against the capitalist system.

Although critics may dismiss the Black unions as a perfunctory cog within the machinery of capitalism in South Africa, arguing that unions merely agitate for more gains from capitalism and essentially legitimate this unjust economic system, the *process of conscientization* through Black worker strikes and organization ineluctably radicalizes Black workers. However, progress is punctured by intermittent trajectories of cooperation and acceptance of capitalist benefits. Since May 1994, when the government gained office, there are signs that the coalescence between the country's largest union, the Congress of South African Trade Unions, and the new government is waning, sparked by worker resentment against such procapitalist legislation as the Draft Bill on Labor of 1994, which provides rights to business employers to dismiss both striking workers and those workers agitating for justice at the workplace.[15]

Another sign of this underlying and latent radicalism is evidenced in the recent commission report of COSATU, which threatens "setting up a new left-wing opposition movement—if the African National Congress does not return to its socialist roots."[16] The organization has also criticized the new government's GEAR program as "right-wing" because it fails to address the underlying causes of poverty and economic inequality and espouses macroeconomic growth at the cost of social development. The report proceeds to assert that though COSATU views alignment with the ANC of strategic importance, it will not support the ANC in the 1999 elections if the ANC abandons socialist principles. COSATU apparently has not followed through on its threats, given the 1999 election results. One point is clear, though, and that is that this large pro-ANC union has not been entirely "co-opted" by the capitalist proclivities of the state and still possesses some radical undercurrents. Evidence of this level of radical politicization is reflected, for instance, in the antiprivatization protest marches involving 5,000 members of NUMSA, the metalworkers' union, on July 19, 1995,[17] and in early 1996, in COSATU's planned twenty-four-hour antiprivatization strikes, reminding the political authorities that it rejected the government's approach of privatizing major sectors of the economy for the lucrative benefit of capital.[18] COSATU has also criticized the government's schemes for massive privatization of major industries as contravention of agreements that the trade union organization had signed with the government prior to the elections of April 1994.[19]

Two unions, the National Union of Metal Workers of South Africa and the South African Clothing and Allied Workers Union, indicated at meetings during 1993–1995 that they cannot endorse capitalism as the economic system for South Africa because capitalism exploits Black workers.[20] Black workers are fully aware that industrial corporations are making handsome profits at the cost of cheap labor, confining the working-class sector to echelons around the poverty line. Further evidence of the increasing radicalization and asseveration of political independence of the Black worker movement is the formation of the South African Independent Trade Unions Confederation, a conglomerate of Black unions including the South African Health and Public Service Workers Union; the Azanian Workers Union; the Black Mining Workers Union; the Education, Hospital, and Church Workers Union; the Metal and Engineering General Workers Union; and the National Association of Civil Service Workers Organization in early August 1995.[21] The combined paid-up membership of this Black-led trade union confederation is 120,000.

Traditional Marxian ideologues would repudiate the piecemeal strategic character of South Africa's Black unions as being "reformist" and generally too "procapitalist." They would therefore dismiss any revolutionary potential emerging from this segment of South African society based on the historical praxes of the union movement, like COSATU. A writer from the Azanian People's Organization provides a cogent rebuttal to this argument, and since it is lucid and persuasive, it needs to be extensively cited here:

> The trade union or labor movement has been seen by some as the most revolutionary movement capable of ushering in a new social order, while others have dismissed it as a reactionary product of a capitalist society. Both views are inaccurate and misleading.
>
> While on the one hand it is true that trade unionism introduces workers to democratic processes of accepting joint responsibility and joint decision making, it also sharpens their consciousness to the relationship that exists between them and the means of production and exchange. This should not be construed as a complete revolutionary process that needs no direction and guidance.
>
> A revolution has been described as "a dialectical progress of historical development" which is "the sum of varied and diverse circumstances of multiplex elements that together added up lead to the solution in a given historical moment of a crisis that has stubborn and deep economic cause." With that in mind, the working class as such is not

synonymous with a revolutionary phenomenon. If this were so, the true picture of society in our country would have been otherwise. . . . As many scholars would have it—and we agree with them—the dominant ideas in any given society are those of the ruling class. Sections of the working class in this country have absconded from their fundamental worker responsibilities and embraced the values of the ruling class. . . . In our country, the labor union movement should be influenced by revolutionary consciousness to transcend its limitations that strait jacket it into assuming a "pressure group" character that concerns itself with the amelioration of working conditions. The movement must discard its reformist character, that is, being solely concerned with factory floor grievances and turning a blind eye to both the existential situation in which workers find themselves and the material conditions that determine their respective background. It should never be divorced from day to day rigors and vagaries of living in a capitalist society; for the direction of its programs it must draw from the ethos and paths of the workers' experiences at both factory floor level and beyond.

As the political writer Sorel has said: "the working class alone—by virtue of its being the most down trodden and oppressed people in our society—has the moral virtues necessary to rejuvenate society and that to perform this mission it must have faith in itself and its purposes." Our duty then is to raise its revolutionary consciousness.[22]

What this critic is propounding here is the view that revolution building among Black workers is a *process*, that workers' radical outlook becomes elevated as they become more conscientized, a task that the radical Black organizations such as AZAPO and the Socialist Party of Azania have full responsibility in assuming. Within the Black Consciousness Movement, proponents are fully aware of the limitations of trade union organizing and tendencies, reflected in this perspicacious critique: "Revolutionary consciousness has to be injected and it necessarily means transcending trade union consciousness. As one comrade put it, 'Unionism is a capitalist trap. Can we use the trap to trap the trapper?'"[23]

The radical Black Consciousness socialist sector of the trade union movement needs to be fully cognizant of the "trap" of capitalist benefits progressively extended to workers so that worker hostility toward the capitalist system is mitigated. Again, Azanian workers' organizations indicate that they recognize the limitations of the trade union movement and its becoming consumed with reformist objectives:

When the trade union movement becomes "big labor," then it is only too easy to fall into the pluralist trap of "balancing" big labor against big business. The myth is that these groups, as they compromise, make deals and pressure each other, will create a natural system of checks and balances.

Pluralism masks the fact that some groups and individuals hold power in this society—specifically the white minority—while the rest of the people are excluded. The union bureaucrat always tries to channel conflict and his work assumes the respectability of the system for the state and the bosses. Listen to Harold Pakendorf in the *Sunday Times* of 16th August, 1987, in an article entitled "Yes, we should be applauding this strike, not regretting it": "Far from looking with distaste on (the) NUM (National Union of Mineworkers) we need to compliment it. There may be doubts as to its bargaining methods, but it is still working with the system. Whatever else may be said, the mine strike is not part of the revolutionary onslaught. It is part of the orderly process of change."[24]

Capital's designs toward co-opting Black union leaders and workers must be forcefully arrested by a militant program of noncooperation, which marks the beginning of the revolutionary "onslaught" that Pakendorf alludes to and clearly repudiates because it threatens the foundations of capitalist civilization, which is seen as normative and necessary for white economic benefit in South Africa.

It is the active collaboration and strategic planning between the radical sector of Black workers and the radical segment of the Black political movement that promises to forge a united revolutionary Black front against white capital and its Black allies. Perhaps the most formidable obstacles to this materialization at this stage are the serious levels of division among Black workers and the Black left and the distancing of the radical movements from the rural peasantry. Notwithstanding COSATU's current threat to disalign itself from the ANC over the latter's abandonment of socialist principles, the union coalition is generally reformist and subscribes to a "nonracial" philosophy as part of this alignment. The program of the organization is thus inclusive of white workers and white leadership, which in actuality translates into a posture of coexistence with capital, since white workers are generally reactionary as a result of their benefiting historically from capitalism. The propensity and desire to overthrow capitalism and settler-colonialism in South Africa is hardly the pervasive tendency among white workers. The outcome is a diluted worker militancy that functions within an unjust existing status quo.

It is imperative that all segments of the Black rural peasantry and working class invest their energies for the long haul and coalesce around the struggle's underlying objectives: to uproot European settler-colonialism and eradicate the capitalist system through a multipronged strategy of ongoing revolutionary educational workshops and classes for Black workers. Such actions are necessary preparation for armed guerrilla combat and for the organization of sustained worker strikes when necessary, which could cumulatively induce paralysis of the capitalist economy entrenched in South Africa. It is also incumbent upon the radical organizations to organize a massive literacy and conscientization campaign among the rural peasantry, which represents a sizable segment of the Black dispossessed, so that these women and men can become empowered toward revolutionary mobilization. The decolonization program must begin with the rural community, with the question of *land* being at the fore, as was the case with the guerrilla movement in rural Zimbabwe.

The talk of "transition to a nonracial democracy" is a bourgeois term that even incisive critics like C.R.D. Halisi still echo, even though they are cognizant of the history and dimensions of Black Consciousness politics.[25] Nonracialism, though efficacious, is a far cry into the distant future, only attainable after Azania and the rest of Africa have become economically independent under a unified socialist Africa, free from the systems of settler-colonialism, capitalism, sexism, and Western imperialism. South Africa will not have any transition to "nonracial" democracy under the existing capitalist and settler-colonialist dispensation. It needs to be liberated from the yoke of "nonracial democracy" as defined by liberal observers, because "nonracial democracy" conceals the contradictions of white settlers' owning almost all of the country's land, resources, production, and wealth. Black Consciousness is still key in the genuine liberation of Azania because it insists on Black unity, depends on coalescing of the oppressed, works tirelessly and defiantly for solidarity among all Black people, and is committed to overthrowing capitalism, while mindful of the fact that there are elements within the Black community, particularly among the bourgeoisie, that function as enemies of Black liberation. It envisages a truly nonracial society *only after* Black working-class power has been achieved politically and economically via socialism and the land has been returned to the indigenous peasantry.

Concomitantly, the Black political left needs to bury personality disagreements, discard egotistical aspirations, dissipate nuanced tactical differences, and converge around solid revolutionary principles. For instance, a major ideological divide that emerged within the radical Black Consciousness Movement in the last year was the question of participation in the country's 1999 elections. The leadership of AZAPO had indicated that it was earnestly considering running as a party in the 1999 elections. This proclivity was viewed as problematic by stalwart adherents of the Black Consciousness persuasion, who viewed such an entrée as participation in bourgeois neocolonial politics and an affirmation of the status quo.

The question that the Black left needs primordially to consider in any response to the existing situation in South Africa is whether revolutionary socialism can be achieved through the ballot. Given the historical and contemporary levels of ignorance on the part of the Black masses, especially in rural areas, regarding the workings of socialism, it is highly unlikely that there will be a populist vote in favor of AZAPO at the polls. A corollary problem is the predicament of intensive indoctrination that the overall Black population suffers from as a result of bourgeois Black and white political fraudulence and obfuscation.

Socialism is hardly a system that bourgeois "democracies" grant to voters. It, like economic power or working-class power, is principally acquired through revolutionary education, organization, and cultural and guerrilla praxes, in essence a full-blown armed insurrection involving Black working-class militant action on political, commercial, economic, and social levels. The designs of the neocolonial regime in South Africa in attempting to convey to the country's population and the world that South Africa is a truly free country where all freedom of expression exists, including revolutionary rejections of capitalism, needs to be understood by all revolutionary cadres. Veritable Black socialist power in Azania can only be acquired through a cultural, political, and armed revolution led by the Black working class. This noble objective is most likely not attainable within the confines of the existing dispensation, primarily because the new South African government is an intentional neocolonial structure orchestrated by Western capital and managed by domestic white capital and its Black allies among the bourgeoisie.

The ANC-dominated government is determined to willfully cater to the needs of the global capitalist system, while enriching a tiny Black bourgeois group at the cost of impoverishing the vast working-class Black majority through the sham philosophy of "economic pragmatism" and its bogus GEAR program—bogus because "growth" has taken the form of capitalist profits and Black working-class poverty. Further, employment for Black urban workers and the rural sector has significantly decreased since 1996, with 500,000 jobs shed over the past five years.[26] The figures in government are fully aware of the repercussions of their duplicitous role in neocolonialism, yet they lust after lucrative personal economic gain and have opted to oppose the interests of the Black working class. Neocolonialist capitalism, like colonialism, is a tenacious and intransigent system that must be uprooted at its foundations. It is thus incapable of being radically transformed through Western bourgeois political avenues.

The revolutionary movement can only be built, advanced, and fructify through firm adherence to revolutionary principles of anticapitalism and prosocialism, antisexism, anticolonialism, and antiracism, all under the auspices of a *conscientized* rural peasantry and Black working class. These are fundamental principles that are nonnegotiable and irrevocable. The Black left needs to stand firmly by these principles notwithstanding their "unpopularity" or seeming anachronistic irrelevance *now*, so that the truth of these solid principles may be borne out in the *future*.

One of the prototypical features of the neocolonial regime in South Africa is its constant wavering of principles, where *everything is negotiable*, contingent upon the extent of one's financial resources. This is reflective of the nature of capitalism, which is distinguished by its sole principle: greed for profit. The news about the multiple scandals of monetary abuse and waste at all levels of government is consistent with the logic of capitalism. The Black left needs to steer clear of the corruptive and predatory system of neocolonialism and instead invest its energies in persistent conscientization of the Black working-class urban and rural masses, in preparation for the consummation of the Azanian socialist revolution, even if it takes the next half century to achieve it. It must resist and refuse to be seduced by the superficial trappings of crass materialism and consumerism, evils that are reproduced at all levels of South African capitalist society and promoted by the corruption of European bourgeois culture.[27] There can be no interclass compromise between capital and Black workers, and the Black left cannot afford to forsake this historical revolutionary principle.[28]

Archie Mafeje contends that in light of the collapse of the Eastern socialist bloc, "there is no strategic value for the few who aspire to socialism to insist on marching ahead of these popular forces and pretending to be speaking on their behalf" and that social democracy is the best strategic objective since socialism is a quixotic ideal at this stage. But he overlooks the fact that capitalism does not permit national development in the underdeveloped world.[29] Unlike the conditions of the liberal capitalist nations of the West, where the material conditions and historical realities are conducive to the establishment of welfare states, such as those in Scandinavia, for example, colonized countries like South Africa do not have that historical luxury since they are under siege by capital. Capital is not benign and does not roll over and play dead in situations like that in South Africa, precisely because of the enormous stakes in the country. Mafeje proposes that in the interim period prior to socialist attainment, "social democratic strategies such as nationalization of essential social services such as health, education, transport, and housing, and of requisite industries could be adopted under the conditions of undeveloped or unevenly developed capitalism." But again, he does not take into account the unmitigated aggressiveness of capitalism in the underdeveloped world, with massive privatization of traditionally state-run industries such as health, transport, and housing taking place at lightning speed throughout Africa.[30]

Transnational capital is on the warpath against the poor of Africa and the underdeveloped world by *taking over ownership* of major national resources of individual countries. The problem is that nations struggling for economic justice are either compelled to brave the storms of imperialism and remain in the struggle for the long haul, like Cuba, or alternatively, accept the entire package of capitalist medicine since capitalism does not come piecemeal in the underdeveloped world. There is no single instance in the underdeveloped world where a benevolent capitalism exists in which social democracy prevails, precisely because Western imperialism's hegemonic mantle is conditioned to destabilize such "benevolence," to project its totalitarian power on the world. Transnational capitalism disallows such models from prevailing in the underdeveloped world because it thrives on the cheap labor of the working classes of these countries and is by nature dedicated to confiscate whatever resources these nations have to offer, leaving a pittance for the indigenous populations.

In the South African context, given the obscenely skewed distribution of resources between white capital and the Black poor and the entrenchment of its colonial trajectory, social democracy as conceived by Mafeje is hardly a "middle-path" option. A precipitous break with the colonial past and the capitalist present is necessitated, as the Chinese discovered in their own history with events like the Long March and the organization of rural cadres. Between April and December 1927, 38,000 radical Communists were killed.[31] The price of sacrifice for revolutionary change is the necessary and the *only* means of eradicating the tenacity of slavery and exploitation.[32]

The attitude of the minority revolutionary segment of the Black working class must be one of building *toward* the revolution, even though the popular "will of the people" is a tacit and diffident coexistence with capitalism. The totalitarianism of capitalism and its projected hegemonic power today paradoxically provide the reason that those who comprehend the nature of the system need to intensify the resistance, even with the unpopularity of radical inclination. The majority is not always right, the United States most poignantly demonstrating this fact. The revolutionaries ought to be the true long-term prophets, those who work tirelessly to *conscientize* the masses about the ephemeral character of profits. They ought not to capitulate and resign themselves to a position of surrender, saying, "We could not win, so we joined them!" The history of the Azanian struggle, specifically that waged by Black Consciousness of the 1960s and 1970s and the movement of Soweto of 1976, and the histories of revolutionary struggles that remain like those in China, Cuba, and Libya, certainly do not testify to this abdication. We would not have been able to advance to this level of struggle today if Steve Biko had opted to coexist with white racism in the 1970s; we are where we are today because of the radicals who perse-vered to struggle for the truth and live out the vision, knowing that the odds and even the masses were against them.

As radicals, we are by no means arguing that socialism is on the near horizon in Azania. We know that it is not imminent given the current political-economic charac-ter of the post-apartheid dispensation and the structures of global capital. We are also fully aware that as Mafeje correctly points out, for instance, in referring to the Chi-nese Revolution, the Communist Party in China was only able to consolidate itself following the peasant cadres' direct actions against the landlord aristocracy and the wealthy, a process that began in the early 1920s and took decades.[33] Socialism only materialized when the workers and peasants were intrinsically involved in radical measures themselves and provided the momentum for the establishment of a socialist status quo. We are under no illusion that socialism, by virtue of its foundations (meaning here an indigenous socialism that emanates from the cultural and social ethos of Azania very much along the lines that Chinese socialism was constructed, in contradistinction to Soviet Marxism-Leninism), is a fundamentally democratic process that like a religious faith, ferments and materializes over the passage of time.

We radicals are fully cognizant, too, that globally, radical movements are not the most popular, whether one beholds the mayoral or gubernatorial elections in the state of Mexico recently or the Black movements in the United States. Putatively radical parties, like the Party of the Democratic Revolution in Mexico or the Sandin-ista Party of Nicaragua, are only able to garner a tiny percentage of the vote.[34] It cer-tainly is not the most sanguine ethos for radical or revolutionary politics, as the 1960s or 1970s were. However, these are passing phases and not the end of history. Capitalism has been able to extricate itself from numerous crises in recent history, supposedly even from the Asian financial crisis. Yet it has also been weakened by the cumulative effect of these crises and is rapidly running out of time and perennial so-lutions to its structural contradictions that precipitate consistent crises. This is nei-ther the end of struggle nor the end of the world, even though capitalism would like the world to believe the former.[35] Movements from Chiapas, Mexico, to New Cale-donia, from Korea to Libya, from Puerto Rico to India and its Dalit Movement indi-cate that not all oppressed people are supine and willing to play dead so that capital-ism can streamroll over them.

Clearly, there is no quick road to socialism in Azania, and any critic who assumes this is myopic and quixotic. What we as radicals are arguing *for* is the cultivation of this revolutionary process, a baking of the indigenous socialist pie, if you will, through a gradual and protracted struggle that starts with conscientization of the Black rural peasantry and the reconscientization of the urban working class, with particular emphasis on the education and mobilization of indigenous women, who possess the greatest potential for the uprooting of colonialism and the overthrow of capitalism and imperialism. We contend that the masses need to understand the es-sential nature of the capitalist system so that they can viscerally reject it in spirit and embodiment, as opposed to learning to work more effectively within it and thus pro-long its tenure.

Creative Cultural Productions and Resistance

Any movement in Africa that is geared toward revolutionary transformation is doomed if it disregards and overlooks the potency of indigenous working-class culture, as Mao Tse-tung and the working classes of China so dramatically illustrated in the historic Chinese revolution. South Africa has a rich history of innovative cultural mediums and expressions that have functioned to give momentum to movements that were waning caused either by pressures exerted by the industrial establishment or by the psychological stress of living in a depressed social environment like the urban township.

Vigorous foot stamping and chanting in unison, called *toyi-toying* in Black parlance, is a popular cultural form of Black working-class resistance. *Toyi-toying* functions as a cultural resistance model that emboldens Black people to face oppressive policies and practices through dancing and chanting in collective formation, coupled with vigorous physical movement. It is an evocation of the historical warrior tradition of African communities that faced the colonial onslaught. *Toyi-toying* is still widely observed in the post-apartheid society at strikes, demonstrations, and protests, as Black workers demand justice and equity. It functions as a cathartic cultural instrument because it allows for the ventilation of pent-up frustrations of women and men who are forced to tolerate the absurdity of capitalist society and the burdens of economic and material deprivation. Additionally, *toyi-toying* functions as an indigenous means of resistance communication and solidarity in which various African languages are collectively used as a protest code that cannot be understood by the colonial oppressor. It is used for empowerment toward mobilization.

Black theater and drama continue to be potent instruments for the education, conscientization, and mobilization of Black workers resisting political repression, social degradation, and economic exploitation. Workers continue to organize cultural projects, utilizing traditional theatrical skills, accompanied by a flamboyant blend of song and dance.[36] The stellar performance of the play *The Long March*, portraying the historic strike of Black workers at BTR Sarmcol, in Howick, Kwazulu-Natal Province, in 1989 captured this power of Black workers' culture. This is just one instance of the interwoven dynamics of culture and political struggle.[37]

The harmonization of the polyrhythms of African freedom songs coupled with vigorous dance and instrumental improvisations seen in worker plays like *The Long March* and *Bambhata's Children*, Maishe Maponya's *Gangsters*, Percy Mtwa and Mbongeni Ngema's *Woza Albert* and *Asinamali*, Matsemela Manaka's *Children of Azazi*,[38] and women's plays like *Wathint' Abafazi, Wathint' Imbokodo* (You Have Struck a Woman, You have Struck a Rock) are polysemic in that they symbolize the resilience of the Black working class and embody the unity and cooperation of workers in challenging exploitation. This cultural coordination of working-class dance and song invigorates Black people in persevering in the struggle against neocolonialism and capitalist hegemony. In the words of the producer of *The Long March*, "Culture is the way we carry our message to the people. It helps to inspire workers so that we can continue even though our company dismisses us."[39]

South Africa possesses a proliferation of creative and imaginative Black artists, painters, sculptors, musicians, poets, dancers, theatrical performers, and populist intellectuals whose skills and talents can function as primary sources for the intensification of the momentum of liberation in the country. Musician Blondie Makhene, who has formulated the classic *Amaqabane* (Comrades) album, the late actor Matsemela Manaka, with his Pan-African plays evoking transformative images from the cultural wellspring of the continent, radical writer Mtutuzeli Mashoba, poet Mzwakhe Mbuli, and so many others all describe the dynamism of Black working-class culture in instructive ways that can be utilized by the people for their own liberation. These artists provide a critical commentary on colonial-capitalist society, often alerting Black people to the excesses and distortions of European capitalist and consumerist culture, warning us to veer away from this destructive path of exploitation and commoditization.

Black music forms like *marabi, mbaqanga, micathamiya,* and *kwela* are musical art forms that empower people to appreciate Black working-class culture, particularly since colonial culture has historically ravaged and deprecated indigenous African culture. *Marabi,* for instance, is a musical form born in the *shebeens* (liquor houses) of the urban Black townships, and it represents an admixture of a variety of African artistic styles. It involves the playing of diverse instruments such as the banjo, guitar, tambourine, and concertina and is a medium of social protest against the alienation wrought by a colonial urban society.[40] *Mbaqanga* is another artistic musical form that combines indigenous African polyphonic rhythms with modern jazz and soul styles that evolved out of the blood, sweat, and tears of Black working-class struggles in the mines and factories in the 1950s. It is a vehicle for the creative articulation of Black resistance to the imposition of the cheap Black-labor system. It functions to conscientize Black workers about their condition of economic and racial subjugation and consists of songs sung during demonstrations against forced removals, such as at Sophiatown in 1955, strikes against rent and bus-fare increases, and at other solidarity events. Songs such as *Bye, Bye Sophiatown* and *Asibadali* ("We won't pay rent") were played by Black disc jockeys on radio stations in the 1950s, 1960s, and 1970s in defiance of government restrictions, in the attempt to inform Black people about the draconian measures of forced removals and living-cost increases promulgated by the apartheid system.[41]

These musical forms are expressions of a defiant working-class movement, rooted in indigenous African historical cultures that opposed oppression. They infused Black people with a sense of cultural power and racial pride, notwithstanding the indoctrination by apartheid that disparaged Black working-class culture. They were viewed as subversive by the previous apartheid state and even led to incidents such as the police shooting of seventeen-year-old Mngcini Mginywa at a funeral in Grahamstown because the police claimed that Black people were "singing in their own language and this causes riots."[42] They will continue to play a role in valorizing the contributions of South Africa's aesthetic pioneers who promote the ethic of Pan-African unity and continental solidarity, nourishing the hopes of the dejected in the working class so that they never abdicate to the psychological taunts of oppression.

Future Revolutionary Transformation in Azania and Africa: The Primacy of Women's Struggles

Azanian and African revolution will never fructify without the centrality and primacy of indigenous women, a point that all radical scholars and activists need to inculcate. The plague of sexism continues to haunt our world today, where though "women account for 67 percent of the world's working hours, they earn only 10 percent of the world's income and constitute the highest percentage of part-time workers."[43] South Africa is no exception to this norm; if anything, it exhibits a rigid conformity. Colonial and capitalist politics have been characterized by staunchly patriarchal structures and practices that have viewed and continue to view women as appendages of labor, valued for sexual exploitation and essentially second-class citizens in a male-dominated world. Ironically and tragically, even many anticolonial and revolutionary movements in Africa have been culpable in perpetuating the inferiorization of women, notably in Zimbabwe, Mozambique, Angola, and South Africa. Josephine Ndiweni, a freedom fighter in the anticolonial guerrilla struggle in Zimbabwe, recalls this dynamic role of women:

> Women played an important role during the struggle because they met with and patiently overcame a lot of hardships. I am not only talking about myself: women worked hard throughout the country. I know men also worked, but sometimes they got tired or fed up, but the women pulled together from the beginning, right until the end of the war—even when it meant going to meetings.[44]

Today, many women in Zimbabwe, including those who participated as coequals in the *chimurenga* (the war of independence), are still awaiting recognition for their historic roles in the struggle.

True, old habits and traditions die hard. However, considering the history of women's involvement in the liberation struggle in South Africa for generations, it is unconscionable and unacceptable that women are relegated to the periphery in vital areas such as education and working-class leadership and as political organizers and policymakers. The trade unions in Azania may demonstrate progressive ideas in areas of the economy and participatory democracy, but most often they fall short when it comes to incorporating women into the leadership and voting for women within the decisionmaking process. This situation of fundamental injustice against women must be uprooted and destroyed in the South African revolutionary process. Che Guevara, the Cuban revolutionary, underscored this difficulty of sexism in anticolonial struggles:

> The part that the women can play in the development of a revolutionary process is of extraordinary importance. It is well to emphasize this, since in all our countries, with their colonial mentality, there is a certain underestimation of the woman which becomes a real discrimination against her.

The woman is capable of performing the most difficult tasks, of fighting beside the men; and despite current belief, she does not create conflicts of sexual type in the troops.[45]

The ANC government has taken comfort because it has inserted a women's rights bill into the new constitution and because there are 100 women out of the 400-seat assembly in Parliament. South Africa now ranks seventh in the world in terms of women representatives in Parliament.[46] It is not well known that "one-third of all candidates who ran on the African National Congress were women."[47] Yet the question of justice for women and the need to address the tenacity of sexism, patriarchy, and capitalism, which dehumanizes all women and exploits Black working-class women in particular, has not been addressed and has since 1994 essentially been rhetorical.[48] The issue of Black *working-class women's power* nevertheless remains, for the majority of women parliamentarians, like their male colleagues, emanate from middle-class backgrounds and generally serve bourgeois women's interests.

The government has retrenched workers in the public service sector, many of whom in the health and education areas were Black women, as part of its plan to downsize the government bureaucracy by 100,000 jobs.[49] In the informal labor sector, the government has failed to recognize the crucial role that Black woman play and provide extended credit facilities and infrastructural supports, substantiating the fact that the commitment of the new regime to Black working-class women is chimerical and devoid of substance.[50]

Constitutional laws promulgated by the government and now boasted about as the corrective to uplifting disadvantaged Black women are woefully inadequate to address structural injustice. Linda Zama, a former secretary of the National Association of Democratic Lawyers, questions the potency of generic statutes in the law books without the correlative transformation of unjust socioeconomic structures that include land and wealth:

> For post-apartheid laws to be significant in the struggle for the emancipation of women they must move in the same direction as other social, political, and economic institutions. If not, the law merely becomes "the law" in the big books without any significance to the real life and toils of women. This is often the case in countries which enact laws decreeing equality of treatment and equality before the law for both sexes. Women remain economically and materially dependent on men because of men's control of, and better access to, productive sources. If the wealth of South Africa is to be shared, we must demand an economic policy that will make space for a percentage of women to prove themselves.[51]

Given Archie Mafeje's contention, with which I concur, that the post-apartheid government is a bourgeois formation essentially protecting the interests of capital and the petit-bourgeois sector of South African society, the struggle for women's liberation assumes formidable proportions: to attain economic justice through socialist revolution, to create participatory democracy and an egalitarian society where working-class women and men are given the power to gain access to needed material, educational, and financial resources.[52] It is imperative, too, that the solid structures be constructed by the revolutionary elements of the working class to ensure that violence against women is prevented through educational and socioeconomic improvement programs. According to the Project for the Study of Violence based at the University of the Witwatersrand, violence against Black women domestic workers by white employers has escalated following Nelson Mandela's release, as part of the ventilation of white rage against the prospects of having a Black leader in South Africa.[53]

The history of the struggle of Black working-class women in Azania as narrated in the first chapter of this discourse illumines the contention that women's struggle does not simply imply adding women into bourgeois patriarchally defined structures, as is the current situation in the South African Parliament and at various levels of government, but wields the potential revolutionization of society by Black female working-class experience and culture.

The Creative Resourcefulness of Indigenous Black Working-Class Women in Revolutionary Struggle

The experience of Black working women cannot afford to be overlooked or underestimated within this equation of radical struggle. It is ironic to see that though over 25 percent of the South African parliamentary composition consists of women, most are urban-based and from elitist backgrounds. This is yet another indication of the embourgeoisiement of women's experience by the post-apartheid regime. Critiquing the dominant ANC–Freedom Charter brand of feminism, Rudo Gaidzanwa, an African womanist critic, in critically reflecting upon the period preceding the post-apartheid era, argues that the "liberalization" ethos extant in South Africa will not produce fundamental changes in gender practice and policy, principally because of the bourgeois orientation of the ANC:

> For women in South Africa, the relative muting of the debate [over] the implications of implementing options other than the social democratic one has meant that the perspectives on change have focused on bourgeois theorization and policy formulation. Very little attention was being paid to the implications of extending the social democratic policies to blacks or to other alternatives to social change.
>
> In this respect, there has been an organic linkage between organizations such as the ANC and the women's movements that they influence and control . . .
>
> For women, if the new order that is produced is a non-racial capitalist one, it follows that women's subordination will be reformed but not challenged substantially. It is also unlikely that the present levels of white security will be maintained while they are extended to blacks at the present level of capital accumulation. In this scenario, the bourgeois black women will join their white sisters on the rungs of privilege that accompany social democratic capitalism but the working and peasant women will have to continue to struggle for better lives than they can achieve in capitalist developing countries.[54]

Given the abysmal situation with regard to liberation of working-class and poor rural women today, five years after the first post-apartheid government was installed, Rudo Gaidzanwa's prognosis is absolutely correct.

Most female representatives in Parliament are drawn from urban communities, often overlooking specifically rural women's concerns. Although Nici Nelson, a development critic, recognizes the formative influences women can play in leadership positions and the sensitivity of women to women's issues, she sounds a caveat in the dynamics of power conditioned by women functioning in elitist and male-biased institutions:

> It would be rash to assert that placing women in positions of power will automatically result in a better deal for rural women. Women appointed to such positions may be from the urban elite (notoriously ignorant about the conditions prevailing on rural areas for both men and women) and will have been trained in universities which reflect the "male bias" of their societies.[55]

This class dimension is etched in sharp relief when one considers that in post-apartheid South African society today, close to 2 million Black women are still forced to work as cooks, maids, gardeners, and nannies for the white community and for elitist Black families, earning an average of R300 ($50) per month for a forty-hour-plus workweek. As Jacquelyn Cock points out in her study *Maids and Madams:*

> Domestic workers are among the most exploited groups in a society marked by extreme inequality. They are situated at the convergence of three lines along which social inequality is generated—class, race, and sex. These inequalities are related to the capitalist system of production in South Africa, which is not unique, but is "perhaps uniquely vicious in its degree of exploitation and repression" . . .
>
> . . . In South Africa, domestic service is predominantly a Black female institution.[56]

Many of these women, young and old, are either illiterate or forced to work for whites to overcome poverty and unemployment. In the rural areas, 42 percent of women are heads of households, and the majority of them live in poverty.[57] If there were opportunities to go to school and engage in productive work, these sisters, mothers, grandmothers, and daughters would surely have taken these options. Yet this institution of indentured slavery persists, in the name of "personal need" and "economic pragmatism." In fact, the new government appears to be using the expression "economic pragmatism" beyond excess. One can understand the tactical use of pragmatism in economic matters but can neither comprehend nor accept the essential immorality of tacitly accepting and incorporating this subhuman institution of domestication in an era of post-apartheid in a "free South Africa."

Sibongile Makhabela, a coordinator of the Advice Centers Association based in Johannesburg and the only woman among those charged in the "Soweto 11" trial of 1976, observes: "Women, along with men, have fought to attain fundamental human rights, equality, justice and peace in South Africa . . . And yet, their contribution goes unrecognized . . . There is still a tendency to regard women as supporters rather than as equals. 'Women' are allowed to debate and discuss certain issues only."[58]

The potency of Black women's thought and activity proffers new paradigms of liberation activity that are qualitatively different from those of men, a factor that serves to edify the entire liberation project. Makhabela contends:

> [T]here is evidence that they [women] utilize resources differently so that the empower-
> ment of women has a ripple effect. In other countries it has been proved that if women
> are conscientized then society becomes conscientized, whether it be about health care or
> civil rights. It has also noticed income earned by women is more likely to be spent on
> improving the quality of life of communities, that improving the literacy education lev-
> els of women has an immediate impact on the educational aspirations of the genera-
> tion.[59]

Speaking from the Tanzanian context of women's struggle, Rose Shayo comments on the multifarious roles that women assume and their innovativeness in executing these roles:

> Besides producing both food and cash crops, women have other roles related to the gen-
> eral development of the society. They are responsible for social production, i.e., ensuring
> the propagation and survival of successive generations. In this role not only do they
> carry the responsibility of child bearing but they are fully charged with duties related to
> caring and bringing up the children until they are able to take care of themselves. They
> also take care of other family members and the extended kin: parents, in-laws, cousins,
> nephews, etc., as well as performing the household chores. Women, especially rural
> women, also participate in community activities such as funerals, weddings, economic
> projects/programmes offered at the community level.[60]

Revolutionary liberation of Black society from colonialism, racism, and capitalism must include emancipation of all sectors of the society from sexist practices and con-ditioning, a point that South African liberation activist Dabi Nkululeko underscores when she asserts:

> As women represent the most oppressed segment of the society (more oppressed than
> even the oppressed men in each society) and constitute the majority, they are an essen-
> tial part of the struggle of the oppressed in each society. For this reason they must resist
> attempts to subordinate them and undermine their role in the struggle against other
> forms of oppression, and in the creation of programs against sexism in their society.[61]

Any movement for authentic transformation from a condition of injustice to one of justice will go nowhere if it either ignores the women of society or relegates women to a secondary status at the back of the movement, particularly working-class women. When discussing the question of women's powerlessness, sexism, and development, it is imperative that the issue of power and class be raised at every turn, a factor that many Western feminists often ignore. Fatima Babiker Mahmoud, director of a three-month residential course for women organized by the Institute for African Alternatives from March through May 1990 in London, reminds us of the need for more sophisticated analysis in feminist critiques, when she contends:

> If the contradiction between men and women is the fundamental contradiction of society, all men must possess power while all women are powerless. But it is clear that some men have no power while some women have power. Powerful men and women derive their power from their class position. But men are superior to women across classes because of the prevailing ideology of patriarchy. Patriarchal power is exercised over women within the family, classes and institutions and strives to maintain inferiority at all levels. This duality of access to power as a member of the dominant class and lack of power as a woman characterises the difference in the forms of oppression experienced by women of different classes.[62]

Should the post-apartheid state become liberated by a new revolutionary movement in the future, the new order ought to make a foremost and primordial commitment to the mothers and sisters of our struggle who have labored and toiled through the muck and filth and squalor of apartheid dispossession and impoverishment. Once the revolutionary state has acquired real political power, it ought to provide free education to all who need and desire it and, if strapped for financial resources, require the private sector (minus the capital monopolies) to donate a monthly stipend of R300 (about $50) to each working-class Black woman so that she can go to school, at night if necessary. Such a program would cost around R7 billion ($1.16 billion), which is still less than what the new government spends on defense, on preparing to fight South Africa's "enemies." One wonders who South Africa's new enemies are that such a large portion of the budget is warranted, when over 35 percent of Black people suffer from illiteracy, with rural women being the most deprived. Since literacy is the key to eradication of ignorance, including political literacy, any revolutionary government needs to invest most substantially within budget allocations toward literacy training and development, as Cuba did after the revolution in 1959, so that Cuba is virtually totally literate, surpassing even the powerful United States in its level of literacy.

Literacy and political education are the preconditions to the emancipation of the dispossessed and disenfranchised and are prerequisites for discarding the blinders of obfuscation and obscurantism imposed by the forces of capitalism and imperialism. In Venezuela, for instance, studies indicate that "there is a direct link between female labour force participation and education," and formally educated women experience more mobility in vocations outside the home.[63] Although such women have fewer children, they do not view children as an obstacle to their active participation in the labor force outside the household. According to Mansa Prah of the University of Ghana in Legon, the education of women is indispensable for fundamental social stability of any nation. He notes:

In the "World Report on Women," Brekke et al. argue that although education—schools, society, and the media—does tend to steer women on a course that sets them down firmly in the home and in badly paid jobs, it also empowers women. They suggest that there is strong evidence that education is one of the most decisive factors that change a woman's life. The World Fertility Survey discovered that women's ability to read and write was more closely related to their fertility, their use of contraception and their children's health than even their income (Brekke et al.). Education, then, is a relevant variable in the struggle for women's self-determination as well as in family planning and maternal and infant health care.[64]

Women, who signify the carriers of the tradition of the African nation and the hope of the future, must be accorded this right of literacy and education, including the right to higher education. In studies on Mozambican education, it was revealed that the enrollment of women in preuniversity level was 28 percent, while that of men was 72 percent.[65] In Somalia, less than one-fourth of girls had access to education at the secondary or tertiary levels. Female participation steadily decreases as one traverses the upper levels of the educational system, so that a tiny percentage of women end up receiving university degrees compared with men.[66] This disparity results in fewer women occupying roles in administrative and decisionmaking positions at various societal levels. Higher education must be challenged to embrace and engage in extraordinary measures to ensure the enrollment and graduation of women. It is not enough to enroll women in school; the quality of education must be such that school graduates leave with a substantial and qualitatively improved knowledge and skills base. In concrete terms, this translates into 3,000 hours of instruction needed to equip students with functional literacy.[67]

Teacher training needs to be revamped in Azania and in most other parts of Africa, so that teachers are qualitatively and efficiently trained through the teacher training curriculum to specialize in teaching in their respective subjects and other related teaching subjects. Holistic curricula, infused with a rounded knowledge base in the physical sciences, social sciences, humanities, the arts, and the primacy of indigenous knowledge systems, are required throughout the African continent. The underlying philosophy of all African educational curricula must be working-class-oriented and indigenous in its orientation, emphasizing that knowledge ought to be utilized for the collective welfare of all, especially the marginalized and deprived communities of society.

In most instances, women are the most enthusiastic literacy learners. When Zimbabwe introduced the Zimbabwe National Literacy Campaign shortly after independence in 1983, 85 percent of the learners were female, in a population of 2.5 million illiterate adults.[68] However, as Blandina Makoni laments in the Zimbabwean context, short of revolutionary transformation, the fundamental structures of inequality, patriarchy, and class privilege still persist, with working-class women confined to subservient and exploited roles: "It appears that Zimbabwe has not been able to capture the essence of a cultural revolution so indispensable in the attempt to carve a national identity. In this vacuum, women who morally raise children, who are the first teachers and custodians of a culture, are left ill-equipped to perform a societal function."[69]

Similarly, in South Africa in particular, gender oppression will not be eliminated without a correlational indigenous cultural, profemale, working-class-led socialist revolution. Arun Naicker, head of the Women's Unit at the UMTAPO Center in Durban, a Black Consciousness affiliate in South Africa, made this point abundantly clear at a meeting of the Institute for African Alternatives entitled "African Women: Transformation and Development" held in Dar es Salaam in 1991, when she asserted: "I believe that the anti-sexism movement in South Africa needs to be under the leadership of Black working class women who are the majority, and that we have to take our place alongside our sisters in the rest of Africa and shed the Eurocentric ideas and beliefs that have been foisted on us over the years."[70]

All of the rhetoric from the post-apartheid government regarding women's rights and social justice rings hollow, since the regime has opted to align itself with Western capital and the forces of neocolonial reaction.

Gender is a liberation *and* a development issue for working-class and indigenous women of the Two-Thirds World, principally because:

> Women are usually responsible for domestic work; the care of children, family health, cooking and providing food and other household services. In most societies they also play a major role in the productive activities of the family; in farming, paid domestic labour, services, industries and income-generating activities. In some societies they also have clear community roles.[71]

It has been verified by groups such as Women in Development in the United States that neocolonialism and capitalism have contributed toward the decline in the living standards of women in the Two-Thirds World.[72] It is for this reason that the equity approach was adopted as the strategy to forge women's development; the capitalist paradigm was rejected. It was a criterion of development priorities that women be brought into the development process as decisionmakers, initiators, and social-change agents as opposed to being passive recipients of development and change paradigms. It is imperative in such approaches to identify specific relations of inequality within familial, cultural, and socioeconomic structures so that such manifestations of inequality and injustice can be confronted and overcome through carefully deliberated policy and institutional practice. The knowledge of indigenous women must be considered at the outset when probing solutions to community or social problems. Miranda Munro, a gender and development researcher at the University of Reading, states: .

> Arriving at a common analysis of a situation has already been mentioned as an important stage in formulating an agreement for action in the community. In situations where women hold particular indigenous skills and knowledge, it is also important to understand the criteria leading to their decisions and preferences regarding productive or or-

ganizational practices. Ranking methodologies have been used with some success to identify local knowledge.⁷³

In Alexandra, a township of over 400,000 people outside Johannesburg, notwithstanding overwhelming economic and sociopolitical obstacles, women have prevailed in working to keep families together as well as being primarily involved in establishing day-care and kindergarten facilities for the community's children, such as administering the Ikaneng Pre-School. Women like Martha Mokgatle, a retired social worker, her daughter Tumi, and her granddaughter Palisa work tirelessly to advance the cause of children's education and nurturing.⁷⁴ Whereas men have generally lost faith in their ability to organize and transform the tenacious structures of oppression, women have seized every opportunity to protect and preserve the children for future generations, so that such children would be able to enjoy the fruits of a liberated Azania.

Another organization that warrants mention is the UMTAPO Center, an educational nongovernmental organization in Durban, Kwazulu-Natal, that is affiliated with the South African Association for Literacy and Education, a coalition of grassroots, nongovernmental organizations. It has made education around women's struggles a central issue in its overall program of Black improvement. UMTAPO held a National Women's Day Conference in August 1997, where various women gathered to discuss the issue of the oppression of women and resolved to intensify steps that challenge the oppression of women. Commissions discussed issues of the empowerment of women, critiquing the government's capitalist GEAR program, issues of violence against women (rape, domestic violence, child abuse), women's health (AIDS, reproductive rights, teenage pregnancy), women's access to land, empowerment of women for educational leadership, self-reliance needs for women, drugs and alcohol abuse, and marriage and dowry systems. An important resolution of the conference was to address the reeducation of the country's magistrates and justices who deal with issues of violence against women and "the setting up of a special unit within the justice system, headed by a woman, to handle cases of violence against women" and "setting up of a movement such as 'Fathers Incorporated'" to educate young men in particular to become members of a more humane society." The question of the conscientization of Black women and radical struggle, as elucidated earlier, was prominently featured at this important Women's Day Conference, attended by women activists from Jamaica and various parts of Africa.⁷⁵

Perhaps the two most important issues facing Black women in the struggle for liberation of Azania today are: the need for education in the broadest sense of the word and the need to empower indigenous rural women to function as change agents in the emancipation of South Africa. Deborah Mfugale, an educator from Tanzania, reflects on the struggles there for educational liberation and justice for women:

Adult education is becoming more important to development in the Third World. Government departments, development planners, economists, adult educators and even illiterate people are now convinced that priorities in planning, especially in rural areas, should feature adult education.

Our aim of education in Tanzania is to provide youth and adults with the basic tools of knowledge: literacy and numeracy . . .

Most Tanzanians engaged in agricultural production, the backbone of our economy, are women. To increase agricultural production through the use of modern methods, women must be taught to read and write through primers reflecting their environment and daily work. Adult education through literacy classes should improve the women's knowledge and skills and make them confident, competent, and aware.

As Paulo Freire points out, literacy "develops the students' consciousness of their rights, along with their critical presence in the real world. Literacy in this perspective, and not that of dominant classes, establishes itself as a process of search and creation."[76]

What is evident is that Black women in South Africa, like their dark-skinned sisters around the world, have generally always been working women and have historically engaged in economic production through agricultural and industrial labor. They have shouldered the responsibilities of providing for their families economically, as well as functioning as the leading protagonists in decolonization and anticapitalist struggles.[77] Black working-class women have often been more vigilant than their male counterparts in the Black resistance movement and have refused to cooperate with structures of racist and colonial repression at times when Black men were weakened by such policies.

At this stage of the Azanian liberation struggle, the principal ideological issue facing Black women is the imposition of economic, political, and cultural capitalism, as part of the strategy of neocolonialist designs for South Africa. Black working-class women particularly need to be most vigilant in rejecting the seductive tentacles of capitalism and consumerism that are being championed by the paltry middle-class Black elite. Since most Black women in South Africa are working class, it is critical that they organize themselves into a revolutionary mass grouping that will defy the assumptions and practices of capitalism in the country and unite with revolutionary Black male workers. It is then that the contradictions of capitalist exploitation in claiming to advance Black workers' interests while paying lip-service and below-livable wages to Black workers and accruing gargantuan profits for white industrial capitalists will be openly exposed. The Azanian Workers Union, with a membership of over 65,000 workers, is currently organizing on this revolutionary Black anticapitalist and prosocialist level.[78]

It is relevant to point out some conflictual philosophical issues that have emerged in the terrain of gender struggle, influenced by the politics of transformation in South Africa. The dominant "nonracial" liberal politics of the ANC, in which bourgeois white women are involved in expounding the feminist agenda often on behalf of Black women or concerning Black women, is at odds with the Black Consciousness position that quintessentially contends, "Black woman, you are on your own!" In other words, the Black women's struggle is one in which Black women alone are responsible for advancing, because of the materiality of

Black women's oppression and the general privileged middle-class status of white women.

Although one does not want to sound racially chauvinistic or essentialist, it is nevertheless ironic that the ANC government initially appointed Sue Lund, a white woman, as head of the Land Reform Pilot Program in 1994.[79] On such a decisive issue as land reform or land reclamation, the portfolio of head of any program pertaining to land reform could have been assigned to a Black person, particularly a Black woman, who is academically knowledgeable and viscerally familiar with concerns of dispossession and disenfranchisement. In fact, according to Mihloti Mathye, head of women's rights in the Land Sub-Directorate in the Department of Land Affairs, "the power relations between men and women are nowhere near as unequal as when it concerns land" and "very few" women have benefited from the land reform program.[80] Mmakgomo Tshatsinde, a rural and development economist, notes that though rural women play a decisive role in agricultural production in South Africa, they still experience shortage of water and land, hindrances of access to markets and finance, lack of support from agricultural authorities, and lack of managerial know-how.[81] She points out that land deprivation is the most burdensome problem for rural women: "Rural women cannot own the land they cultivate because they are black and female. Land which is available to them is usually officially owned by their husbands and is communal. Thus they cannot make improvements without the husband's consent and because it is communal, everyone is entitled to it."[82]

Until racism is obliterated from the South African socioeconomic map and the nation is truly a race and color-blind society, race will continue to be an issue in such areas. It is not just a question of gender consciousness that is necessary in areas of land reform, as Shamim Meer asserts, but issues of capitalism, culture, and power.[83] Black women are unequivocally closest to such issues and struggles. As Mihloti Mathye argues: "A key factor which contributes to the failure of women to overcome poverty is a lack of access to and rights to own and control land . . . It is only when women are active participants in land reform that they can influence and share control over development initiatives, resources and the decisions that affect them."[84] Mathye is referring to African women in this instance, specifically rural women.

The solidarity often sought by white feminists in South Africa with their "Black sisters" in the struggle for national liberation may be a noble goal, but it generally falls short in terms of substance. A key problem is that most white feminists have a stereotypical and colonial image of African culture that is ahistorical, fixed in time and space, and essentially "antifemale," one that views African culture as possessing no potential for indigenous transformation of its male-dominated structures. Valerie Amos and Pratiba Parmar, reflecting on Black women's experience in Britain, confirm this tendency by Eurofeminists:

> By adopting the research methods and frameworks of white male academics much academic feminist writing fails to challenge their assumptions, repeats their racial chauvinism and is consequently of less use to us.

One such assumption is that pre-capitalist economies equal backwardness in both a cultural and ideological sense and in fact are responsible for the continued oppression of women in these societies. It is further implied that it is only when Third World women enter into capitalist relations will they have any hope of liberation. . .

Furthermore, when Black and Third World women are being told that imperialism is good for us, it should be of no great surprise to anyone when we reject a feminism which uses Western social and economic systems to judge and make pronouncements about how Third World women can become emancipated. Feminist theories which examine our cultural practices as "feudal residues" or label us "traditional," also portray us as politically immature women who need to be versed and schooled in the ethos of Western feminism. They need to be continually challenged, exposed for their racism and denied any legitimacy as authentic feminists.[85]

These feminists often assume that Western culture possesses more capacity for structural gender transformation, as if to say that European culture is basically pro-female! It is important that indigenous women's cultures in Africa that are empowering for women and were suppressed by colonial ideology be reclaimed in the struggle to assert the primacy of indigeneity. Ifi Amadiume has described the resilience of indigenous Nnobi culture in Nigeria and its empowerment of women in that part of the continent.[86] Musi Katerere, from the Zimbabwe Association of Community Theatre, while pointing out the oppressive elements of Shona culture with regard to women, asserts that concomitantly, there are elements that demonstrate the power of indigenous women. She explains:

That a person's spirit, especially a woman's, is feared most after her death is quite interesting. It really shows how powerful women are in our society. The Shonas strongly believe that the mother's spirit can kill you if they leave you unguarded. They say, "Mudzimu yaamai yadimbura mbereko," when something bad happens, meaning that the spirits of one's mother have untied the knot. Families who consult traditional healers for any family problem are usually told that they might not have paid "mombe youmai." This is a cow which is paid during marriage to the mother of the bride. Bad luck usually surrounds families of those who have not paid the cow to the mother/grandmother. This all proves that without women society would be nothing. There would be no love, no reproduction, no continuity and culture. In a nutshell there is no life and culture without a woman.[87]

It is instructive to note that Shona women possessed intricate scientific and botanical knowledge that was used in agricultural production prior to European colonization, and they engaged in taxonomic classification of seeds and plants, so that they were able to determine exactly when the best time and what the best place would be for seed cultivation. Ntombie Gata, from the Department of Research and Special Services in Zimbabwe, explains:

The onset of the cropping season heralded the crucial role of women in the selection of seed with preferred characteristics. Women selected seeds of characteristics such as color, grain size, agronomic stability, suitability to different types of soils and terrain, drought and disease tolerance, palatability, grain storage and processing, etc. Therefore, women needed to be familiar with these crop types and varieties and possessed invaluable knowledge and experience about plant taxonomy and morphology and the different uses of plant genetic resource material. In fact, a working classification of crops by gender, shows that crops with multiple uses, in both form and function, such as pumpkins, cowpeas, groundnuts, and millet were considered women's crops while those with less diversified uses like tobacco and cotton were considered as men's crops [you can see who carries the day in indigenous agriculture]. This rough classification of crops along gender lines is illustrative of the central role played by women in maintaining and conserving bio-diversity which ensured reasonable options for responding to changing environmental conditions (Mushita, 1992; Chidsonga, 1992) . . .

. . . Women were excellent sources of both genetic and cultural information about species because each had to meet multiple needs within their household mandated sphere of activities.[88]

In *vlei* (hilly) areas of Zimbabwe, women used tillage systems that conserved water while preventing soil erosion. Practices were employed, such as *mateka,* to tap precious water resources, conserve them, and prevent further water loss through soil erosion. Terracing was also practiced by women, particularly in the Eastern Highlands. All of these practices functioned to maximize production, conserve natural resources, harmonize modes of cultivation with the natural environment, and expend low-energy levels during cultivation. Sara Mvududu writes:

Women's knowledge about the environment is often more comprehensive because of the diversity of their tasks. The main responsibility for sustaining the family is usually assigned to the women, increasingly so because of male migration away from degraded rural areas. This makes women's knowledge an important issue in environmental management and rehabilitation.[89]

In the indigenous context of North America, Native women also lament the imposition of patriarchal colonialism, most classically distorted by Eurocentric ethnographers with the depiction of the male "chief" as the undisputed leader of Indian nations. In Africa and the Americas, the "chief" was a colonial creation to facilitate European male communication with indigenous nations for the explicit purpose of dispossession, predicated on a European patriarchal cultural model. Jeff Haynes, a British political historian, explains that "[a]lthough many communities did not have a 'chief' to rule them, because Europeans believed that *all* societies have at their head a leader, many chief-less communities produced a de facto leader when required."[90] Speaking from the Iroquois nation in Canada, Martha Mantour asserts:

When the Europeans came to the North American continent, the Iroquois woman had great economic, political and legal power enshrined in the Iroquois Constitution (Gayanerekowa—this is also known as the Great Law of Peace of the Constitution of the Iroquois Confederacy). Whereas the European patriarchy had successfully suppressed the power of their own women (the European woman upon marriage came under the total bondage and control of her husband who then owned her property, social life, body and children), the Iroquois Constitution organized Iroquois society by large kinship groups or clans.

The decision making power was exercised equally by both the male and female councils in each clan. Since women were considered the progenitors of the Nations, they held the traditional lands for the preservation of the future generations and hereditary titles of chieftainship, war chief and clan membership (Spittal, 1990).[91]

Euro-American feminists need to pay tribute to the Iroquois nations for the institutionalization of all human rights, including women's rights in the United States, since the U.S. Constitution was patterned upon the Iroquois Confederacy, a fact little known to the general public.[92]

Much more of this kind of research needs to be pursued on indigenous cultures and traditions that demonstrate the inherent power of women's roles, particularly within Black cultures in South Africa, so that these cultural trajectories can be re-tapped and reappropriated in the movement for radical transformation. We need to understand that both matriarchal and male-biased cultures existed in Africa in the precolonial era, without assuming that European patriarchy was also normative for Africa.

Musi Katerere's critical observations are important in regard to the question of African women and cultural transformation, intrinsic to any gender and social analysis on the continent, what can be described as female ancestral wisdom. Although simple, they are vitally important for liberation struggle. She incisively declares, with challenges for all radical critics to ponder:

> We should know who we are and where we came from.
> We should know our history and culture;
> What are we doing to preserve our culture and heritage?
> Are we doing it for the future generation or just for our present survival?
> Are we proud to be black?
> Do we have our customs and traditions which bind us together as a people?
> What are we doing to resuscitate and protect the positive aspects of our culture?
> Do we imitate foreign culture to the extent of forgetting where we came from?
> Are we genuinely fighting the oppressive aspects of our culture without the intervention of donors and with the participation of all women?
> Are the criticisms of our oppressive traditions coming from within us or are they imported and instigated by people who have not lived our lives but might have lived with us in order to make judgements?
> Are we liberating ourselves and being exemplary to our future generation?
> What are we going to do to revive our oral traditions, ngano/narratives, praise poetry, proverbs, riddles, expressive songs and dances?
> As mothers/sisters/aunts, what are we doing to de-colonise our minds as well as our children's?[93]

Ruth Meena, an African womanist critic, points out that the perspectives and viewpoints of indigenous African women are still perceived as "raw-data" producers for the "intellectual" factories of the West, compelling the dependence of women from the South on the publishers from the North.[94] Unfortunately, there is little evidence in the post-apartheid era to suggest that many white feminists' behaviors have been fundamentally different from their role in the 1970s and 1980s, particularly in their failing to advocate for and support socialist-led revolution waged by the Black working class. Personal economic interest is still extremely difficult to sacrifice for the advancement and liberation of working-class people, as most middle-class people discover. In this regard, bourgeois feminist critics in South Africa do need to consider Oyeronke Oyewumi's critique when she forthrightly asseverates:

> Western writings . . . on Africa have been racist and ethnocentric, projecting Africans among other things, as savage, sub-human, hypersexed. I contend that such images are represented in Western feminist discourse on Africa today. I am suggesting that despite the professed radical nature of feminism with regard to African Studies, it is not distinct from traditional Africanist scholarship . . . it appears as if a second nineteenth century is beginning all over again.[95]

Oyewumi, like other indigenous-oriented scholars, is sounding the clarion call for African studies, like women's studies in Azania and the rest of the Black world, to become *indigenized*, so that they both are freed from the shackles of Western epistemological paradigms and metaphysical systems that have unconscionably imposed themselves on indigenous ways of knowing and doing.[96] Appolonia Kerenge, an organizer with the Pan African Institute for Development, based in Yaounde, Cameroon, echoes this need for indigenous intellectual movement in the decolonization of Africa and the empowerment of women, given that colonialism institutionalized and entrenched the oppression of women, when she contends that "social science parameters derived from industrialised Western societies tend to lack the intellectual concepts and tools with which to describe and analyze present day transformations in Africa."[97]

The axiological question within the discourse on gender transformation in South Africa is that of Black working-class female servitude, which borders on the level of contemporary slavery, in which everyone, particularly whites, middle-class "Coloreds," "Indians," and Africans expect that all menial chores ought to be performed primarily by poor African women. Capitalism exacerbates this situation, especially in the current post-apartheid context, because Black women are particularly susceptible to economic vulnerability. The national monetary currency plummets daily, and the ravages of unemployment for both skilled and unskilled Black workers is astronomical, at 40 percent in some urban areas. It is this ethos of servility that needs to be radically disrupted, prefaced by a revolutionary ideology that eradicates internalized oppression where Black women view themselves as "destined to do the dirty work for others" and are made to feel powerless. This perverse condition must be supplanted by one in which Black working-class women refuse to be inferior and are transmuted into freedom fighters who strike out at those who demand their deference and servility. They must refuse to cooperate with the male masters or female mistresses of oppression, the modern day slave owner or slave exploiter, signifying the launching of the Black women's working-class revolution.

A revolutionary consciousness is necessitated on all levels of Black working-class women's culture, mediated by revolutionary organizations such as the Azanian Workers Union, the Black Allied Workers Union, the Socialist Party of Azania, and other such radical formations. These organizations need to become instruments of empowerment, rooted in the indigenous African precept of collective sharing of economic resources, meager as these are in many instances. Yet this is necessary to break the yoke of dependence of Black women forced into serving whites and others because of economic vulnerability. They need to have women at the helm and in solid representation in their rank and file.

Concomitantly, as part of the revolutionary conscientization and Black worker praxis of resistance, Black working-class organizations need to challenge the neocolonial state by protracted demonstrations, strikes, and other resistance actions, demanding that the state subsidize Black women's education so that they are not prevented from attaining at least high-school educational levels. The state needs to be confronted and forced to create viable and productive employment for Black women in areas of industry, housing, agriculture, and education and to provide support services in areas like social welfare and child care.

The budget for this fitting tribute to the Black grandmothers, mothers, and sisters of the liberation struggle ought to come from the R40 billion ($6.7 billion) that the new government continues to pay each year to international banks and lending institutions for debts accruing from apartheid.[98] On the side of diplomacy and resistance, the ANC-led government needs to be constantly reminded through daily strikes, protests, sit-ins, and demonstrations that it is betraying the interests of Black workers by meeting these global economic obligations and that it has sold the Black workers out to the market of predatory imperialism. Should such actions go unheeded and the regime demonstrate increased intransigence (as it probably will, given the recent historical and current responses to Black worker demands), it may unfortunately become necessary to launch guerrilla attacks on symbolic industrial and financial installations of the post-apartheid economy and engage in a subsequent guerrilla war of attrition.

Further, it becomes imperative for those revolutionary organizations like the newly formed Socialist Party of Azania and AZAWTU to adopt a revolutionary understanding of gender and economic concepts. No longer can women's labor at home be relegated to "marginal" work with no compensatory value. By the same token, all Black working-class men need to assume roles of parent, father, and, on occasion, housekeeper, depending on individual circumstances. The days of men frowning upon child care as subordinate "women's work" and viewing such work with derision are long gone in this revolutionary culture. R. P. Ngcongo explains that for working-class Black women and for women in rural areas of South Africa, strictures by male counterparts are pronounced and sexist attitudes are more obdurate. She states:

It is my observation that the dilemma of many working class women, particularly rural women, is worse. However, rural working-class women talk about it less openly. This may be because of their limited access to education. These women are also more dependent financially on men or their families. Thus they may not feel sufficiently equipped to challenge men, and may not want to lest they lose desperately needed financial support.

The traditional role of men as superior and as decision makers in rural settings is thus deeply entrenched and is likely to be seen as normal by some women and men. Women's vulnerability encourages compliance to men's perceived superiority. Such compliance is seen as synonymous with respect. . .

Women, especially working class women, sometimes see men's dominance as normal. They either do not notice oppression or believe they are powerless and helpless, thus resigning themselves to the situation. Indeed, some African women believe men are predestined to be superior to women and to have women at their service. They refer to Biblical texts, such as Ephesians 5:22–33 and Deuteronomy 21:11–13 to support their thinking.

The perceived inferior position of women is reinforced by what Nene (1987) calls games of powerlessness. These include pretending to be weak in the presence of men or pretending to agree with them. The alleged inferiority is also made real by the women's self-victimization. Some women believe that without men they are nothing, have no one to appreciate or to live for.[99]

Sexist conditioning of both Black women and men needs to be eviscerated so that women will be liberated from enslavement to the internalized ideology of inferiorization and men will be liberated from the regressive and spurious sense of maleness possessing superiority over women. In any revolutionary project in establishing a liberated Azania, this pathological conditioning requires magnanimous uprooting from South African urban and rural society, together with the triple intrinsic injustices of colonialism, racism, and capitalism.

Women and men ought to complement one other in terms of the division of labor, with men performing domestic or household chores just as women may be called on to be bricklayers and carpenters; this is envisaged as part of the revolutionary transformation of Black working-class culture. Understandably, this does not mean that men's and women's roles are totally obscured; but it does imply that regardless of what they are and the transformation of certain male-only and female-only vocations, both women and men are accorded gender complementarity and mutual equality and reciprocity. In this context, sexism in the Black community is critiqued from an indigenous cultural standpoint, maintaining the essence of African working-class culture while eschewing its oppressive and negative practices, prototypical of all cultures.

Although Black working-class women are forced by the pressure of historical, political, and economic structures under the colonial-capitalist system to work as the lowest-paid factory workers and as maids, cooks, and nannies for white families, they have never accepted their domination and subjugation as a permanent part of the South African sociopolitical landscape.[100] They have borne the brunt of the apartheid whip, working six to seven days a week at white households and taking care of white children while being forced to leave their loved ones behind, a supreme sacrifice made just to feed, clothe, and educate their children.[101] Once these women are conscientized, they will be potential guerrilla freedom fighters in the intensification of the struggle toward the future Black working-class revolution.[102]

Black working-class women are the heroines of the South African liberation movement yet are still deprecated in post-apartheid South Africa because of the sexist conditioning of society and the state. Revolutionary liberation by Black working-class culture must include emancipation of all sectors of society from this legacy of the repressive condition of misogynist behavior and practices.[103] Black working-class women have a distinctive contribution to make toward the materialization of the future liberated South Africa, not only because they exemplify cultures of resistance but because the irruption of Black women's culture possesses certain distinctive features in marshaling resources for the consolidation of the revolutionary forces. They are no longer diffident about their abuse by patriarchy and capitalism and are demanding transformation of these dehumanizing socioeconomic structures in concert with working-class ethics of justice, complementarity of gender divisions of labor, and conservation of scarcely available material resources. R. P. Ngcongo's recommendations are also instructive here, when she powerfully declares:

> There is an urgent need for education for all African men and women about what women's liberation means. Women's liberation is linked to the liberation of all Africans and all South Africans. It is about the freedom of humankind. If women and men are allowed an opportunity to be what they can be and are given necessary support and opportunity, an awesome potential could be unleashed. Women's liberation is not about being the same as men: it is about the removal of all factors, laws, norms, attitudes and rules which prevent women from attaining the growth and liberation of their potential as human beings of equal dignity with others.
>
> Education about liberation is essential because it clears myths about what women's liberation is not. Whereas women's liberation is seen by some Africans to contribute to the destabilization of families or which encourages women to challenge men or to be non-compliant, with education it will be clear that women's liberation is likely to enable women to be more resourceful family members. This strength will emanate from the opportunities for mutual satisfaction of needs which are important to men and women. Viewed in this way, women's liberation suggests men's liberation.[104]

Revolutionary Black working-class transformation must include the expunging of the evil of sexism at every level so that the vocational goals and aspirations of Black women and men workers, as R. P. Ngcongo argues, can be holistically realized. At all levels of education, beginning with preschool and running through elementary, high school, and college and university, educational philosophies and curricula are pivotal in redressing issues of gender inequality and the roles and statuses of women.[105] The colonial approach and missionary mentality that preferred the education of boys over girls, albeit a tiny segment of the colonized African population, has resulted in two of every three illiterates in Africa being women.[106] Appolonia Kerenge, an organizer for radical social change, underscores the gravity of illiteracy when it comes to women:

> Illiteracy is of course a major stumbling block to a person's development. The illiterate woman will be dysfunctional in situations where literacy skills are needed; she is likely to be less employable, suffer from isolation in the sense that she cannot read and therefore has no contact with the outside world beyond her family circle and neighbourhood. Having little exposure to the outside world, women tend to be more conservative and low risk takers. This state of affairs makes them less visible not only to planners and policy-makers but also to researchers, field enumerators, and extension workers of whom the majority are men, and who are likely to by-pass women due to cultural and language barriers.[107]

The erstwhile and contemporary attitude of viewing women as ill-suited for the sciences, for instance, mathematics and engineering, needs to be eradicated so that these oppressive distortions do not linger in educational institutions. Transformation of sexist attitudes in the construction of a revolutionary Azanian society must be accompanied by a heightened awareness of the need to consciously insist that women be represented at levels of leadership of all societal, educational, economic, political, and cultural restructuring processes in South Africa. In this manner, the revolutionary order being proposed will conscientize working-class Black women at its core. In this regard, the grassroots organization called the African Association for Literacy and Education (AALAE), founded in 1984, is relevant since it has been engaged in positive educational and consciousness-raising efforts geared toward reaching diverse and deprived sections of the population throughout the African continent. The AALAE program aims to implement the following measures as part of a Pan African literacy-conscientization work plan:

A. Building self-reliant structures and organizational frameworks
B. Creating program networks
C. Strengthening South-South exchange solidarity
D. Promoting innovative initiatives and undertaking anticipatory actions to strengthen and popularise ideas, approaches, methods or techniques that may for the time being appear unimportant but whose development potential is tremendous
E. Helping to attract and mobilize more funds and resources for adult education work
F. Enhancing members' professional and technical capacities, including the ability to influence public policy
G. Creating environments for the promotion of literacy
H. Legitimizing work at the grassroots[108]

The program that AALAE has initiated functions as a model for developing literacy skills and conscientization awareness, with both women and men assuming important leadership roles. It has launched three-year plans as part of its modus operandi and is involved in the following areas: building institutional capacities in the adult education movement in Africa; developing program networks in the spheres of literacy, women in education and development, community education, participatory research and training, environmental education, artists for development, and university adult and continuing education; catalytic initiatives, which entail developing materials for workers' education in certain target areas in Africa such as South Africa, Namibia, Ethiopia, Mozambique, Angola, Guinea Bissau, Sudan, Egypt, and Somalia; exchange programs which involve worker representatives conferring across the continent and abroad; peace education, human and people's rights, which attempts to understand the causes of war and generate strategies toward peace and against militarism; research, training, and evaluation, which entails training teams for educational training and program review; and information, publication, and communication, which involves disseminating innovative ideas and information, distributing relevant educational literature, and decentralizing mechanisms for the production and popularization of educational material.[109] One of the important facets of AALAE is its emphasis on decentralized structures and dissemination of knowledge, as opposed to a centralized, inward-looking, and narrowly focused operation. This posture is imperative for effective grassroots organizing in Africa, particularly among rural communities.

The philosophical emphasis on self-reliance is vital for the success of any radical transformative movement, given that dependence on exogenous agencies and organizations for indigenous program development perpetuates the cycle of dependence and paternalism from the colonial era. Azanians need to embody this principle in its crystallized form, as a logical consequence of the Black Consciousness philosophy of Black self-reliance and self-sustenance: to struggle to achieve material transformation as opposed to depending on anticipatory handouts from benevolent institutions. Furthermore, it is critical that South-South solidarity formations and networks transpire, as an extension of the Pan-African network proposed by this revolutionary program. The demon of imperialism that induces underdevelopment, poverty, illiteracy, and cultural stagnation has its roots in the West and has afflicted the entire colonized world. Working-class and rural peasant women, men, and youth, in Africa, Asia, Latin America, North America, the Caribbean, and the Pacific are all struggling against the same monstrosity of oppression.

We can learn much from our sisters and brothers in Cuba, who continue to defy the largest superpower in the world, the United States, notwithstanding the economic blockade against the struggling island nation. Yet Cuba continues to stave off imperialist attempts to subvert its revolution and maintains an educated and progressive population, with free education and health care for all.[110] Cuba trains doctors from other Latin American countries despite its own economic hardships and when Hurricane Mitch tore down the plantations of Honduras and Nicaragua, causing untold human suffering, it was Cuban doctors who were the first to establish medical aid camps for the injured.[111]

So, too, we can learn from our friends in China who are involved in maintaining a socialist revolution in the world's most populous country, facing economic pressures from the West. Although China has vacillated in terms of its foreign policy and has supported dictatorial regimes in Indonesia and South Korea out of its own cynical geopolitical and economic self-interest, China has developed a successful social model and has been able to feed its domestic population, provide mass health care, and support its educational institutions, so that the Chinese working classes, though of modest means, are economically and socially independent. Axiological in China's success is its wide-scale program for the enhancement, education, and training of women in the rural sector, where vigorous efforts were made to train women at all levels in "agriculture, industry, handicraft, commerce, education, health, and other subjects" with the assistance of investment funds and rural policies geared toward women and agriculture in particular.[112]

Azanian revolutionary organizations need to connect and develop solidarity networks with organizations representing the Sandinistas in Nicaragua, the indigenous Mayan movements of Guatemala, the indigenous revolutionary movements of Chiapas, the movements for self-determination and independence among the Australian Aboriginal people, the Maoris of Ao Te Roa (New Zealand), the movements fighting for independence from colonialism and self-determination in the United States like the American Indian movement, the Committee of Five Hundred Years of Dignity and Resistance based in Cleveland, Ohio, the Black Radical Congress, and the Chicano resistance movements, and those engaged in struggling for independence from French neocolonialism in New Caledonia and Palau in the Pacific.

Ultimately, the struggle for Azanian independence and socialism led by Black workers is an indigenous struggle for continental independence and socialism for the African Diaspora, as well as an intrinsic part of the global struggle for socialism and the liberation for all oppressed and colonized indigenous peoples, from Iraq to Yugoslavia, from Australia to Alaska, and from Mexico to Chile. The revolutionary struggle is *one*, since we are all battling the same oppressive beast, trying to survive *genocide*. Revolution moves us into the mode of not just surviving and accepting the dictates of imperialism and capitalism but empowers us to demand what is rightfully ours as colonized classes, to live in the way the Creator destined us to be: thriving, living in justice and in harmony with each other and all of creation.

Summary and Conclusion

Black working-class women represent the most dynamic and powerful force of African society. They are the tireless and undaunted carriers of Africa's cultures of resistance and represent the epicenter for the assumption of the new decolonized Azanian national identity and independent continental African identity. Any truly Black working-class-oriented government in South Africa would do well to keep these points in the forefront of its thinking on public, economic, social, and educational policy. Mao Tse-tung, the late revolutionary leader from China, once declared that the condition of women indicates the extent to which a society is authentically liberated.[113] The future of a liberated Socialist Republic of Azania and a unified and integrated socialist Africa indeed lies with the revolutionary organization and mobilization of indigenous working-class women. A revolutionary Azania may just be the linchpin for the launching of a continentwide revolution that materializes in a united and socialist Africa.

Colonialism and capitalism in the South African situation have resulted in the recreation of a European ruling-class culture, at the cost of humiliating Black people and forcing us to serve this oppressive culture. South Africa, after all, is neither in Europe nor is it an "island of Europe in Africa," as Steve Biko argued.[114] Radical Black working-class cultures of resistance wield the greatest potential for the materialization of a de-Europeanized and re-Africanized South Africa/Azania so the nation may be restored to its authentic African national *working-class* and *indigenous socialistic* character, in terms of its economy, land, polity, language, symbols, names, educational pedagogy, identity, and most important, ownership of resources. There are prospects for revolution led by a *conscientized* rural peasantry and by radical Black workers, as well as limitations considering the elusive and overbearing nature of capitalist hegemony. The most decisive factor is the ability of radical Black peasants and workers to organize themselves more effectively against the capitalist and neocolonialist odds and the ability to mobilize working-class culture in the forging of revolutionary transformation through ceaseless self-sacrificial militant praxis and armed struggle. The resolve of radical Black workers in Azania in this direction remains to be seen.

Notes

1. Carl Brecker raises such questions in his article "The ANC After Mandela," *International Viewpoint,* June 1999, and it appears that he wrings his hands in frustration and wonders how the "left" could assume a more formidable role in determining the change process in post-apartheid society in the face of the apparently unopposed "neo-market" forces.

2. Archie Mafeje expresses his profound frustration at the Eurocentrism and even condescendingly racist attitudes of white members of the South African Communist Party, such as Dan O'Meara, Duncan Innes, Ben Turok, and the late Joe Slovo, who insist that they know what is best for South Africa based on their whiteness and knowledge of revolution; see Mafeje's *The National Question in Southern African Settler Societies* (Harare, Zimbabwe: SAPES, 1997), 1–12.

3. Nigel Harris, *The Mandate of Heaven: Marx and Mao in China* (London: Quartet Books, 1978), 158–159.

4. C.R.D. Halisi, "The Political Role of the Trade Union Movement," in Anthony Freeman and Diane Bendahlmane, eds., *Black Labor Unions in South Africa: Report of a Symposium* (Washington, D.C.: Foreign Service Institute, U.S. Department of State, 1987), 42.

5. Although domestic television channels broadcast local news concerns, the preponderance of Western films and productions, particularly those from the United States, acts as a brainwashing medium that vitiates the minds of Black people, including the Black poor, into thinking that the Western capitalist system is the ideal condition for all of humanity.

6. Diego Ribadeneira, "National Grass-Roots Campaign Tackles Widening Wealth Gap," *Arizona Daily Star,* July 3, 1999.

7. Joerg Rieger, introduction to *Liberating the Future: God, Mammon, and Theology,* ed. Joerg Rieger (Minneapolis: Fortress Press, 1998), 6.

8. Ibid., 7.

9. The Jubilee Movement refers to the Biblical Jubilee, where debts would be forgiven every fifty years. Pope John Paul and the World Council of Churches have called for 2000 to be a Jubilee year, marking the two-thousandth anniversary of Jesus' birth. See Robert Naiman, "G7: Drop the Debt or Stand Aside," *Sunday Journal* (Washington, D.C.), June 13, 1999. These religious leaders and organizations fail to excoriate the immoral basis under which these debts were accumulated and the fact that it is the West which ought to be making reparations to the underdeveloped world for its pillaging of resources from the very countries paying these debts and for perpetrating genocide against the peoples of Africa, Asia, and the Americas for the past five centuries.

10. A recent book that substantively documents the savagery of genocide by European settler colonization in North America is James Wilson's *The Earth Shall Weep: A History of Native America* (New York: Atlantic Monthly Press, 1999).

11. George Casalis has an incisive critique of the systemic nature of ideology and its orchestration by oppressive systems for purposes of subjugation in *Ideas Don't Fall from The Sky* (Maryknoll, N.Y.: Orbis Books, 1985). See also Richard Weaver's critique of the development of crass materialism and exploitation by oppressive ideologies and the argument for the need to assert individual freedom in the face of ruling-class control in his *Ideas Have Consequences* (Chicago and London: University of Chicago Press, 1948).

12. Telephone interview with Barney Mokgatle, June 29, 1999.

13. See, for instance, the Kwazulu-Natal provincial publication *UMXOXI,* February–March 1995, which explicitly states, "You have been warned—no work, no pay." Staff members are warned that they will not be paid for participating in illegal or legal strikes, according to Resolution 261 of the Kwazulu-Natal cabinet. Similar kinds of legislation are being considered in other provincial and industrial circles.

14. The author has witnessed Patrick Mkhize's effectiveness in organizing Black youth into revolutionary cadres in rural areas outside Durban in Kwazulu-Natal. Even though the scope of the organizing is small, it is precisely the cultivation of small, decentralized revolutionary cells around the country that will cumulatively make the revolutionary difference.

15. *South African Labor News*, June 6, 1995.

16. *Mail and Guardian* (Johannesburg), August 8–14, 1997.

17. *Business Day* (Johannesburg), July 20, 1995.

18. *Star* (Johannesburg), January 4, 1996.

19. *Pretoria News*, December 8, 1995.

20. SACAWU, for instance, issued statements about its skepticism of the entire negotiations process between the apartheid regime and the ANC at a congress meeting held at the University of Natal in Durban in July 1993.

21. *Sowetan* (Johannesburg), August 3, 1995.

22. *Azanian Labor Journal* (Johannesburg), Vol. 1, No. 2, 1989.

23. *Azanian Labor Journal* (Johannesburg), Vol. 1, No. 1, 1988.

24. Ibid.

25. C.R.D. Halisi, "Racial Proletarianization and Some Contemporary Dimensions of Black Consciousness," in R. Hunt Davis Jr., ed., *Apartheid Unravels* (Gainesville: Center for African Studies, University of Florida Press, 1997).

26. Jordi Martell, "South Africa ANC Victory: Masses Expect Action," article disseminated by the Black Radical Congress listserv, June 16, 1999.

27. See all of the latest reports emerging from South African newspapers, magazines, and tabloids that emphasize commercialization and consumerism, appealing particularly to the

Black consumer. *Drum*, February 1995, and the June 1993 issue of *Tribute* magazine are examples.

28. Solomon Mlambo, a COSATU activist, cautioned against the superficial interclass alliances being forged in the desperation to topple the apartheid regime, coalitions that overlooked conflicting class allegiances and aspirations. See *South African Labor Bulletin*, Vol. 13, No. 8, 1989.

29. Mafeje, *The National Question in Southern African Settler Societies*, 18.

30. Ibid.

31. Harris, *Mandate of Heaven*, 13.

32. This struggle and extraodinary sacrifice is described, for instance, in ibid., pt. 1.

33. Ibid., 28.

34. *New York Times*, July 5, 1999.

35. A recent interview by David Barsamian with Eduardo Galeano, a Latin American writer, gives us a sense of the world of struggle, quite different from the somber picture painted by diffident bourgeois historians and social critics. See the *Progressive*, July 1999.

36. See, for instance, a collection of innovative Black cultural resistance plays, in Duma Ndlovu, ed., *Woza Afrika: An Anthology of South African Plays* (New York: G. Braziller, 1986). P. V. Shava's *A People's Voice: Black South African Writing in the Twentieth Century* (London, and Athens, Ohio: Zed Books and Ohio University, 1989) is another informative source on Black cultural resistance literary expression in novels, poetry, and drama.

37. Debbie Bonin and Ari Sitas, "Lessons from the Sarmcol Strike," in W. Cobbnett and R. Cohen, eds., *Popular Struggles in South Africa* (Trenton, N.J.: Africa World Press, 1988), 54.

38. These plays are found in Ndlovu, *Woza Afrika*.

39. From a pamphlet advertising the *Long March*, produced by SAWCO (Sarmcol Workers' Collective), October 1988.

40. Jonas Gwangwa and Fulco van Aurich, "The Melody of Freedom: A Reflection on Music," in W. Campschreur and J. Divendaal, eds., *Culture in Another South Africa* (New York: Olive Branch Press, 1989), 148.

41. Ibid., 151.

42. Ibid., 146.

43. Adepoju G. Onibokun, Ajibola J. Kumuyi, and Comfort Akinsette, *Women and Leadership in Nigeria: An Analysis of Factors Determining the Participation of Women in Development Process* (Ibadan, Nigeria: Centre for African Settlement Studies and Development [CASSAD], 1995), 1.

44. Cited in Irene Staunton, comp. and ed., *Mothers of the Revolution: The War Experiences of Thirty Women* (London: James Currey, and Bloomington and Indianapolis: Indiana University Press, 1991), 213.

45. Che Guevara, "The Role of the Woman," in *Guerrilla Warfare* (New York: Vintage Books, 1969), 86.

46. Tania Flood, Miriam Hoosain, and Natasha Primo, *Beyond Inequalities: Women in South Africa* (Bellville: University of Western Cape Gender Equity Unit and Harare, Zimbabwe: Southern African Research and Documentation Center [SARDC], 1997), 13.

47. Rachel Kagan, "South Africa's Other Majority," *Lies of Our Times: A Magazine to Correct the Record*, August 1994.

48. Brigitte Mabandla, "Promoting Gender Equality in South Africa," in Susan Bazilli, ed., *Putting Women on the Agenda* (Johannesburg: Ravan Press, 1991), 76.

49. Neva Seidman Makgetla, "Women and Economy: Slow Pace of Change," *Agenda: Empowering Women for Gender Equity,* No. 24, 1995, 16.

50. Ibid., 17.

51. Linda Zama, "Theories of Equality: Some Thoughts for South Africa," in Bazilli, ed., *Putting Women on the Agenda,* 59.

52. Archie Mafeje, *In Search of an Alternative: A Collection of Essays on Revolutionary Theory and Politics* (Harare, Zimbabwe: SAPES Books, 1992), 64.

53. Suzane Williams, "From 'Mothers of the Nation' to Women in their Own Right: South African Women in the Transition to Democracy," in Tina Wallace with Candida March, eds., *Changing Perceptions: Writings on Gender and Development* (Oxford: Oxfam, 1991), 128.

54. Rudo Gaidzanwa, "Bourgeois Theories of Gender and Feminism and Their Shortcomings with Reference to Southern African Countries," in Ruth Meena, ed., *Gender in Southern Africa: Conceptual and Theoretical Issues* (Harare, Zimbabwe: SAPES Books, 1992), 106.

55. Nici Nelson, "Mobilising Village Women: Some Organizational and Management Considerations," in *African Women in the Development Process,* ed. Nici Nelson (London: Frank Cass, 1981), 49–50.

56. Jacquelyn Cock, *Maids and Madams: A Study in the Politics of Exploitation* (Johannesburg: Ravan Press, 1980), 5, 70.

57. Cited by Flood, Hoosain, and Primo in *Beyond Inequalities,* 21.

58. Sibongile Makhabela, "Women and Liberation," *UMTAPO Focus,* August 1988, 7.

59. Ibid., 8.

60. Rose Shayo, "Integrating Women in the Mainstream: The Case of Tanzania," in Fatima Babiker Mahmoud, ed., *African Women: Transformation and Development* (London: Institute for African Alternatives, 1991), 11.

61. Dabi Nkululeko, "The Right to Self-Determination in Research: Azania and Azanian Women," in Christine N. Qunta, ed., *Women in Southern Africa* (London and New York: Allison and Busby, in association with Johannesburg: Skotaville, 1987), 104.

62. Fatima Babiker Mahmoud, introduction to *African Women,* 3.

63. Abena Oduro, "Women's Economic Roles" in *Gender Analysis Workshop Report,* Proceedings of the Gender Analysis Workshop, organized by the Development and Women's Studies Programme of the Institute of African Studies, University of Ghana, Legon, 14–16 July, 1992, 98.

64. Mansa Prah, "Women and Education," in *Gender Analysis Workshop Report,* 89.

65. Pilar Aguilar-Retamal, "Community Involvement in Basic Education for Life Skills: The Girl-Child in Eastern and Southern Africa," in UNESCO: Bi-Annual of UNESCO Office, Dakar, No. 12, March 1996, 31.

66. Ibid.

67. Ibid.

68. Blandina M. Makoni, *Crisis in Education and Culture and Its Social Reflection on Women: A Case Study of Zimbabwe* (Harare: Zimbabwe Institute of Development Studies, 1991), 18.

69. Ibid., 10.

70. Arun Naicker, "Women in the Struggle Against Apartheid in South Africa," in Ruth Bedsha, ed., *African Women: Our Burdens and Struggles* (Johannesburg: Institute for African Alternatives, 1991), 97.

71. April Brett, "Introduction: Why Gender Is a Development Issue," in Wallace, *Changing Perceptions,* 4.

72. Caroline Moser, "Gender Planning in the Third World: Meeting Practical and Strategic Gender Needs," in Wallace, *Changing Perceptions*, 163.

73. Miranda Munro, "Ensuring Gender Awareness in the Planning of Projects," in Wallace, *Changing Perceptions*, 176.

74. The author had the opportunity to stay with the Mokgatle family in Alexandra in January and July–August 1998 and in July–August 1999 and saw the role that women play in leadership and social action organizing firsthand.

75. The proceedings of this conference are published in UMTAPO's recent publication, "Reinforcing Women's Solidarity for the 21st century," Durban, October 1997.

76. Deborah Andrew Mfugale, "Women's Participation in Literacy Classes in Rural Areas," in Bedsha, *African Women*, 76. The quote from Freire is found in his *The Politics of Education: Culture, Power, and Liberation* (1985).

77. Molara Ogundipe-Leslie, a Nigerian womanist scholar, discusses the issues of African women's experiences of oppression and struggle for emancipation in historical and contemporary African societies in a comprehensive manner in *Recreating Ourselves: African Women and Critical Transformations* (Trenton, N.J.: Africa World Press, 1994). See especially the section "African Women, Culture, and Another Development."

78. The author spent various periods of time from 1991 to 1999 with the general secretary of AZAWU, Patrick Mkhize, who continues to be a role model for committed and principled revolutionary activists in the country. Mkhize is also general secretary of the newly formed Socialist Party of Azania, a splinter tendency from the Azanian People's Organization.

79. Noted in Shamim Meer, "Giving Back the Land," *Agenda: A Journal About Women and Gender*, No. 24, 1995, Agenda Women's Collective, Durban, South Africa.

80. Cited by Flood, Hoosain, and Primo in *Beyond Inequalities*, 15.

81. Mmakgomo Tshatsinde, "Rural Women in Development: Issues and Policies," *Agenda: A Journal About Women and Gender*, No. 18, 1993, 66.

82. Ibid., 67.

83. Shamim Meer, "Giving Back The Land" in *Agenda: A Journal About Women and Gender*, No. 24, 1995, Agenda Women's Collective, Durban, South Africa.

84. Cited by Flood, Hoosain, and Primo in *Beyond Inequalities*, 27.

85. Valerie Amos and Pratibha Parmar, "Challenging Imperial Feminism," *Feminist Review: Many Voices, One Chant, Black Feminist Perspectives*, No. 17, Autumn 1984, London, 6–7. Amos and Parmar cite the work of Maxine Molyneux, author of "Socialist Societies Old and New: Progress Towards Women's Emancipation (*Feminist Review*, No. 8, 1–34) as a classic example of this kind of Western feminist arrogance toward women and societies from the Two-Thirds World.

86. See, for instance, the works by Ifi Amadiume, such as *African Matriarchal Foundations: The Igbo Case* (London and Lawrenceville, N.J.: Karnak House, dist. Red Sea Press, 1995), and *Reinventing Africa: Matriarchy, Religion and Culture* (London and New York: Zed Books, 1997). Amadiume has done some superb work in unearthing indigenous traditions of West Africa that have matriarchal foundations but that were brutalized and fractured by colonialism and patriarchy. Research on other cultures of the continent need to be pursued in this regard. This exploration by no means diminishes the fact that there are indigenous patriarchal traditions in Africa that have oppressed women prior to colonialism and that continue to subjugate women.

87. Musi Katerere, "Women in the Context of African Culture," in National Seminar Series, "Women Organizing for the Future," Ranche House College, August 7–9, 1995, Pan African Women Liberation Organization, 34.

88. Ntombie Gata, "Indigenous Science and Technologies for Sustainable Agriculture, Food Systems and Natural Resource Management with Special Reference to Zimbabwe," cited in J. Matowanyika, V. Garibaldi, and E. Musimwa, eds., *Indigenous Knowledge Systems and Natural Resource Management in Southern Africa*, Report of the Southern African Regional Workshop, Harare, Zimbabwe, 20–22 April, 1994 (World Conservation Union, 1995), 87.

89. Sara C. Mvududu, "Gender Concerns in Indigenous Woodland Management," in Calvin Nhira, comp. and ed., *Proceedings of the Conference on "Indigenous Woodland Management,"* (Harare: Center for Applied Social Sciences, University of Zimbabwe, January 1995), 43.

90. Jeff Haynes, *Third World Politics; A Concise Introduction* (Oxford and Cambridge, Mass.: Blackwell, 1996), 113.

91. Martha Mantour, "Matriarchy and the Canadian Constitution: A Double-Barrelled Threat to Indian Women," *Agenda: A Journal About Women and Gender,* No. 13, 1992, 60.

92. See Bruce E. Johansen's book *Forgotten Founders: How the American Indian Helped Shape Democracy* (Boston: Harvard Common Press, 1982) for a detailed treatment of the way the Iroquios Confederacy and other indigeneous Natives shaped the democratic principles of the U.S. Constitution.

93. Musi Katerere, "Women in the Context of African Culture," National Seminar Series, "Women Organizing for the Future," Ranche House College, 7–9 August, 1995, Pan African Women Liberation Organization, 36.

94. Ruth Meena, "Gender Research/Studies in Southern Africa," in Ruth Meena, ed., *Gender in Southern Africa: Conceptual and Theoretical Issues* (Harare, Zimbabwe: SAPES Books, 1992), 3.

95. See Oyeronke Oyewumi, "Mothers Not Women: Making an African Sense of Western Gender," Ph.D. diss., University of California, Berkeley, 1993.

96. Oyeronke Oyewumi's 1997 work, *The Invention of Gender: Making an African Sense of Gender Discourses* (Minneapolis: University of Minnesota Press), is an important piece of critical literature in African Studies and Women's Studies, in which Oyewumi argues that gender discourses as defined by Western feminists ought not to be used as universal paradigms for understanding society's functioning, with particular regard to the precolonial Yoruba tradition, which she claims is predicated on seniority categories and not on gender. She also persuasively argues that most studies of Africa, including those by African scholars, are Western-produced and -directed and that African scholars need to begin to investigate African societies from the perspective of indigenous histories, cultures, and questions, a legitimate and much-needed task for African scholarship today. Her critique, trenchant as it is, however, fails to consider the question of *power* that underlies all societal relations, including that of precolonial Africa, within which the experiences of different classes of women and men were couched. Certainly, groups of individuals such as members of the royalty and the commercial elites possessed more power and privilege than the peasantry. Precolonial Yoruba society, like all other societies, was constituted by conditions of undulating power relations within which women from different classes featured and functioned. A second question that needs to be raised is: How does one move from the precolonial to the existential experience of women where material oppression and exploitation is a reality? The final critical point is that Oyewumi's work needs to differentiate between what is normative and what is exceptional in diverse indigenous African societies.

Notwithstanding these critical observations, *The Invention of Gender* is an extremely valuable work that all scholars of African Studies and Women's Studies should read, particularly with regard to the call for constructing indigenous discourses free from the tutelage of Western hegemony, a struggle of import equal to that of Africa's economic and political independence in the world.

97. Apollonia Kerenge, *Women in Development: Training Experiences in Sub-Saharan Africa,* Women and Development Series, Collection Femme et Development, PAID Report No. 14 (Yaounde, Cameroon: Pan African Institute for Development, 1992), 12.

98. In 1997–1998, R40 billion of the R186-billion national budget was spent on debt servicing. Cited in the *Sunday Times* (Johannesburg), January 25, 1998.

99. R. P. Ngcongo, "Power, Culture, and the African Woman," *Agenda Women's Collective,* No. 19, 1993, 7–8.

100. See, for instance, Jan Barrett et al., eds., *Vukani Makhosikazi: South African Women Speak* (London: CIIR, 1985), for an illumination of the expressions of Black working-class women on their struggles and frustrations in fighting apartheid capitalism and male domination. See also June Goodwin's *Cry Amandla! South African Women and the Question of Power* (New York: Africana Publishing Company, 1984), as well as *Working Women* (Johannesburg: South African Council on Higher Education and Ravan Press, 1985) and *Factory and Family: The Divided Lives of South Africa's Women Workers/Efetro Nase-Khaya: Impilo Yabesifazane Base-Mzansi Afrike* (Durban: 992 Women Workers and the Institute for Black Research, 1984), for a further treatment of the theme of Black working women's subjugation by the apartheid system. Ellen Kuzwayo's *Call Me Woman* (San Francisco: Spinster's Press, 1985), which describes the telling story of a Black working-class woman's resolve to overcome oppression in the face of overwhelming odds, and Julianne Malveaux's article, "You Have Struck a Rock," in Margaret Sims and J. Malveaux, eds., *Slipping Through the Cracks: The Status of Black Women,* (New Brunswick, N.J.: Transaction Books, 1986) are also informative works on this subject. For projections on the role of women in a liberated South African/Azanian society, see Ivy Matsepe Casaburri's "On the Question of Women in South Africa," in B. Magubane and Ibbo Mandaza, eds., *Whither South Africa?* (Trenton, N.J.: Africa World Press, 1988).

101. Cock's *Maids and Madams* provides detailed documentation of this inhumane and horrific institution of the domesticity of Black women.

102. This view is expressed toward the conclusion of the United Nations film documentary *Generations of Resistance,* 1978.

103. See Dabi Nkululeko, "The Right to Self-Determination in Research: Azania and Azanian Women," in Qunta, ed., *Women in Southern Africa,* 104.

104. Ngcongo, "Power, Culture, and the African Woman," 9.

105. For an elaboration of the status of Black women at South African universities, see Nasima Badsha and Piyushi Kotecha, "University Access: The Gender Factor," *Agenda: A Journal About Women and Gender,* Durban, No. 21, 1994, Agenda Women's Collective, Durban, South Africa.

106. *Within Human Reach: A Future for Africa's Children* (New York: UNICEF, 1985).

107. Kerenge, *Women in Development,* 12.

108. "New Dynamism in the African Adult Education Movement," African Association for Literacy and Adult Education, Report of the Secretary General to the First General Assembly, July 22–24, 1987, Nairobi, Kenya, 6.

109. Much more detail is discussed in the *AALAE Three-Year Programme, 1990–1992,* African Association for Literacy and Adult Education, Nairobi, August 1990, 1–24.

110. See, for instance, Andrew Zimbalist and Claes Brundenius, *The Cuban Economy: Measurement and Analysis of Socialist Performance* (Baltimore: Johns Hopkins University Press,

1989), which discusses the arduous and torturous course of struggle that Cuba was subject to following the collapse of the Soviet Union, which it depended heavily on for trade and aid.

111. *Arizona Daily Star,* July 4, 1999.

112. All-China Women's Federation, *The Impact of Economic Development on Rural Women in China: A Report of the United Nations University Household, Gender, and Age Project* (Tokyo: The United Nations University, 1993), 62–64. This research project provides concrete details of the changes in women's status since the 1949 revolution, the emergence of "New China," and the subsequent role that women played in advancing the revolution, particularly in areas of the agrarian economy. The study also identifies obstacles and difficulties faced by women in China's socioeconomic development such as obdurate male-oriented attitudes and the persistence of traditional attitudes that have stifled women's progress.

113. The liberation of women was reiterated by Mao Tse-tung throughout his career, reflected in his assertion, for instance, that "whatever men comrades can accomplish, women comrades can accomplish too," from *Peking Review,* Vol. 15, No. 10, October 6, 1972, cited in J. Bryan Starr, *Continuing the Revolution: The Political Thought of Mao* (Princeton: Princeton University Press, 1979), 218. For an educational understanding of the relevance of China's cultural revolution to Azania's struggle for revolutionary decolonization, see *The Selected Works of Mao Tse-tung* (Peking: Foreign Languages Press, 1971); Al Imfield, *China as a Model of Development* (Maryknoll, N.Y.: Orbis Books, 1977); David Bonavia, *The Chinese: A Portrait by David Bonavia* (New York: Penguin Books, 1980); and Andres D. Onate, *Chairman Mao and the Chinese Communist Party* (Chicago: Nelson-Hall, 1979).

114. Steve Biko, *I Write What I Like* (London: Bowerdean Publishing, 1996), 145.

Epilogue

This book has discussed many ideas about the current situation in South Africa/Azania and the rest of Africa as understood from the vantage point of Black working-class women and men. Many theorists on South Africa or Africa may view the revolutionary option proposed here as "unrealistic" or "unlikely." They may argue that the Black working classes will not desire to pursue the revolutionary alternative. Certainly, one cannot prognosticate as to what will occur in South Africa/Azania or on the continent over the next decade or two. For us as South Africans/Azanians, the important question is: How can those of us who are committed to reshaping the future of South Africa/Azania and the rest of Africa ensure that the vast majority of the continent's people, the rural peasantry working classes, become uplifted and liberated, able to equitably own the resources of the continent and enjoy the fruits of their labor? Africa's working classes are entitled to enjoy the rights of adequate food, decent shelter, gainful employment, a sound and enlightened education, proper health care, social security, and development of their full human potential, as *inalienable human rights*. If there is another path to the attainment of these goals outside of the revolutionary path, we certainly are willing to enter the discussion of such avenues. However, given the historical and contemporary objective conditions in South Africa/Azania in particular, and Africa in general, we see such possibilities as extremely slim.

So, too, there have been tremendous changes in the rest of Africa and the world, some positive and others not, some substantial, most not. The summer of 1997 saw the final demise of the regime of Mobutu Sese Seko in Zaire and the installation of a new government under the leadership of Laurent Kabila, who renamed the country the Democratic Republic of the Congo. The military regime of Johnny Koroma in Sierra Leone was removed by the ECOMOG forces led by Nigeria in February 1998, and Ahmed Tejon Kabbah was reinstated in March 1998, ironically by a nefarious military regime. Cuba continues to defy international odds and maintain a socialist system, notwithstanding a visit from the Catholic pontiff in early 1998 and the continued economic blockade imposed by the United States. In occupied Palestine, the Palestine Authority, led by Yasser Arafat, continues to hope for a just and fair brokered agreement with Israel that would result in Israel withdrawing from the occupied territories and would eventuate in a Palestinian state. In Northern Ireland, Sinn Fein has been drawn into a negotiated settlement with the British government as part of a solution to the eight-century occupation of Northern Ireland by the British. Meanwhile, in Asia, the nations hailed as capitalist and free-market success stories—Thailand, Indonesia, and most surprisingly, South Korea—all saw their economies teetering on the verge of collapse, even though now there is talk about some recovery. These "dragon economies" had to be rescued by the institutions of Western capital such as the IMF and the World Bank, to the tune of scores of billions of dollars. Of course, the IMF is primarily concerned that the monies that it lent to these capitalist governments and institutions and dictators like the former Suharto in Indonesia *are repaid*, not that economic stability or empowerment of the impoverished Asian masses be instituted.

Capitalism is not as strong, stable, and secure as the rhetoricians and champions of the market claim. Overall, there has been a consolidation of global financial capital so that North America, Europe, and East Asia now generate virtually 100 percent of world financial flows, and the disparities between rich and poor nations and classes are more pronounced than ever before. Satoshi Ikeda, a research associate at the Fernand Braudel Center, notes in this regard:

> [T]he integration of the world economy progressed primarily as an intra-core affair in the second half of the twentieth century in contrast to the period of colonial imperialism in the previous era. This, however, does not mean to deny the crucial and expanding role of the periphery and semi-periphery as the suppliers of low-cost, labour-intensive manufactured goods as well as underpaid workers for the core. Furthermore, core-centered expansion resulted in a widening absolute income gap between the populations in the core and those in the periphery and the semiperiphery.[1]

The global capitalist system prevails, certainly today and perhaps for a short time tomorrow. We continue to live in what Cornel West refers to as "the age of Europe." Will the new millennium become an "age of the indigenous people?" We certainly hope so. No singular continent or people can dominate our beautiful but strife-torn world for ever. The question, though, is: When? Only time will tell.

Notes

1. Satoshi Ikeda, "World Production," in Terence Hopkins and Immanuel Wallerstein, eds., *The Age of Transition: Trajectory of the World System, 1945–2025* (London: Zed Press, and Leichhardt, Australia: Pluto Press, 1996), 47.

2. Cornel West, "Decentering Europe," in *Prophetic Thought in Postmodern Times*, (Monroe, Maine: Common Courage Press, 1993), 124.

Index